Assumptions Inhibiting Progress in Comparative Biology

Assumptions Inhibiting Progress in Comparative Biology

Edited by
Brian I. Crother
Lynne R. Parenti

CRC Press
Taylor & Francis Group
Boca Raton London New York

CRC Press is an imprint of the
Taylor & Francis Group, an **informa** business

The cover image is modified from Donn Rosen's copy of Hennig's 1966 *Phylogenetic Systematics* (see Chapter 3). In the margin of page 90, where Hennig argues that only synapomorphies—not symplesiomorphies—diagnose monophyletic groups, Donn penciled, his objection: "A gross and silly distortion of a well-established and accepted concept." Later he changed his mind, erased the comment, and covered his erasure with a bold check. This graphically symbolizes an important take-home message: *erase assumptions to make progress in comparative biology.*

CRC Press
Taylor & Francis Group
6000 Broken Sound Parkway NW, Suite 300
Boca Raton, FL 33487-2742

First issued in paperback 2020

© 2017 by Taylor & Francis Group, LLC
CRC Press is an imprint of Taylor & Francis Group, an Informa business

No claim to original U.S. Government works

ISBN-13: 978-1-4987-4127-9 (hbk)
ISBN-13: 978-0-367-65832-8 (pbk)

Library of Congress Cataloging-in-Publication Data

Names: Crother, Brian I., editor. | Parenti, Lynne R., editor.
Title: Assumptions inhibiting progress in comparative biology / [edited by]
Brian I. Crother and Lynne R. Parenti.
Description: Boca Raton : Taylor & Francis, 2016.
Identifiers: LCCN 2016027008| ISBN 9781498741279 (hardback : alk. paper) |
ISBN 9781498741286 (e-book)
Subjects: LCSH: Evolution (Biology)--Research--Methodology. |
Biology--Classification--Research--Methodology. |
Biogeography--Research--Methodology.
Classification: LCC QH362 .A87 2016 | DDC 570.72--dc23
LC record available at https://lccn.loc.gov/2016027008

Visit the Taylor & Francis Web site at
http://www.taylorandfrancis.com

and the CRC Press Web site at
http://www.crcpress.com

Contents

Foreword

Donn Rosen and Inhibiting Assumptions

Norman I. Platnick

In the 1970s, the American Museum of Natural History was an amazing place for a systematist to work, and Donn Rosen was one of the main reasons. A kind and affable biologist, Donn was always happy to interact with new students (and even new AMNH curators such as myself). As the leader of the museum's Systematics Discussion Group, Donn facilitated and fostered exciting and exhilarating grappling with a wide variety of issues in comparative biology. Those monthly meetings invariably resulted in spirited exchanges between the new guard (proponents of the ideas of Hennig, Croizat, and Popper) and the Mayr-Simpsonites who had ruled the roost there for some decades. Those spirited exchanges usually continued on late into the night at the after parties (because my wife and I lived in a brownstone apartment only a block from the museum, many of those parties took place in our living room and—a rarity for a Manhattan apartment—back yard!). Often the arguments continued for days afterward as well, at lunch and at our usual after-lunch coffee gathering in the staff lounge. Because Donn had begun as a thorough Mayr-Simpsonite, he was probably the only person in the museum widely enough respected by both sides to keep the arguments civil (at least for the most part!).

Donn's effectiveness as a teacher can easily be grasped by perusing the 1985 list he created for students at the University of Miami that serves as the unifying theme for the varied contributions in this volume. Titled "Assumptions that *inhibit* scientific progress in comparative biology," this list of 33 assumptions demonstrates the wide range of Donn's interests, and reminds me of the "95 Theses" that Martin Luther nailed to the doors of the Wittenberg Castle church five centuries ago, starting the Reformation (with the twist that the 33 assumptions are posed instead in the sardonic, devilish style adopted by C. S. Lewis in *The Screwtape Letters*).

As with many of his published writings, Donn's list includes items that are likely to attract the attention of any student. Perhaps my favorites of his "General" assumptions are the last two:

> "6. Discovering you were wrong is bad in some sense (never publish until you're convinced you've found the truth).
> 7. Your graduate advisor and/or your distinguished visiting professor is probably right most of the time."

For me, these two assumptions are intimately interrelated, as I learned only one thing of lasting value from my graduate advisor (the late Herb Levi, an arachnologist at Harvard's Museum of Comparative Zoology): "The only way to make no mistakes is to do no work." Herb's pithy statement removed any perceived pressure to publish only results of which I was "certain." Later on, I learned from Karl Popper that scientific progress hinges not on our ability to be error-free (much less "certain"), but rather on conjectures and refutations: our ability to pose testable hypotheses so that errors can be found, and corrected, as quickly and efficiently as possible.

Thinking about Donn and his accomplishments, I can't help but wonder what changes he might have made to his list of "Assumptions" had he lived long enough to see some of today's approaches and controversies. I suspect that his list for "Taxonomy" might have started very differently:

1. Cladistics = parsimony.

For those of us who, like Donn, became cladists only after we first became practicing systematists, cladistics = clustering by synapomorphy, not parsimony. By this I don't mean that if one constructs a matrix of putatively synapomorphic characters, and then analyzes that matrix parsimoniously, one necessarily arrives at anything other than a Hennigian synapomorphy scheme. But today's analyses typically involve what I call "gigomatrices" (gigantic, garbage-in, garbage-out matrices). Many or most of the so-called "characters" included in these gigomatrices do not actually represent anyone's hypotheses of synapomorphy; rather, they are mere observations of features that may, or may not, contain any phylogenetic information whatever. As a result, sensible results from these matrices are often attainable only by adopting character weighting (cf. Donn's original assumption 1: "Character weighting can be avoided by the use of numerical procedures"!). Applying today's sophisticated numerical methods of character weighting often amounts to merely an exercise in waste management (i.e., can we find a way to minimize the highly deleterious effects of the plethora of non-synapomorphic information included in this gigomatrix?). Thus, it is hardly surprising that some of the advocates of gigomatrices indicate that they purposefully "no longer cluster by synapomorphy"; in my view, such workers are no longer cladists.

It is as a biogeographer that I remember Donn best, particularly for the technique he used to compare the distributions of taxa. Donn eschewed all mathematical, grid-based attempts to describe taxon distributions, preferring instead to generate outline maps of those distributions on plastic transparencies and then overlay the transparencies on a map to detect shared patterns among them. This approach reflected his general, artistic approach to life; Donn's brother was a concert pianist, and Donn's hobby was painting (with birds as his favorite subject). For me, it is the last of Donn's eight biogeographic assumptions that is the most telling:

"8. Some organisms are better indicators of biotic history than others depending on their ecology and means of dispersal."

Donn may have intended his wording to cast doubt on the first half of that sentence as well as the second, but I would argue that some groups of organisms are indeed better indicators of biotic history than others. The reason, however, has nothing to do with their ecology or dispersal abilities. As Donn's fifth biogeographic assumption indicates, widespread taxa are uninformative, as are relationship statements involving fewer than three taxa or areas. So if, like Donn, we are interested in the biogeography of Central America, species that occur throughout that area are uninformative, as are the distributions of species about the relationships of which we cannot make at least a three-taxon statement. For any given region, the best indicators of biotic history are those groups that show the highest species-level diversity within the area, and whose species show the smallest average distribution ranges. The smaller the areas of endemism that are suggested by a group, the greater the information that group can potentially offer about the biotic history of that region. Unfortunately, this result dramatically highlights the pitiable state of our current knowledge of earth's biodiversity. The taxa whose distributions we know best (i.e., vertebrate animals and green plants) seldom rank highly on either relevant scale—within any given region, they tend to have relatively few species, and those species tend to have relatively large range sizes. It is among the other 97% of the world's species, about which vastly less is known, that the groups with the greatest diversity and smallest average range size are to be found. Despite his affection for fishes (the potentially most informative vertebrates), I think Donn would agree that we need to study that other 97%!

Preface

In February 1985, the eminent evolutionary biologist, biogeographer, systematist, and ichthyologist Donn Eric Rosen (1929–1986) accepted the invitation of Jay M. Savage, Professor of Biology and a herpetologist, to be a Distinguished Visiting Professor of Biology at the University of Miami, Florida. Donn Rosen was an affable, erudite, and gregarious man who surrounded himself with colleagues and students with whom he liked to debate any idea in biology. He was a gifted teacher who inspired a generation of systematic biologists. At Miami, Donn taught a mini-course on "Evolving Evolutionary Thought" which covered a wide range of topics across the broad fields of evolution, systematics, and biogeography. At the end of the course he gave the students a list of what he argued were assumptions that *inhibit* scientific progress in comparative biology which we publish here as Chapter 1. A perusal of the assumptions revealed, to us, a list with continued relevance. They range from the general ("Scientists are more objective than other people," "Ultimate causes are knowable") to much more specific, addressing topics in Evolutionary Theory ("Competition Theory is important and well formulated" and "Adaptation scenarios have important general explanatory powers"), Taxonomy ("Deliberately formulated paraphyletic groups are analytically useful," "Convergence really occurs"), and Biogeography ("Geographic hybridization and biotic mixing make vicariance analysis impossible," "Some organisms are better indicators of biotic history than others depending on their ecology and means of dispersal").

The year 2015 marked the 30th anniversary of Rosen's list, which seemed like an appropriate time to revisit his eclectic, thought-provoking ideas in comparative biology. We wondered what, if any, progress has been made in the theory and methods of comparative biology in the past three decades. We, a herpetologist and student of Savage (Crother), and an ichthyologist and student of Rosen (Parenti), planned to confront Rosen's bold ideas in a symposium at the Joint Meetings of Ichthyologists and Herpetologists (JMIH) held in Reno, Nevada, in July 2015. The setting was ideal. Rosen was devoted to the American Society of Ichthyologists and Herpetologists and looked forward each year to the annual meetings where he could engage with both colleagues and students. How would our colleagues react to a typed list that was yellow with age and scarred from the thumbtacks and multiple layers of tape added every time the list was taken down and moved? The response to our invitation to participate in the symposium and revisit Rosen's list was overwhelmingly positive. Brazilian ichthyologist Mario de Pinna expressed the thoughts of many: "Great idea! The Rosen document is a marvelous piece of history and freethinking. It is fascinating how some of the topics have aged somehow, others are more modern than ever, and others still point to the future even from the standpoint of 35 years beyond."

We convened the JMIH symposium always with the idea that we would publish the proceedings in a book. It is appropriate that not all of the topics on Rosen's list are covered here and some, such as biogeography, get much attention. Although

Donn Rosen (center) at an annual meeting of the American Society of Ichthyologists and Herpetologists talks to Robert Inger of the Field Museum as the late Robert McDowell, New Zealand ichthyologist and biogeographer (left), waits his turn. Photographer and date unknown; possibly 1981.

several prior publications include a dedication to Donn's memory, there was no proper Rosen *Festschrift*, an oversight that we are happy to correct. We know that Donn would be pleased.

Brian I. Crother
Lynne R. Parenti

Acknowledgments

We gratefully acknowledge support for the 2015 JMIH symposium from the American Society of Ichthyologists and Herpetologists, the Herpetologists' League and the Herbert R. and Evelyn Axelrod Chair in Systematic Ichthyology and our late colleague, Richard P. Vari, in the Division of Fishes, National Museum of Natural History, Smithsonian Institution. We thank the speakers who all exceeded our expectations to present a vibrant, sometimes controversial, well-attended symposium. Chuck Crumly provided encouragement throughout development of the symposium and this edited volume.

Our following colleagues graciously provided presubmission reviews of chapters in this book: James Albert, Frank Burbrink, Don Buth, Raul Diaz, Malte Ebach, Richard Glor, Gene Hunt, Joe Mendelson, Juan Morrone, Chris Murray, Gareth Nelson, and David Williams.

LRP thanks Tina and Ella for all the jokes.

Contributors

Prosanta Chakrabarty, PhD
Department of Biological Sciences and
    Museum of Natural Science
Louisiana State University
Baton Rouge, Louisiana

Brian I. Crother, PhD
Department of Biological Sciences
Southeastern Louisiana University
Hammond, Louisiana

Maureen A. Donnelly, PhD
Department of Biology
Florida International University
Miami, Florida

Mallory E. Eckstut, PhD
Department of Biological Sciences
Southeastern Louisiana University
Hammond, Louisiana

Lance Grande, PhD
Integrative Research Center
The Field Museum
Chicago, Illinois

Craig Guyer, PhD
Department of Biological Sciences
Auburn University
Auburn, Alabama

David Kizirian, PhD
Department of Herpetology
American Museum of Natural History
New York, New York

Caleb D. McMahan, PhD
Department of Zoology
The Field Museum
Chicago, Illinois

Randall D. Mooi, PhD
Zoology
The Manitoba Museum
Winnipeg, Manitoba, Canada

Christopher M. Murray, PhD
Department of Biology
Tennessee Technological University
Cookeville, Tennessee

Gareth Nelson, PhD
School of BioSciences (Botany)
University of Melbourne
Melbourne, Victoria, Australia

Kirsten E. Nicholson, PhD
Department of Biology
Central Michigan University
Mt. Pleasant, Michigan

Lynne R. Parenti, PhD
Department of Vertebrate Zoology
National Museum of Natural History
Smithsonian Institution
Washington, D.C.

John G. Phillips, BS
Department of Biological Sciences
University of Tulsa
Tulsa, Oklahoma

Brett R. Riddle, PhD
School of Life Sciences
University of Nevada, Las Vegas
Las Vegas, Nevada

Donn E. Rosen, PhD (deceased)
Department of Ichthyology
American Museum of Natural History
New York, New York

Subir Shakya
Department of Biological Sciences and
   Museum of Natural Science
Louisiana State University
Baton Rouge, Louisiana

Mary E. White, PhD
Department of Biological Sciences
Southeastern Louisiana University
Hammond, Louisiana

1 Assumptions That *Inhibit* Scientific Progress in Comparative Biology

*Donn E. Rosen**
American Museum of Natural History

General

1. Ultimate causes are knowable.
2. Bridge principles are useful.
3. The *ceteris paribus* is generally applicable and helpful.
4. Scientists are more objective than other people.
5. Redefining the problem can solve it.
6. Discovering that you were wrong is bad in some sense (never publish until you're convinced you've found the truth).
7. Your graduate advisor and/or your distinguished visiting professor is probably right most of the time.

Evolutionary Theory

1. Competition Theory is important and well formulated.
2. Random mutation, natural selection, and microevolution combine as a progressive research program to explain the hierarchy of organisms.
3. Pure paleontology provides essential information about the history of life.
4. Punctuated equilibrium rescues the imperfections of the fossil record from criticism and is testable.
5. It is important to search for ancestors and/or to formulate archetypes.
6. The stratigraphic sequence never lies.
7. Fossils specify the age of their including taxon.
8. Adaptation scenarios have important general explanatory powers.
9. The Biological Species concept is an essential concept.
10. Goldschmidt's ideas must be wrong because they conflict with neodarwinism.
11. Ditto for neolamarkism.

---

* Deceased 1986.

Taxonomy

1. Character weighting can be avoided by the use of numerical procedures.
2. Cladograms and classifications are decoupled because they express different and equally useful notions.
3. Deliberately formulated paraphyletic groups are analytically useful.
4. Primitive characters are informative.
5. Convergence really occurs.
6. A knowledge of evolutionary processes and genetics is necessary to do taxonomy.
7. When comparative anatomy and ontogeny are fully informative they can disagree.

Biogeography

1. All distribution patterns result from vicariance.
2. All distribution patterns result from dispersal.
3. Centers of origin can be found.
4. Historical geology tests can reject biological theories of area relationships.
5. Widespread taxa and two taxon systems are informative about the relationships of areas.
6. Geographic hybridization and biotic mixing make vicariance analysis impossible.
7. A knowledge of geologic processes is necessary for a vicariance analysis of areas of endemism.
8. Some organisms are better indicators of biotic history than others depending on their ecology and means of dispersal.

February 8, 1985

DONN ERIC ROSEN
Distinguished Visiting
Professor of Biology
University of Miami

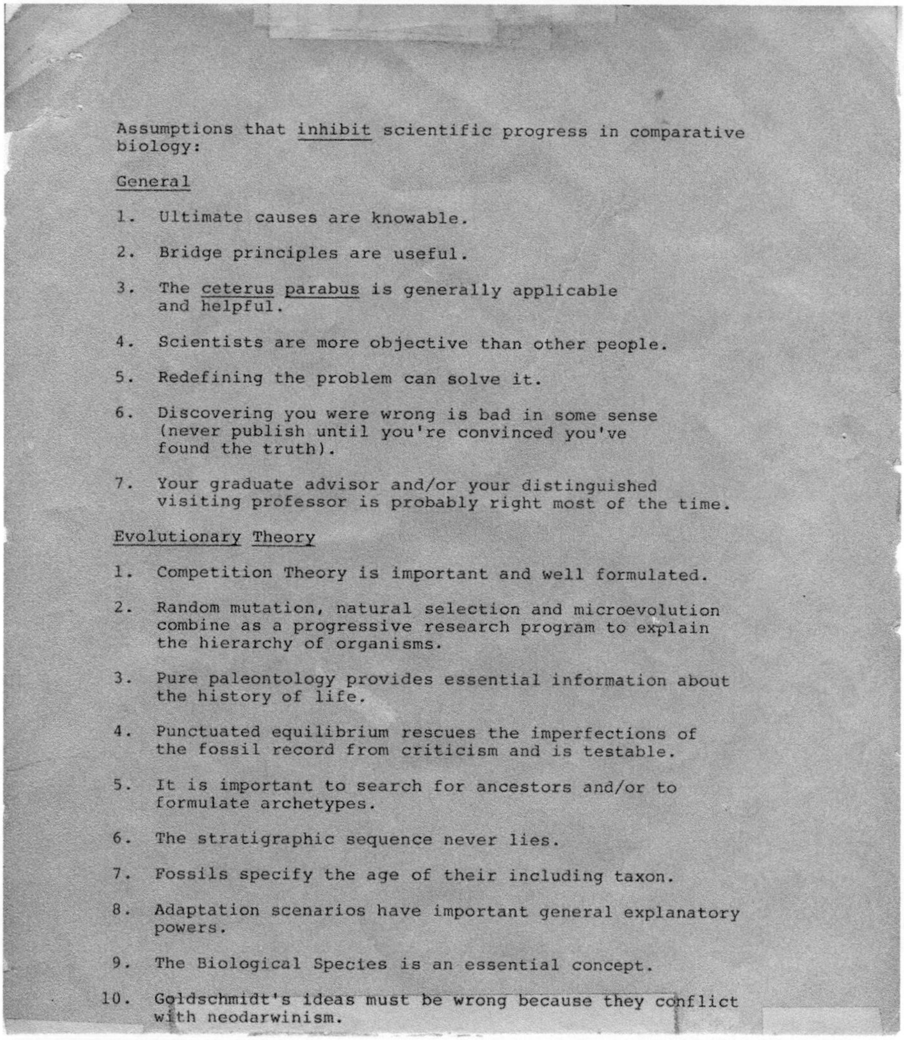

Assumptions that <u>inhibit</u> scientific progress in comparative biology:

<u>General</u>

1. Ultimate causes are knowable.

2. Bridge principles are useful.

3. The <u>ceterus parabus</u> is generally applicable and helpful.

4. Scientists are more objective than other people.

5. Redefining the problem can solve it.

6. Discovering you were wrong is bad in some sense (never publish until you're convinced you've found the truth).

7. Your graduate advisor and/or your distinguished visiting professor is probably right most of the time.

<u>Evolutionary</u> <u>Theory</u>

1. Competition Theory is important and well formulated.

2. Random mutation, natural selection and microevolution combine as a progressive research program to explain the hierarchy of organisms.

3. Pure paleontology provides essential information about the history of life.

4. Punctuated equilibrium rescues the imperfections of the fossil record from criticism and is testable.

5. It is important to search for ancestors and/or to formulate archetypes.

6. The stratigraphic sequence never lies.

7. Fossils specify the age of their including taxon.

8. Adaptation scenarios have important general explanatory powers.

9. The Biological Species is an essential concept.

10. Goldschmidt's ideas must be wrong because they conflict with neodarwinism.

FIGURE 1.1 (a, b) The original list that Donn Rosen typed for students at the University of Miami. Note the holes from thumbtacks and multiple layers of tape added every time the list was taken down and moved.

(Continued)

11.   Ditto for neolamarkism.

Taxonomy

1.   Character weighting can be avoided by the use of numerical procedures.

2.   Cladograms and classifications are decoupled because they express different and equally useful notions.

3.   Deliberately formulated paraphyletic groups are analytically useful.

4.   Primitive characters are informative .

5.   Convergence really occurs.

6.   A knowledge of evolutionary processes and genetics is necessary to do taxonomy.

7.   When comparative anatomy and ontogeny are fully informative they can disagree.

Biogeography

1.   All distribution patterns result from vicariance.

2.   All distribution patterns result form dispersal.

3.   Centers of origin can be found.

4.   Historical geology tests and can reject biological theories of area relationships.

5.   Widespread taxa and two taxon systems are informative about the relationships of areas.

6.   Geographic hybridization and biotic mixing make vicariance analysis impossible.

7.   A knowledge of geologic processes is necessary for a vicariance analysis of areas of endemism.

8.   Some organisms are better indicators of biotic history than others depending on their ecology and means of dispresal.

February 8, 1985   DONN ERIC ROSEN
                                          Distinguished Visiting
                                          Professor of Biology
                                          University of Miami

FIGURE 1.1 (CONTINUED)   (a, b) The original list that Donn Rosen typed for students at the University of Miami. Note the holes from thumbtacks and multiple layers of tape added every time the list was taken down and moved.

2 Donald Eric Rosen (1929–1986)

Gareth Nelson
University of Melbourne

CONTENTS

2.1 INTRODUCTION

Donn's family home was at 101 West 78th Street in Manhattan. This address is in the 1940 Manhattan telephone book under the name of Anita Rosen, his mother. It is mentioned also in obituary notices of his brother Charles (Tannenbaum 2012):

> Charles Rosen, the pianist, polymath and author, who died on Sunday [Dec 9 2012], had lived here with his parents, Anita and Irwin, and younger brother, Donald. In 1940 Charles was 13 and Donald was 11.

The building is on the corner with Columbus Avenue across from the American Museum of Natural History. Donn was a volunteer there from the age of eight, destined to find employment as a curator in its Department of Ichthyology. Early on he absorbed, as his own, the ideas about evolution that prevailed in the Museum—the New Synthesis. In his own research on fishes, he later came to reject this Synthetic Theory because of its "assumptions that inhibit scientific progress in comparative biology" (best summarized in his 1985 course given at the University of Miami). His changed outlook—his progress—resulted from his studies in biogeography. This field has remained contentious ever since, but is still explored in the hope of finding illumination, one might even say enlightenment. There Donn became sensitive to the conflict between discovery and assumption, or in his terms "empirical evolutionary research versus neo-Darwinian speculation" (title of Rosen and Buth 1980). For him the outcome approached the motto of the Royal Society of London, *Nullius in verba*, or in plainspeak: "We take nobody's word for it."

5

2.2   DONN'S BACKGROUND

In his primary school (PS 87), then located at 77th St. and Amsterdam Ave., teachers did not accept an apparent nickname for classroom use. So he added a terminal "n," perhaps inspired by a Columbia University PhD student at the Museum, Bobb Schaeffer. Bobb was the family name of his mother, Mary Mabel Bobb, who married Jacob Parsons Schaeffer in 1903. In 1961 when Donn was hired as an Assistant Curator by the Museum, his first publication was by Bobb and Donn (Schaeffer and Rosen 1961).

When in high school, Donn's "volunteer work shifted to the Fish Genetics Laboratory of the New York Zoological Society, then headed by Myron Gordon and located on the sixth floor of the Whitney Wing of the Museum" (Nelson et al. 1987:542). From there eventually came Donn's first scientific publication (Gordon and Rosen 1951).

2.3   DONN AND MAYR

In the Museum from 1931 to 1953, Ernst Mayr (1904–2005) was Curator of the Whitney-Rothchild Collections of birds. In 1979, Donn recollected (Nelson and Rosen 1981:xi):

> Over 30 years ago, however, a small group was formed by Ernst Mayr, then a curator on the museum's staff, in which graduate and undergraduate students working in or associated with the museum's collections discussed their own research or reviewed items in the current scientific literature. The group met at irregular intervals, sometimes as often as each week, and initially included about a dozen students—most of whom now are professionally engaged in systematic research.

In early 1961, Donn was hired as Assistant Curator by the Museum, and he immediately revived Mayr's discussion group. He recollected (Nelson and Rosen 1981:xi):

> When Mayr left the American Museum of Natural History [1953] to assume a position at the Museum of Comparative Zoology, Harvard University, the group remained active because of the students' interest. In the early 1960s the group named itself the Systematics Discussion Group (SDG), adopted a program of monthly meetings for the academic year, was assigned a meeting room at the museum, and was officially acknowledged by the museum's administration.

In 1953, the book *Methods and Principles of Systematic Zoology* (Mayr et al. 1953), one of a series in "McGraw-Hill Publications in the Zoological Sciences," was published. By the late 1960s, its success prompted a revised edition written by Mayr alone (Mayr 1969). Donn was asked to read and comment on the manuscript. He did so, and his comment reveals his state of mind (*in litt.* to Mayr, 21 July 1967, quoted in Nelson 1991:306):

> In regard to Hennig, I very much like your discussion of the pitfalls of the "cladistic approach." I place the foregoing in quotes only because I do not regard the said approach as anything more than bad biology. For that reason I think Throckmorton's [e.g., 1965] ideas should be elaborated. Many people are impressed by Hennig—and by Brundin's interpretation of Hennig—because of what probably seems to them a

simple way out of having to think for themselves. "make it all nice and arbitrary and according to the numbers" is more or less what has been suggested, and with the new English version of Hennig's work the problem of undoing this type of thinking will I am sure become more acute. There is one particular point of Hennig's and Brundin's that I really don't remember your having discussed in the manuscript and it concerns the use of primitive characters in phylogenetic analysis. Brundin in particular suggests that they are of no use whatever, and this is, of course, an insane point of view.

"An insane point of view"—a seemingly harsh judgement, but all the more signifi-cant because it was the reflection of the collective opinion of the Museum staff of that time. Donn was merely delivering the message.

In time, and to his credit, Donn changed his mind about the "cladistic approach." After a few years he once asked me into his office to show me something. He sat at his desk and I sat opposite. From the desk drawer he passed me a sheet of paper. It was a letter to Mayr, explaining why Donn was abandoning the New Synthesis and adopting the ideas of Willi Hennig. I commented favourably. Years later, when I was reviewing Mayr's recent book (Nelson 1991), I unsuccessfully searched for a copy of this letter in Donn's files of correspondence, created over many years by Department Secretary Victoria Pelton, and now archived in the Museum Library. I wrote to Mayr enquiring if he ever received such a letter; he denied receiving any such. I concluded that the letter was never sent and destroyed.

When Donn became President-Elect of the Society of Systematic Zoology, he suc-cessfully promoted the "Ernst Mayr Student Award in Systematics," "given at the Annual Meeting of the Society of Systematic Zoology for an outstanding presentation by a student" (Anon 1976). The award was first given in 1976 and nearly every year since.

I believe Donn changed his mind about cladistics through his studies of bioge-ography. General interest in cladistics and biogeography grew exponentially in the years since, as shown in Figures 2.1 and 2.2.

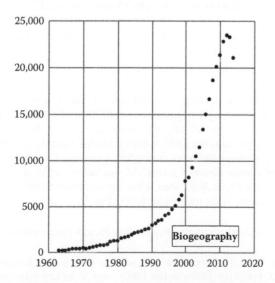

FIGURE 2.1   "Biogeography": Number of citations per year (Google Scholar 2015).

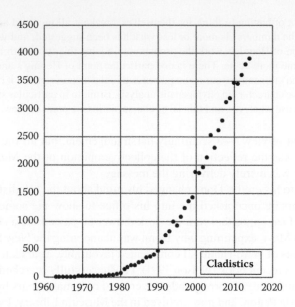

FIGURE 2.2 "Cladistics": Number of citations per year (Google Scholar 2015).

2.4 RECENT DEVELOPMENTS

In a recent book (de Queiroz 2014:278), fig. 11.2 sourced from "a search of the Web of Science database":

> Graphing an intellectual sea change: scientific studies using or discussing molecular clock analyses have taken off in the past twenty years, almost certainly driving the parallel rise in studies that use the term "long-distance dispersal."

The import is much the same as that of Figure 2.3, which extends over a longer time, and shows an earlier and independent history of "long-distance dispersal."

A review of this book states (Heads 2014:282):

> The main theme of Alan de Queiroz's (2014) book is that a flawed approach, vicariance theory, dominated biogeography from the 1970s to the 1990s, but that a more reliable theory, chance dispersal, has since claimed the field. However the reality is different; chance dispersal is not a recent theory, but has been the dominant paradigm in biogeography ever since the rise of the modern synthesis in the early 1940s. Despite the dominance of chance dispersal theory, the significance of vicariance began to be taken seriously in the 1970s. Since then, it has become more widely accepted, despite its radical undermining of the traditional theory (Fig. 1).

Heads' "Fig. 1" is much the same as Figure 2.4. Heads comments (p. 300):

> Since the 1970s, several components of vicariance theory have become much more widely accepted. From the 1940s to the 1980s, one of its key concepts, vicariance, was almost completely suppressed by authors such as Mayr (1965, 1982), Stebbins

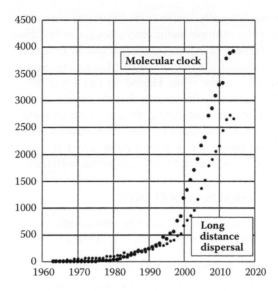

FIGURE 2.3 "Molecular clock" and "long distance dispersal": Number of citations per year (Google Scholar 2015).

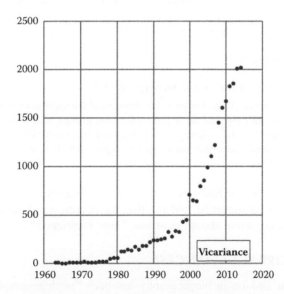

FIGURE 2.4 "Vicariance": Number of citations per year (Google Scholar 2015).

(1966) and Grant (1971, 1981) in their widely used text books. Thanks to the work of Croizat et al. (1974), vicariance was introduced to the mainstream, and by now it is well established (Fig. 1).

Croizat et al. (1974) appeared in *Systematic Zoology* when I was editor of this journal, published by the Society of Systematic Zoology. It had seemed to me that the idea of vicariance was virtually absent from discussion current at the time. I rarely encountered it, except for example, in the entomological publications of PhD students of George Ball, University of Alberta (retrospective in Erwin et al. 1979). I knew that the idea was fundamental to the outlook of Leon Croizat, and I invited Croizat to submit a manuscript on the topic. He did so and I had the manuscript widely reviewed. From the reviews it was evident that the reviewers did not meaningfully connect with the manuscript, and their comments were unconstructive. Eventually, I asked Croizat to allow Donn (then President-Elect of the society) and me to revise the manuscript. Donn added a new introduction, and I revised the footnotes and references. The resulting manuscript was re-reviewed, with a more positive result.

2.5   VICARIISM VERSUS VICARIANCE

Because George Ball and his students were in entomology, I had assumed that their interest in vicariance stemmed from Hennig (e.g., 1950:220):

1. Die biogeographische Methode (Vikarianztypenlehre)

Wenn im vorstehenden und in früheren Abschnitten festgestellt wurde, daß die Vikarianztypen höherer Ordnung mit dem verschiedenen Entstehungsalter der taxonomischen Gruppen höherer Ordnung in der Weisein Verbindung zu bringen seien, daß jedem Vikarianztypus ein bestimmtesEntwicklungsalter zugeordnet sei, und daß daher Artengruppen, die dengleichen Vikarianztypus zeigen, gleiches Entstehungsalter haben müßten, so gilt das nur innerhalb gewisser Grenzen.

I translate this passage as:

1. The biogeographic method (Vicariance type theory)

As determined in the above and in previous sections, if the vicariance type be related to the various ages of origin of the higher taxonomic groups, and if the age of each specific vicariance type be assigned, then groups of species which show the same vicariance type should have the same age of origin—which is probably true only within certain limits.

Years later I asked George Ball where his ideas of vicariance came from—in the sense of the history of ideas; he told me that the source was Stanley Cain (Cain 1944). However, Cain wrote about "vicariism," not "vicariance."

2.6   STILL MORE SEA CHANGES?

The most popular fashion in biogeography has been "phylogeography," as shown in Figure 2.5, the word appearing first in the 1980s (Avise 2009). Other candidates show a similar, but less exuberant, development, running to some hundreds rather than thousands of citations per year: "vicariance biogeography" (Nelson and Rosen 1981), "cladistic biogeography" (Humphries and Parenti 1999) and "comparative biogeography" (Parenti and Ebach 2009). In this respect they are similar to "panbiogeography," shown in Figure 2.6. Briggs (2007, abstract) commented (cf., Figure 2.6), belatedly reflecting the viewpoint of the "New York School" (Nelson and Ladiges 2001):

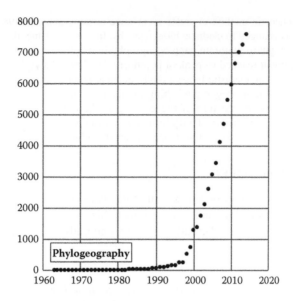

FIGURE 2.5 "Phylogeography": Number of citations per year (Google Scholar 2015).

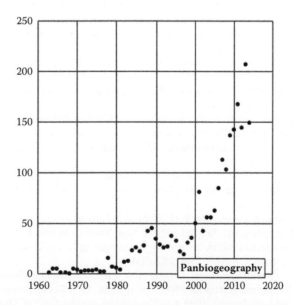

FIGURE 2.6 "Panbiogeography": Number of citations per year (Google Scholar 2015).

From the viewpoint of 2007, one can trace the history of an interesting and contentious trend in biogeography and evolution that began with Croizat's concept of panbiogeography in 1958. After a quiescent period of about 16 years, some young biologists in New York and in New Zealand read Croizat's books and became enthusiastic supporters. In New York, in the early 1970s, panbiogeography was combined with a part of

Hennig's phylogenetic method to create vicariance biogeography. In 1986, the name of the latter was changed to cladistic biogeography. In the meantime, the followers in New Zealand sought to maintain panbiogeography in its original form without reference to phylogeny. It reached its peak of popularity in 1989–90 and then began to die down. In comparison, cladistic biogeography became much more widespread, especially when its followers began publishing laudatory books and papers. Its decline became noticeable after the turn of the century as the dispersal counterrevolution began to have its effect. It served a useful purpose by engaging the interest of young biologists who otherwise may not have become aware of biogeography.

2.7 AND THE FUTURE?

Donn did not survive into the present age of molecular systematics, which, one might imagine, would furnish additional assumptions to his list. One obvious source is the current enthusiasm for molecular dating (Nelson and Ladiges 2009).

REFERENCES

Anon. 1976. Ernst Mayr student award in systematics. *Systematic Zoology* 24(4):513.
Avise, J. C. 2009. Phylogeography: Retrospect and prospect. *Journal of Biogeography* 36(1):3–15.
Briggs, J. C. 2007. Panbiogeography: Its origin, metamorphosis and decline. *Biologiya Morya* 33(5):323–328.
Cain, S. A. 1944. *Foundations of plant geography*. New York: Harper and Brothers. Reprint 1974, New York: Hafner Press.
Croizat, L., G. Nelson and D. E. Rosen. 1974. Centers of origin and related concepts. *Systematic Zoology* 23(2):265–287.
de Queiroz, A. 2014. *The monkey's voyage: How improbable journeys shaped the history of life*. New York: Basic Books.
Erwin, T. L., G. E. Ball and D. R. Whitehead, eds. 1979. Carabid beetles: Their evolution, natural history, and classification. *Proceedings of the first international symposium of carabidology*, Smithsonian Institution, Washington, D.C., August 21, 23, and 25, 1976. The Hague: Dr. W. Junk.
Gordon, M. and D. E. Rosen 1951. Genetics of species differences in the morphology of the male genitalia of xiphophorin fishes. *Bulletin of the American Museum of Natural History* 95(7):409–464.
Grant, V. 1971. *Plant speciation*. New York: Columbia University Press.
Grant, V. 1981. *Plant speciation*. Second edition. New York: Columbia University Press.
Heads, M. 2014. Biogeography by revelation: Investigating a world shaped by miracles. *Australian Systematic Botany* 27(4):282–304.
Hennig, W. 1950. *Grundzüge einer Theorie der Phylogenetischen Systematik*. Berlin: Deutscher Zentralverlag. Reprint 1980, Koenigstein: Otto Koeltz Science Publishers.
Humphries, C. J. and L. R. Parenti. 1999. *Cladistic biogeography*. Oxford: Oxford University Press.
Mayr, E. 1965. Summary. In: *The genetics of colonizing species*, eds H. G. Baker and G. L. Stebbins, 553–562. New York: Academic Press.
Mayr, E. 1969. *Principles of systematic zoology*. New York: McGraw-Hill Book Company.
Mayr, E. 1982. *The growth of biological thought: Diversity, evolution, and inheritance*. Cambridge: Harvard University Press.
Mayr, E., E. G. Linsley and R. L. Usinger. 1953. *Methods and principles of systematic zoology*. New York: McGraw-Hill Book Company.

Nelson, G. 1991. Principles of systematic zoology. Second edition [review of Mayr and Ashlock 1991]. *Cladistics* 7(3):305–307.

Nelson, G. and D. E. Rosen. 1981. Preface. In: *Vicariance biogeography: A critique. Symposium of the Systematics Discussion Group of the American Museum of Natural History May 2–4, 1979*, eds G. Nelson and D. E. Rosen, xi–xii. New York: Columbia University Press.

Nelson, G., J. W. Atz, K. D. Kallman and C. L. Smith. 1987. Donn Eric Rosen (1929–1986). *Copeia* 1987(2):541–547.

Nelson, G. and P. Y. Ladiges. 2001. Gondwana, vicariance biogeography and the New York School revisited. *Australian Journal of Botany* 49(3):389–409.

Nelson, G. and P. Y. Ladiges. 2009. Biogeography and the molecular dating game: A futile revival of phenetics? *Bulletin de la Société Géologique de France* 180(1):39–43.

Parenti, L. R. and M. C. Ebach. 2009. *Comparative biogeography.* Berkeley: University of California Press.

Rosen, D. E. and D. G. Buth. 1980. Empirical evolutionary research versus neo-Darwinian speculation. *Systematic Zoology* 29(3):300–308.

Schaeffer, B. and D. E. Rosen. 1961. Major adaptive levels in the evolution of the actinopterygian feeding mechanism. *American Zoologist* 1(2):187–204.

Stebbins, G. L. 1966. *Processes of organic evolution.* Englewood Cliffs, NJ: Prentice-Hall.

Tannenbaum, A. 2012. [Charles Rosen, obituary]. www.directme.nypl.org.

Throckmorton, L. H. 1965. Similarity versus relationship in *Drosophila. Systematic Zoology* 14(3):221–236.

3 Rosen Decomposing

Lynne R. Parenti
National Museum of Natural History,
Smithsonian Institution

Brian I. Crother
Southeastern Louisiana University

CONTENTS

3.1 INTRODUCTION

Donn Rosen (1929–1986), curator of ichthyology at the American Museum of Natural History, and British Museum (Natural History) paleoichthyologist and comparative morphologist Colin Patterson (1933–1998), were close colleagues who worked together with some frequency in the 1960s and 1970s. During the height of their collaboration, almost annually either one or the other took a trans-Atlantic flight to New York or London so they could study specimens, debate ideas or write manuscripts together. They found much to agree on, especially the disruption of traditional teleost classification. Together they wrote several influential papers, notably two monographs published in the *Bulletin of the American Museum of Natural History*: one on the structure and relationships of the paracanthopterygian fishes (Rosen and Patterson 1969) and another on Mesozoic teleost fishes and the theory and practice of classifying fossils (Patterson and Rosen 1977).

Patterson and Rosen became known for their bold implementation of Willi Hennig's (1966) principles of cladistics. Colin told the story of his quick adoption of cladistics in a lecture that he presented to the Systematics Association in London in late 1995. David Williams and Anthony Gill edited Colin's detailed lecture notes and published them in a special volume dedicated to a critique of the ongoing conflict between the use of molecular versus morphological data in systematics (de Carvalho and Craig 2011). Although Colin readily became a cladist,

Donn needed some persuading before he would abandon traditional systematic methods:

> "I learned that Hennig's book, published in German in 1950 (Hennig 1950), had come out in a new English version in the States in 1966 (Hennig 1966), but I couldn't find a copy anywhere in London in the summer of 1967 so I asked Donn Rosen to bring me one when he came over in September to work with me. Donn brought the book, but read it first and scribbled all over it in pencil—things like "nonsense" and "a misunderstanding of well-known principles." He thought the book was rubbish; I read it and thought the opposite. So while Donn and I spent a couple of years working on paracanthopts, I was a committed Hennigian and he wasn't." (Patterson 1995, published 2011).

3.2 ROSEN'S COPY OF HENNIG (1966)

One of us (LRP) was a doctoral student of Donn who visited his wife, Carmela (Mel) Rosen, in their New Jersey home soon after he died in 1986. Mel was distributing Donn's personal library and offered several books, including his copy of Hennig (1966) and Léon Croizat's *Space, Time, Form: The Biological Synthesis*, among others. Donn's Hennig (1966) is worn, but has held up well over 50 years.

Donn wrote a list, in pencil, on the back of the paper book jacket: p. 79–80 as marked, p. 90, p. 155, p. 157, p. 190. On p. 79, he drew a horizontal line in pencil under the last line of text and on p. 80 drew curly brackets on either side of the first six lines of text, thus highlighting this passage (Hennig 1966:79–80):

> "In the phylogenetic system the categories at all levels are determined by genetic relations that exist among their subcategories. Knowledge of these relations is a prerequisite for constructing the categories, but the relations exist whether they are recognized or not. Consequently here the morphological characters have a completely different significance than in the logical and morphological systems. They are not themselves ingredients of the definition of the higher categories, but aids used to apprehend the genetic criteria that lie behind them."

Donn drew a squiggly line to the left of the text on p. 190 in which Hennig argued for the superiority of the phylogenetic over the typological system of relationships and classification. In neither case can we know how Donn interpreted these passages or why they drew his attention. That is not true for p. 90 and pp. 155 and 157.

3.3 ROSEN'S ERASURES

On p. 90, Hennig (1966) made the argument that only synapomorphies, not symplesiomorphies, diagnose monophyletic groups. In the left-hand column, there is a bold curly bracket, in pencil, that marks the second full paragraph and to the left of it a check mark, perhaps for emphasis or to signify acceptance of the text. Underneath these symbols is an erased comment that we have scanned and enhanced for legibility in Figure 3.1. It reads: "A gross and silly dis-tortion of a well-established and accepted concept" (see Chapter 4 this volume p. 35 for further comment on this erasure).

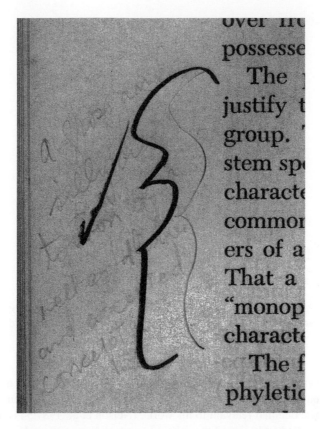

FIGURE 3.1   Portion of p. 90 from Rosen's personal copy of Hennig (1966). The erased text in pencil reads: "A gross and silly dis-tortion of a well-established and accepted concept." The word distortion is hyphenated in the original.

On p. 157, Donn drew curly brackets on either side of the middle paragraph in which Hennig argued in favor of a phylogenetic rather than syncretic system. In the left-hand column, Donn wrote "see p. 155".

On p. 155, Donn drew a bracket to the right of Hennig's (1966) paragraph that reads:

> "In principle there is nothing against determining the absolute rank of taxa on the basis of degree of morphological divergence, provided this is done within the limits set by phylogenetic systematics—that sister groups be coordinate, and thus have the same absolute rank."

To the right of the bracket is a bold check mark over an erased comment that we have scanned and enhanced for legibility in Figure 3.2. Donn had written, then erased, this rejoinder:

> "What about evolutionary rate and new adaptive thresholds?"

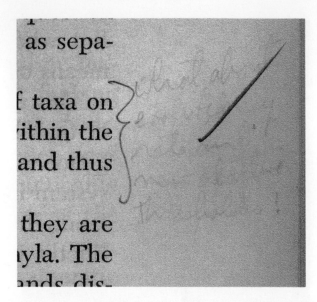

FIGURE 3.2   Portion of p. 155 from Rosen's personal copy of Hennig (1966). The erased text in pencil reads: "What about evolutionary rate and new adaptive thresholds?"

Persuading Donn Rosen to leave evolutionary systematics behind must not have been easy, but the outcome was inevitable: "Donn always strove to be current in all areas of theory that he deemed relevant to his work" (Nelson et al. 1987:544). In less than a decade, Rosen had fully joined the debate in systematics on the side of the cladists (e.g., Rosen 1974).

3.4   END NOTE

The title of our essay, Rosen Decomposing, suggests an old joke or cartoon of uncertain attribution that was one of Donn's favorites. In one version of the cartoon, German composer Ludwig von Beethoven has died and is sitting up in his grave erasing his musical notation. The caption reads "Beethoven decomposing."

ACKNOWLEDGMENTS

We thank Gary Nelson for highlighting Colin Patterson's comments on Rosen's erasures in his presentation at the JMIH symposium in Reno, Nevada, July 2015.

REFERENCES

de Carvalho, M. R. and M. T. Craig (Eds). 2011. Morphological and molecular approaches to the phylogeny of fishes: Integration or conflict? *Zootaxa* 2946:1–142.
Hennig, W. 1950. *Grundzüge einer Theorie der phylogenetischen Systematik*. Deutscher Zentralverlag, Berlin.

Hennig, W. 1966. *Phylogenetic systematics*. University of Illinois Press, Urbana.

Nelson, G., J. W. Atz, K. D. Kallman and C. L. Smith. 1987. Donn Eric Rosen 1929–1986. *Copeia* 1987:541–547.

Patterson, C. 2011. Adventures in the fish trade. 29th Annual Lecture, Systematics Association presented December 6, 1995. *Zootaxa* 2946:118–136. [Edited and with an introduction by D. M. Williams and A. C. Gill.]

Patterson, C. and D. E. Rosen. 1977. Review of ichthyodectiform and other Mesozoic teleost fishes and the theory and practice of classifying fossils. *Bulletin of the American Museum of Natural History* 158:81–172.

Rosen, D. E. 1974. Cladism or gradism?: A reply to Ernst Mayr. *Systematic Zoology* 23:446–451.

Rosen, D. E. and C. Patterson. 1969. The structure and relationships of the paracanthopterygian fishes. *Bulletin of the American Museum of Natural History* 141:357–474.

4 Donn E. Rosen, Skepticism, and Evolution

Brian I. Crother
Southeastern Louisiana University

CONTENTS

Two men say they're Jesus, one of them must be wrong.

"Industrial Disease," Dire Straits

4.1 INTRODUCTION

Making a list of assumptions that *inhibit* scientific progress (Rosen 2016), especially when the assumptions include beliefs long held and cherished by many in comparative biology, identified Donn Rosen as a skeptic. Skepticism has a long, storied, and varied history, which I will not recount, but I can probably be safe to define skepticism as a philosophical approach that poses questions about the nature of evidence in terms of sufficiency or probity that may be presented to justify propositions. The kind of doubts, the kind of sufficiency, the kind of evidence, and the kind of justification associated with schools of skepticism vary widely, but suffice to say the general feel of the definition covers most forms. Thus, the simple act of questioning existing dogma or knowledge ties the schools of skepticism together. The goal of this chapter is to review and reveal Rosen the skeptic, specifically with regard to his list of inhibiting assumptions. And revealing it is, with clear instances of growth of thought as his skeptical side developed.

I propose no new thoughts or angles of skepticism and how these may relate to or were hinted by Rosen. My own perspective that all knowledge claims are tentative, that there is no capital T in truth, can be traced back to the Hellenistic period, to the Platonic Academy, where the skeptics declared that no knowledge was possible. My view is not quite that harsh though, and is softened by Humean thought which says that some things can be known such as egocentric knowledge (knowledge peculiar to the individual such as thoughts, feelings, and perceptions), and logically deduced truths. However, knowledge about everything else cannot be claimed with certainty (e.g., Watkins 1984). Therefore at best our knowledge claims are tentative propositions, and no more. While Truth in knowledge is unattainable, I argue that some claims are closer to Truth than are other claims, but not in an inductive sense. Least rejected and more highly corroborated claims are to be more highly regarded than more often rejected claims (e.g. Popper 1959). My skepticism about knowledge is a blend of the strict Academic school, coupled with Humean sensibilities, and the more optimistic metaphysical view of Popper, which was part of his answer to Hume's skepticism on knowledge.

Given Rosen's attraction to Popperian thought (Platnick and Rosen 1987) coupled with his obvious skeptic streak, I offer that Rosen's own skepticism was similar to mine. Rosen was not a formal skeptic in that he did not devise approaches or arguments to detect when evidences or beliefs did not support knowledge claims. I argue he was a skeptic in that he voiced concerns, doubts, about apparent evidence that supposedly supported particular beliefs, ideas, or dogmas. So his skepticism was based on evidence, on his conclusions that observation did not match theory, regardless of how dearly the theory was held by others. What Rosen certainly was not, was a denier, as in the individual who simply denies something without regard to evidence, an "I don't believe in evolution" type of denier. Rosen was thrown into the pattern cladist school that was characterized by Beatty (1982) as being antievolution (Carpenter 1987) and thus possibly considered a denier by some. Brady (1982), among many others (e.g. Patterson 1982; Platnick 1982; Carpenter 1987) rescued pattern cladistics from the misinterpretation of Beatty. Regardless, if we are stuck with the dictum that the evolution of life followed a strictly neo-Darwinian process, then what other ideas are left to explore when inconsistencies crop up? That was Rosen's position as a skeptic; to ask can we do cladistics without a neo-Darwinian covering theory? And so, we see that as his inhibiting assumption 2 under Evolutionary Theory: Random mutation, natural selection, and microevolution combine as a progressive research program to explain the hierarchy of organisms and inhibiting assumption 6 under Taxonomy—A knowledge of evolutionary processes and genetics is necessary to do taxonomy. He was a skeptic indeed.

4.2   THE ASSUMPTIONS IN ROSEN'S WRITINGS

I was interested to see where, in his writings, the assumptions (Rosen 2016) first appeared and how they were used. While analyzing Rosen's works, I found many, many examples of where assumptions in the list appeared in his writing. I easily found 27 of the 32 assumptions but I will not share every example. There are simply too many. What I share, I hope will be interesting and instructive. Rosen learned from his skepticism, and hopefully the reader can too.

A. General
1. *Ultimate causes are knowable.*

In the search for explanations regarding features of organisms, biologists are often not satisfied with *what or how did it happen* answers, but seek to extend to *why did it happen* answers. For example, we may know what happens to make snakes limbless, but why it happened is purely a matter of speculation and that is what

Rosen addressed here. This inhibiting assumption was well-developed in Rosen (1982:270–271) over a couple of dense pages. There is a lot there (including a whiff of General inhibiting assumption 3—The *ceteris paribus* is generally applicable and helpful), and I have removed some of it, but retain the context and flavor. The final sentence, besides being a clever play on the words proximate and ultimate, is certainly an example of General inhibiting assumption 4—scientists are more objective than other people:

> "Two kinds of statements have been made about the origin of this structure that have to do with proximate and ultimate causes. The proximate cause is the ontogenetic migration forward of the anterior dorsal fin spines onto the head ... Not long ago Lewontin (1978) admitted the futility of searching for the ultimate causes, these 'why did it happen' questions, but pleaded nonetheless that some things cry out for explanation: '... it must be feasible to make adaptive arguments ... [and] this in turn means that in nature the ceteris paribus assumption must be workable.' These things that cry out in the night, can be given plausible, or at least non-mysterious and therefore soothing, causal interpretations. *What* happens to the anglerfish dorsal spine during ontogeny can, and has been, described in some detail, but plausible, soothing stories about why it happened only serve to conceal that the singular historical process, this mystery, cannot be justified empirically ... I imagine that the empirical limitations imposed by systematists will continue to go unnoticed in the evolutionism of some functional anatomists. If so, what will constrain their explanations of how and why? There seem to be only two limiting factors: proximately, their own inventiveness, and ultimately, the gullibility of their audience."

2. *Bridge principles are useful.*

Rosen (1984) railed against neo-Darwinian explanations for the diversity and hierarchy of life. He decried the use of bridge principles as a way to prop up what he considered a faulty evolutionary explanation. The following quotes are quite explicit about his thinking on this inhibiting assumption.

First (Rosen 1984:91):

> "One might conclude from the history of the neo-Darwinian period that a good reason for research in population genetics is to enlarge our understanding of the genetics of populations, not to discover a mechanism that drives evolution by the gradual accumulation of micro-variations that add up to macro-differences (speciation). Nevertheless, it was the absence of evidence to connect changes in gene frequencies with speciation that seemed to reinforce the continued use of some version of Darwin's theory. 'Adaptation by natural selection' is the bridge principle used to connect the two. The use of bridge principles is a technique for combining in a coherent theory two or more kinds of data for which an empirical connection has never been demonstrated and, in this case, is undemonstrable because of an immunizing stricture that evolution by this principle is too slow to observe. The embarrassments of this rationalization were eliminated by a stroke of the pen (Michod, 1981, pp. 3, 4, and 20): 'gene frequency changes *are* evolution'! A bridge principle is what has been called a canonical statement (one constructed to agree perfectly with known facts). Since it is *made* to agree with the data, and since any number of other canonical statements are possible, a given bridge principle must be accepted as true axiomatically, i.e. cannot be questioned by those who play by the rules."

Rosen (1984:96) ended with the following paragraph, which included mention of General inhibiting assumption 5—Redefining the problem can solve it:

> "I conclude that something new and constructive can be learned about the evolution of ontogenies and their genetic control when such theories are constrained by the empirical data of natural order. This conclusion acknowledges the importance of progress to date on the nature of genetic systems, ontogeny, and the genetics of populations as general statements. It also recognizes that progress to date to relate this general biology to the evolution of life as we know it is largely wishful thinking couched in the obscure language of bridge principles and thinly veiled extrapolations. Or it evades the question of how they are related by redefining the problem: 'gene frequency changes *are* evolution'."

3. *The ceteris paribus is generally applicable and helpful.*

Under General Assumption 1, the *ceteris paribus* is mentioned in a quote from Lewontin (1978), but is not directly commented on by Rosen. It is worth speculating on where Rosen may have begun to think about problems with *ceteris paribus*. Brady (1979) published a fascinating paper titled "Natural selection and the criteria by which a theory is judged" and Rosen is thanked in the acknowledgments for suggesting the topic. Brady (1979) harshly criticized the use of the *ceteris paribus*, and one might wonder if Rosen was influential in developing the criticism or if Rosen realized the beauty of Brady's arguments and latched onto them. To understand what Rosen meant in his inhibiting assumption about the *ceteris paribus*, we may turn to quotes from Brady (1979). There is much more on *ceteris paribus* in Brady (1979) than what I quote here (p. 611), but the quotes help us understand why Rosen would be skeptical of the "all things being equal" argument:

> "General principles do not predict specific events in themselves, but only specify a contribution to those events. When a particular application of a principle is made, it must always be accompanied by a *ceteris paribus* assumption, since in any concrete situation other causal parameters could interfere with the result. But when we see that in any particular test the congruence or incongruence of the anticipation with the result may be due to parameters external to our theory, it becomes clear that tests neither prove nor disprove theories."
>
> "Faced with the reality of a false prediction, the researcher must suppose either: (1) that the basic theory has been falsified, or (2) that the *ceteris paribus* clause attendant to the application has been falsified."
>
> "Of course, since the *ceteris paribus* clause of any application is potentially inexhaustible–one could never know how many parameters are actually hiding under its blanket inclusion–if the new prediction also fails we are still not finished with our investigation. We may continue to adjust our hypothesis as long as possible causal parameters occur to us."

6. *Discovering that you were wrong is bad in some sense (never publish until you're convinced you've found the truth).*

This inhibiting assumption perhaps represents the best advice for budding scientists and speaks directly to current problems in systematics in the age of molecular phylogenetics, in which workers publish their molecular phylogenies but are reticent to finish the job by recognizing unique lineages as species. In the following quote, Rosen (1985a:55) is explicitly referring to hypotheses of homology, but the point is easily extrapolated to hypotheses of lineages:

> "But ambiguous or inconsistently distributed characters still plague fish systematics and are used again and again to support some proposed relationship. Having been guilty of the crime myself, I am naturally tolerant of others who commit it, allowing that it is a far better thing to have proposed and been rejected on grounds of nonhomology than never to have proposed at all."

B. Evolutionary Theory
1. *Competition Theory is important and well formulated.*

Darwin (1859) argued that competition was universal across life (p. 115), "all organic beings are exposed to severe competition" and was important to his theory of natural selection to explain the origin of species. One needs only to peruse *The Origin of Species* to realize the strong connection Darwin made between his process of natural selection and competition. Nowhere is he more direct than here (p. 445), "As natural selection acts by competition ..." This became standard thinking of adherents of the neo-Darwinian process.

Rosen's inhibiting assumption about competition was based on his skepticism that competition could account for the diversity of life. Because of the close association between competition and natural selection, his skepticism spilled over into the next inhibiting assumption in evolutionary theory (see below) that argued for the emptiness of neo-Darwinian theory as explanation for the diversity and hierarchy of life. Donnelly (2016), expounds on the issues with competition; Rosen provided choice quotes for this assumption. There is a hint, ever so slight, that Rosen was not always a skeptic about the power of competition. In their paper on adaptive levels and evolution of actinopterygian feeding mechanisms, Schaeffer and Rosen (1961:197) wrote:

> "There was thus a lag in the teleost radiation throughout the Jurassic, perhaps related to holostean competition."

The implication was that competition drove the evolution of teleosts, which subsequently with the development of new features, was able to out-compete the holosteans and thus led to increased rates of speciation. Rosen clearly changed his thoughts on the ability of competition to drive speciation. For example, from Rosen and Buth (1980:304) we have:

> "That is, competition theory, based on observations of phenotype frequencies, may refer to a trivial aspect of organic diversity—the noise superimposed upon the natural hierarchy of organisms as represented by a hierarchy of epigenetic programs."

and

> "Competition is not an issue."

and

> "Lamarck, therefore, might, in one sense, have been closer to the bone than Darwin,
> for, if the theory is correct, the environment is primarily influential in 1) promoting
> variability and 2) releasing previously hidden, or changing the expression of, geno-
> types, whereas the idea of environmentally mandated competition could only be
> viewed as influential in bringing the process of change to a stop (i.e., reducing and
> stabilizing phenotype or genotype frequencies)."

2. *Random mutation, natural selection, and microevolution combine as a
progressive research program to explain the hierarchy of organisms.*

As noted above, this inhibiting assumption is tied closely to the importance of competition to explain species diversity. This is perhaps Rosen's boldest skeptical position, explicitly attacking the neo-Darwinian paradigm so dearly held by the vast
majority of biologists. As noted in the introduction, Rosen was not a denier, and
his skepticism of the neo-Darwinian explanation for the diversity of life was based
on his view that observation did not match theory. His skepticism was not empty.
He proposed alternative ideas to explain diversity and the hierarchy of life (e.g. see
Evolutionary Theory assumptions 10 and 11).

Rosen did not always take this skeptical position, instead his writings exhibit clear
growth of thought. For example (Hinegardner and Rosen 1972:632):

> "However, it is not necessarily high DNA content that leads to extinction or to the rela-
> tive paucity of primitive generalized organisms in large and diversified groups, but the
> process of natural selection itself that favors new adaptations and, by definition, the
> relatively more specialized forms."

And especially:

> "A combination of mutation, recombination, and natural selection is probably the basis
> for speciation."

Rosen's views of neo-Darwinism shifted dramatically between 1972 and 1978, when
he published a review of a volume on natural selection (Rosen 1978a). There he
laid out his opposition to neo-Darwinism, especially natural selection, as a creative
mechanism to explain the diversity of life. He posed the question (p. 371):

> "If the concepts of population genetics fail to explain the evolution of the world's
> diverse taxa, and Darwin's original idea lacks the necessary independent enabling
> criteria of a testable theory, how can one account for the persistence of the argument
> about natural selection as the driving force behind evolutionary history?"

This 1978 paper was as strong a damning publication against neo-Darwinism as any
he wrote. Rosen and Buth (1980:307) pushed beyond neo-Darwinian explanations
and in the process condemned those explanations:

> "But with regard to mechanism, a half-century of neo-Darwinism has focused only
> on small-scale changes in phenotype frequency in a population and has contributed no
> evidence about how there came to be 'horses and tigers and things.'"

In the introduction of the book *Vicariance Biogeography: A Critique* (Nelson and Rosen 1981), Rosen (1981:1–2) continued his assault on Darwinian explanation:

"These axioms, concerned with competition, fitness, and adaptation, are considered by present-day evolutionists–as they were by Darwin over a century ago–a collective explanation of organic change through time and the observed diversity of the modern world."

"During the century of their application to the organic world, these axioms have taken on a life of their own, and the study of organisms in space and time has been richly adorned with Darwinian fabric–the emperor's clothes."

And further, from Rosen's "Do current theories of evolution satisfy the basic requirements of explanation?" paper (1982:83–84):

"Selection theories succeed only in explaining why some potentially viable developmental programs might not survive. They include no predictions about patterns that agree with the hierarchical patterns of life."

"So what we are left with after more than a century of conceptual struggle are three kinds of theories to explain the diversity of life. The Darwinian-neodarwinian kind has turned out to be a theory of ecological interaction whose parameters are sometimes behavioral, populational, genetic, or fanciful."

"The relevance of neodarwinism to either phylogeny or ontogeny, in being founded on notions of genetic and ecological chance, appears to me to be simply undecipherable."

As a final example, here Rosen (1984:90) kept his foot on the neck of neo-Darwinism:

"Sensitive to criticisms that population genetics has made no contributions to understanding causal mechanisms underlying the structure of life's hierarchy or even the origin of its humblest species Michod insists nonetheless that all is well with neo-Darwinism, that the empirically empty concept of fitness (the fit are those that survive; those that survive are fit) can be 'fixed' by adopting an ecological bridge principle (i.e. an ecological extrapolation) about ecological constraints that are independent of survival but for which there are at present no data and no known way of acquiring them."

3. *Pure paleontology provides essential information about the history of life.*

Rosen et al. (1981) addressed the relationships of lungfishes and tetrapods and in the process exposed the perils and problems of paleontology and plesiomorphy (the latter is addressed in an assumption 4 under Taxonomy). The point is made quite strongly in this paper and in the following quote that reliance on pure paleontology significantly inhibited the authors' own progress in comparative biology. The frustration is palpable and jumps from the quote. As Rosen et al. (1981:264) put it:

"We record our own fascination with these likenesses for a particular reason–the conviction that, had the study of lungfish relationships developed directly from comparisons among living gnathostomes without interruption by futile paleontological searches for ancestors, there would have been no "rhipidistian barrier" to break through, this review would have been written decades ago, and we would all have been spared a generation of anatomical misconstruction, dispensable scenarios about the fish-amphibian transition, and hapless appeals to plesiomorphy. Finally, it should be obvious from the nature of this review that our argument is not with the use of fossils, but with the use of the traditional paleontological method."

5. *It is important to search for ancestors and/or to formulate archetypes.*

One of the major changes that occurred with the Hennigian revolution was the recognition that ancestors were not empirically discoverable. Internal nodes of clades became HTUs, hypothetical taxonomic units, and with that, these once sought after ancestors became interesting points of speculation, but nothing more. All taxa, extinct and extant, were included as terminal taxa in the analyses and thus the search for sister relationships took over the search for ancestors. Patterson and Rosen (1977) explain some of the problems associated with the search for ancestors, including the notion that although ancestry and descent are assumed in cladogram inference, one cannot then claim they have discovered ancestors from the cladogram (1977:154):

> "We find it hard to believe that many of our colleagues would entertain this tautology as a sufficient basis for their work."

These quotes (Patterson and Rosen 1977:154) further explain their dissatisfaction with the importance of searching for ancestors:

> "The theoretical difficulties in classifying fossils with extant organisms that have been pointed out by Hennig ... among others, arise from two sources: the impossibility of classifying ancestral species in the same hierarchy as their descendants (since the ancestral species is equivalent to the higher taxon containing its descendants), and the impossibility of ranking ancient fossil species or groups in a Linnaean hierarchy in which absolute age is the ranking criterion (where categories are age classes, in Griffiths's terminology). In our opinion, both these theoretical problems are circumvented by treating fossil species as terminal taxa, i.e., as sister-groups of other taxa."
>
> "Concerning only the practice of identifying ancestors, as distinct from the theoretical justification for such an undertaking, it is evident, therefore, that looking for ancestors involves a secondary, and negative, concept, i.e., after finding a taxon to have membership in a more inclusive taxon its ancestral status is decided secondarily on the basis of what it does not have. Hence, any species found to be more primitive with respect to all known or analyzable characters than other members of its group is a priori ancestral to the other included members of its group. Even allowing this procedure, with its inevitable consequence, one may ask: What is learned by discovering that a taxon, assigned to a group on the basis of one or more synapomorphies with all members of the group, shares no other apomorphies with any part of the group? The answer manifestly is that the taxon deserves membership in the group and that its relationship to any part of the group is equivalent to its relationship to the whole group, i.e., it is not more closely related to one than to any other member. That is a statement of cladistic relationship which cannot distinguish whether the taxon in question shared a common ancestor with the other members of the group or was ancestral to them. How, then, does one go from such statements to conclusions about ancestry? This question has never been answered, and little wonder, for the fact is, in phylogenetic investigation, we are simply assuming the appropriateness of the concept of ancestry and descent (or genealogy) as an axiom for the interpretation of a pattern of character distributions."

7. *Fossils specify the age of their including taxon.*

"When we see a species first appearing in the middle of any formation, it would be rash in the extreme to infer that it had not elsewhere previously existed." (Darwin 1859:302).

Given Rosen's skepticism with Darwinian/neo-Darwinian explanations of diversity, inclusion of the above quote from Darwin may seem ironic. Perhaps so, but I cite it to demonstrate the intuitiveness of this inhibiting assumption. It is clearly puzzling to Rosen that somehow biologists overlook the simple logic that fossils only give minimum ages as well as the implications for ignoring this logic.

From Rosen's (1975:458) Caribbean vicariance paper:

"So accustomed are we to interpreting the occurrence of Pleistocene fossils (which are relatively abundant in the Caribbean region) as indicating that the regional history is tied to Pleistocene events, that we have forgotten that fossils give us a minimum rather than a maximum age for the groups of which they are members."

Patterson and Rosen (1977:154) wrote:

"It also entails arguments consistent with the ideas that specimen age cannot settle questions about age of the taxa to which fossils are assigned and that specimen age can never falsify theories of relationship based on biological evidence."

And from Rosen's (1985b:636) exquisite geological hierarchies paper:

"Other biologists (Darlington, 1965; McDowall, 1971; Briggs, 1984) tried and still try to rescue the past by agreeing that the geography did in fact move but that the timing of these great events was wrong in relation to the ages of the biotas. Such attitudes might invoke the ages of fossils to show that all the taxa are too young to have been influenced by the geographic cataclysms. This view involves two assumptions, both wrong at some level: 1) that fossils can tell us how old a taxon is and 2) that the ages of the geologic events have been correctly assigned. The first assumption is wrong because fossils give a minimum rather than maximum age of a taxon, and the second assumption is put into question by recent age reassignments."

8. *Adaptation scenarios have important general explanatory power.*

In the 1985 Miami course, Rosen conducted an exercise on adaptation, and it's an exercise I still use in some of my courses. Across the top of a long chalkboard, he listed flatfish asymmetry, long legs of giraffes, poor eyesight of rhinoceroses, snout asymmetry in freshwater porpoises, and the coiled/uncoiled shell morphology variation in oyster drill mollusks. He then asked us to consider the adaptive advantages of each one. We constructed long lists of possibilities for each one and then Rosen asked us which one was correct for each example. We were flummoxed by the question, because it seemed simple. You just test them one at a time until you can't falsify one. Discussion ensued and it became clear that the study of adaptation has the problem of uncertainty associated with it. In fact, why must a feature, any feature, be an adaptation at all (e.g., Gould and Lewontin 1979)? Of course, if the adaptive explanations don't work out, one could always resort to modifications of the explanations by invoking another variable, which cleverly brought us right back to the *ceteris paribus* (see inhibiting assumption 3 under General). It was a terrific lesson that made me a skeptic of adaptive explanations for features of organisms.

Rosen did not always think this way. As with neo-Darwinian explanations for biological diversity, Rosen first accepted the dogma, then grew skeptical, and challenged it with vigor. Early on, he clearly pondered adaptive explanations. Schaeffer and Rosen (1961:187, 193, 197, respectively), wrote:

> "As this biting mechanism was modified and perfected through time, the potentiality for adaptive radiation in the entire feeding mechanism increased."
> "The adaptive significance of the coronoid process, ... is perhaps best explained in terms of torque."
> "The adaptive modification of the premaxilla created a mechanical problem for the movable maxilla."

Rosen and Bailey (1963:20–21) titled a section: "Adaptive Significance of Gonopodial Structures" and this was not with regard to skepticism, but explanation. In summarizing the work Rosen and Bailey (1963:151) wrote:

> "It is suggested that what matters from the view point of mechanical control of the gonopodium is not the embryological origin of the different suspensorial structures, or even the precise point at which these structures arise, but the total adaptation for the job of suspension."

And

> "The view expressed by some authors that the specialized terminal structures of the gonopodium are of no adaptive significance is contested."

Sometime after 1963 Rosen seems to have begun to wrestle with the problems of adaptive scenarios because the word "adaptation" and variants of it vanish from his writings (at least from what I read) until 1974, where it starts to come under attack.

Rosen (1974a) is a significant attack on evolutionary taxonomy and gradism, and much of the paper concerns criticisms of adaptive zones, adaptive peaks, and other such criteria for recognizing the elusive grades. However, he does take a moment to go after the identification of adaptations (1974a:449):

> "Operationally here, as in trying to estimate genetic content, adaptive novelties are identified by evolutionary taxonomists with some amalgam of taste, intuition, experience, and tradition."

In their work in which they reimagined the origin of diversity, Rosen and Buth (1980:305) threw adaptation into the mix:

> "The importance of these observations is that they are examples of what evolutionists have called adaptations but have been caused by the environment acting on adaptable

epigenetic systems. When such "convergent" adaptations are considered as a whole, it becomes clear that the environment is the constant and the different genealogical histories of the organisms are the variables. It is hardly surprising, therefore, that Darwin invented "selection of random variations," a notion to explain adaptation that is rendered unnecessary ..."

In a treatment of evolutionary novelties through the philosophy of Popper, Platnick and Rosen (1987:16) mused:

"Perhaps it is not selection but the ability of epigenetic systems (genotypes plus ontogeny) to solve problems posed by new environments (and to create new environments, with new problems) that generates major evolutionary novelties."

The quote from Platnick and Rosen (1987) refers to the origin of things we would call adaptations, the things we would then concoct scenarios about. It is easy to see that an adaptation would be considered by some as a solution to a problem posed by the environment. That would be a standard neo-Darwinian interpretation. But Platnick and Rosen were getting to the alternative, what if selection is not the key?

I am not sure if Rosen read the works of Gabriel Dover, or vice versa, who also presented alternative ideas to neo-Darwinian explanation of diversity, but Platnick and Rosen's (1987) quote is similar in thought to something from Dover (1986:164): "This type of exploitation and the reasons behind it can be called 'adoption,' in contrast to the Darwinian process of adaptation. In the latter, the environment technically sets the 'problem' and the 'solution' is provided by selection from a subset of existing variant individuals. In the former, the 'problem' is posed by the internally driven changes in a population of phenotypes with the environment providing a 'solution'".

The compelling questions these skeptics asked about neo-Darwinian adaptation were: Is the feature a solution to a problem posed by the environment, or is the environment a solution to a problem posed by the feature? The symmetry is beautiful and the vectors of the explanations are opposite, which reveal a significant gap in neo-Darwinian explanation.

9. The Biological Species concept is an essential concept.

As Rosen developed his ideas on cladistics and vicariance biogeography, he ran into a conceptual problem with the biological species concept. How can the simple ability to interbreed dictate taxa, especially in the face of conflicting evidence from inferred phylogenies and distributions? The quotes below show how he handled this problem, ultimately rejecting the concept and in the process, subspecies as well.

Rosen (1978b:17) recognized the problem:

"To the extent that biogeography searches for patterns and their historical explanation, the choice of a species concept by the practicing biogeographer is crucially important. This is so only because conventional taxonomic practice regards the species as a fundamental evolutionary unit. A concept such as the biological species, however, appears to be inapplicable to observable nature, because it incorporates criteria that are generally undiscoverable. It requires that we identify noninterbreeding sympatric units (I have discussed the reproductive criterion elsewhere: Rosen, in press). But in practice what are these units—crabs and fishes, lions and zebras?"

Rosen (1979:276–277) solved his conceptual dilemma:

"Within the framework of the 'biological species' concept, zones of secondary inter-gradation (hybridization) in nature between recognizably different natural populations have been taken by some taxonomists as prima facie evidence that the two popula-tions represent only a single species (Mayr, 1969, p.195).... To so argue, however, requires that reproductive compatibility is evidence of relationship-and, moreover, evidence that transcends all other criteria of relationship in its biological importance. But, within the history of any lineage, reproductive compatibility is an attribute of the members of the ancestral species of that lineage, an attribute which is gradually dimin-ished and ultimately lost in its descendants during geographic differentiation. In other words, reproductive compatibility is a primitive attribute for the members of a lineage and has, therefore, no power to specify relationship within a genealogical framework."

"The only other such criterion of which I am aware is the potentially nonarbitrary reproductive property of 'biological species' in nature, the search for which, however, is logically flawed (Rosen, 1978) and which implies, as an underlying premise, the use of a primitive character to specify relationships. I am, thus, compelled to reject both the 'biological species' as a conceptual tool and the 'subspecies' as a methodological one, and this argument constitutes my reason for now recognizing as species forms that were hitherto recognized as subspecies."

It wasn't until 1983 that Rosen felt he had finally broken free from the biological species concept (and revealed his fine sense of humor). In the obituary by Nelson et al. (1987:544), they told a story about Rosen with regard to his surgery and the biological species concept:

"After his first surgery in 1983 he proclaimed the operation completely successful in removing from his brain the last vestiges of the biological species concept."

10. *Goldschmidt's ideas must be wrong because they conflict with neo-Darwinism.*
11. *Ditto for Neo-Lamarckism.*

With neo-Darwinian mechanisms not adequate to explain the diversity and hier-archy of life, Rosen looked for alternatives. In Rosen and Buth (1980) he began to espouse epigenetic theories, far before they were popular, to explain his observa-tions. He considered new ideas and evidence on the inheritance of acquired char-acteristics (neo-Lamarckism, IAC) and the idea that change need not be gradual. Goldschmidt (1940) was the early champion of the latter, of rapid phenotypic change, therefore macroevolution and the concept of the hopeful monster (this inhibiting assumption is covered by White, this volume). Waddington (see 1975 for a com-pendium of his works; Rosen introduced Waddington's works to me in 1985) also pushed for alternative mechanisms to explain the diversity of life, using canalization of environmentally induced changes to explain evolution. Gould (see 2002 for his masterful overview) was an advocate of the importance of development in producing radical novel phenotypes, essentially hopeful monsters. All of these epigenetic phenomena that Rosen thought held the secrets to diversity have become accepted by many (e.g. Landman 1991; Raff 1996; Theißen 2009) and recent empirical work has even demonstrated specific mechanisms for how inheritance can be accomplished through epigenesis (e.g. Ashe et al. 2012). Rosen tried to raise the ghosts of Lamarck and Goldschmidt for good reasons, and frankly, I think he was right.

First up is a simple, direct statement (Rosen 1982:82) about Goldschmidt's most famous creation:

"What, in fact, does it take to make a bird? In a real sense, birds, and at least a great many if not most other taxa, are hopeful monsters."

The first quote from Rosen and Buth (1980:304) is a part of a larger quote under competition, but is worth reusing here:

"Lamarck, therefore, might, in one sense, have been closer to the bone than Darwin, for, if the theory is correct, the environment is primarily influential in (1) promoting variability and (2) releasing previously hidden, or changing the expression of, genotypes …"

As noted above, Rosen and Buth (1980:307) argued for epigenetic processes:

"Four independent kinds of observations and theories about genetic and epigenetic systems are highly consistent with one another and are combined to derive an alternative approach to the study of mechanism…" "Our purpose, however, is decidedly not to convince Hairston or anyone else that a new explanation has been found, only that investigation of epigenetic systems is an alternative to the studies in population genetics which have been notably unproductive in their attempts to discover clear links between genotypes, genetic equilibria, 'natural selection' and the origin of taxa."

C. Taxonomy
2. *Cladograms and classifications are decoupled because they express different and equally useful notions.*

Rosen's battle against grade expectations in classifications, grades being the hallmark of the evolutionary school of thought (contra cladistic, phenetic schools) no doubt led to this inhibiting assumption. There is no question that Rosen thought classifications should be derived from cladograms. By 1969, Rosen (Rosen and Patterson 1969) had already concluded that classifications should be based on phylogenetic relationships and monophyly, even though they suggested that grades might be "useful." The notion of the usefulness of grades had vanished five years later with his (Rosen 1974a) "Cladism or Gradism?: A reply to Ernst Mayr" paper, which could be quoted in entirety as a paean to the eventual (it was still years away, Hull 1988) decisive victory of cladistics over gradistics. Rosen and Greenwood (1976) and Patterson and Rosen (1977) further criticized evolutionary taxonomy and perhaps felt by then the problems of grades in evolutionary classifications were well understood because the issue did not reappear.

From Rosen and Patterson (1969:362):

"Before presenting the detailed evidence on the structure and interrelationships of the paracanthopterygians, we should state that our aim is to recognize phylogenetic relationships and to produce a classification containing monophyletic groups, with consequent emphasis on 'vertical' divisions. The concept of 'horizontal' grades or 'levels of organization,' although useful, appears to us to be a preliminary and transient stage in deciphering the relationships of organisms– the recognition of grades is essentially the recognition that the representation in those levels of organization is or may be polyphyletic. In the phylogenetic arrangements we have attempted to emphasize groups united by shared specialized characters, and to avoid definition of groups characterized by shared primitive features. The classification of the acanthopterygian fishes as it stands at present is an example of the ultimate futility of using assemblages of primitive characters to express relationship, with large, catch-all basal groups in effect defined arbitrarily by specializations that they do not have (Percoidei), leaving many derived and specialized groups of high rank and of uncertain relationships with one another and with the basal groups."

Even though the entire Cladism or Gradism paper concerns this inhibiting assumption, I present only a single, explicit statement (Rosen 1974a:451):

"It seems clear enough that, if the cladistic method is accepted as an operational tool, and not encumbered with the insubstantial ideas of estimated genetic content and degrees of divergence, the only recourse in the construction of classifications is the direct translation of the cladogram into a cladistically ranked and ordered hierarchy."

In a review of synbranchid eel classification, Rosen and Greenwood (1976:6) criticized evolutionary taxonomy:

"The few striking specializations uncovered by anatomical investigations have been applied according to that uncritical dictum of evolutionary taxonomy that hierarchically separates species with respect to the magnitude of their differences but without regard to their genealogical relationships."

Finally, Patterson and Rosen (1977:162) appear to give evolutionary taxonomy a compliment, only to snatch it away by calling the information inserted into such a classification as irretrievable:

"It is evident to us and to those familiar with the controversy between cladal versus gradal schools of phylogenetic reconstruction and classification that the various questions raised above are problems of real concern only to the precise requirements of the cladistic method. Traditional evolutionary classifications, in contrast, are capable of absorbing all manner of grade categories, such as the 'Pholidophoridae' and 'Leptolepididae,' multiple furcations as final statements of relationship, ancestor-descendent relationships (as opposed to sister-group relationships), and nonphyletic measures such as 'degrees of adaptational divergence' and 'adaptive zones' as valid representations of evolutionary history. Although evolutionary classifications are simpler to construct, because they depend so heavily on authoritative judgments and opinions, little of the informational input is retrievable from the formal hierarchy."

3. *Deliberately formulated paraphyletic groups are analytically useful.*

If Rosen were a comedian, he might have quipped, "Take my paraphyletic groups, please!" (with apologies to Henny Youngman). That sums up what he thought about the usefulness of paraphyletic groups. In "Vicariance Models and Caribbean Biogeography" (1975:433), Rosen expressed his view of the utility of such groups in biogeographic investigation:

> "For example, both Simpson and Gilmore (see Simpson, 1961) defend the traditional usage of groups that are non-monophyletic—a nonmonophyletic group (e.g., a para-phyletic group) being one that does not include all the descendants of an ancestral taxon. From the standpoint of biogeography such subjective practice (i.e., selectively discarding known data from a system) leads to the formulation of distribution patterns that are incomplete with respect to the occurrence of species or lineages that are not classified with their genealogically closest relatives. Likewise classifications of organisms that are based on the occasional or frequent use of non-monophyletic groups—the stock-in-trade of 'evolutionary' classifications—inhibit general biogeographic studies by making it difficult or impossible for workers in one discipline to know which if any of the hierarchical categories of another discipline are monophyletic."

And Rosen et al. (1981:178) noted dryly:

> "Yet the sequences consist of nothing more than abstractions from paraphyletic groups such as rhipidistians, osteolepiforms, and labyrinthodonts." "What use can these abstractions have in analysis?"

4. *Primitive characters are informative.*

It is difficult to tell when Rosen altered his thinking on characters, but based on Rosen and Parenti's (1981) discussion of Rosen (1964), in which Rosen did not explicitly differentiate between apomorphies and plesiomorphies, and Rosen and Patterson (1969), where they were explicit, I offer the speculation that in between 1964 and 1969 he read Hennig (1966). As support of my speculation, see Parenti and Crother (2016). They present a scanned erasure (their Figure 1) from Rosen's personal copy of Hennig (1966:90). A paragraph is bracketed and the erased note says, "a gross and silly distortion of a well established and accepted concept." The bracketed paragraph is where Hennig (1966:90) initiated his arguments to object to the use of primitive characters to define monophyletic groups, "The possession of plesiomorphous characters (symplesiomorphy) does not justify the conclusion that the bearers of these characters form a monophyletic group." Even stronger evidence that Rosen read Hennig between 1966 and 1969 comes from Patterson (2011; which renders my speculation moot) in which he relates a story asking Donn to bring a copy of Hennig to London because there was not a copy to be found. I'll let Patterson (2011:124) tell the tale:

> "Donn brought the book, but read it first and scribbled all over it in pencil—things like 'nonsense' and 'a gross misunderstanding of well-known principles.' He thought the book was rubbish; I read it and thought the opposite. So while Donn and I spent a

couple of years working on paracanthopts, I was a committed Hennigian and he wasn't. But Gary Nelson joined Donn Rosen in New York in October 1967, and began the campaign of argument and persuasion that eventually turned the American Museum of Natural History into the world's leading institute of systematics, or a hotbed of crazy cladists, depending on your point of view. You can find the detail of Gary's campaign in David Hull's book Science as a Process (Hull 1988). Donn Rosen soon came round, to become one of the leaders in cladistics, particularly in developing vicariance biogeography during the late 1970s."

G. Nelson said (pers. comm.) that when Patterson later visited New York he noted that Rosen's penciled comments in the book had been erased (Fig. 1). A coincidence? I think not. After Hennig (1966), everything changed. For example, from Rosen (1964:257), we have:

"If one reasonably assumes that the great similarity of the more generalized members of each group is a measure of their common heritage ..."

whereas Rosen and Patterson (1969:362) explained:

"In the phylogenetic arrangements we have attempted to emphasize groups united by shared specialized characters, and to avoid definition of groups characterized by shared primitive features."

Here and elsewhere Rosen denounced the utility of primitive characters in phylogenetic inference, including Rosen (1974a,b), Patterson and Rosen (1977), and Rosen and Parenti (1981). Below are two examples from Rosen et al. (1981:163, 264, respectively):

"We attribute the century of confusion about the structure and position of lungfishes to the traditional paleontological preoccupation with the search for ancestors, to the interpretation of *Eusthenopteron* in the light of tetrapods and the reciprocal interpretation of fossil amphibians in the light of *Eusthenopteron*, and to the paleontological predilection for using plesiomorphous characters to formulate schemes of relationships."

And:

"We record our own fascination with these likenesses for a particular reason-the conviction that, had the study of lungfish relationships developed directly from comparisons among living gnathostomes without interruption by futile paleontological searches for ancestors, there would have been no 'rhipidistian barrier' to break through, this review would have been written decades ago, and we would all have been spared a generation of anatomical misconstruction, dispensable scenarios about the fish-amphibian transition, and hapless appeals to plesiomorphy."

6. *A knowledge of evolutionary processes and genetics is necessary to do taxonomy.*

This inhibiting assumption takes us back to the final paragraph of the introduction, the idea of pattern cladistics and the notion that evolutionary process did not need to be invoked to infer phylogeny and the classification that derived from it. At a minimum, and maybe maximum, for that matter, which Patterson and Rosen (1977:155) noted, is the necessary assumption of descent with modification, or as they put it "concept of ancestry and descent." Some would still agree with this assumption as necessary and sufficient, but many others, especially in the stastico-phylogenetics schools, see the necessity of inserting process in the form of models in the inference of phylogeny (I presume taxonomy follows from phylogeny). Models are required in Bayesian and likelihood methods, thus some knowledge of process is required. Whereas model selection for DNA bases is reasonable (to a degree), the push to model process for the evolution of morphological characters enters the realm of science fiction, regardless of the model comparison outcomes (e.g., Wright and Hillis 2014). Based on Rosen's writings, I do not think he would have been an advocate of this school of phylogenetic inference.

Rosen's (1974a:450) initial problems with process came with regard to evolutionary taxonomy:

"Many systematists have by now recognized that their open dissatisfaction with the ambiguous and doubtfully useful estimates of genetic content and divergence (morphological gaps) prods evolutionary taxonomists to the supposition that phylogenetic systematics is unconcerned with the processes and consequences of evolution."

Patterson and Rosen (1977:154) gave the necessary and sufficient assumption:

"We are simply assuming the appropriateness of the concept of ancestry and descent (or genealogy) as an axiom for the interpretation of a pattern of character distributions. There is good reason to use this genealogical axiom, because it gives order and direction to comparative biological data and because genetic studies have revealed no other mechanisms for genetic continuity and change (evolution) than that provided by the consequences of ancestry and descent (genealogy)."

Rosen and Buth (1980:307) were succinct and explicit:

"We know that testable cladistic theories about evolutionary relationships (what we call phylogeny) can be made without invoking a specific evolutionary mechanism."

7. *When comparative anatomy and ontogeny are fully informative, they can disagree.*

Rosen pondered the phylogenetic information relationship between comparative anatomy and ontogeny early, as seen here (1964:246):

"It is not now possible to say whether the early developmental differences between the caudal skeletons of mugiloids and those of atherinoids are more important phylogenetically than their final adult similarities, but it seems probable that both are important. The developmental differences may mean, as advocated here, that atherinoids and mugiloids are members of different lineages. The similarities may mean that those lineages are intimately related. This interpretation leads again to the view expressed above that the halfbeaks, killifishes, and silversides are the descendants of a transitional group that stood in the ancestry of the perch-like fishes."

And here (1964:253):

> "Is the addition of this single highly distinctive feature during an intermediate stage of development really indicative of wide phyletic separation from fishes that lack this detail?"

The above quotes were not about character state transformation series information, which is probably what Rosen was referring to in this inhibiting assumption, but nonetheless the information content of ontogeny was on his mind. The theme of this assumption runs through Patterson and Rosen (1977) and appears in explicit form in Rosen (1982) where he became much more specific about his view of the relationship of ontogeny and comparative anatomy. For example (1982:76):

> "I will maintain that the underlying justification for the comparative method should be the expectation that studying patterns of hierarchical character distribution among organisms will enable us indirectly to recover the information about the hierarchy that is provided directly by ontogeny"

And (1982:81):

> "The discovery of such congruence enlarges the data base pertaining to group membership, minimizes the chance of error inherent in single-character comparative anatomical argumentation, and makes an attested assertion about character transformation. The asserted transformation should agree with ontogeny. And ontogeny is taken as an arbiter of such assertions because, empirically, unambiguous character-state reversals during development (types 1I-VI, above) are unknown. That is the essence of von Baer's Law."

The latter quote leaves little doubt about Rosen's position on the superiority of ontogenetic data in the inference of transformation series. As explicit as Rosen was in 1982, a later paper, "Hierarchies and History" (Rosen 1984), pushed the power of ontogeny with even more ardor. Consider the opening paragraph:

> "Ontogeny teaches that each organism is ordered hierarchically during its transformation from zygote to adult. A comparison of different ontogenies teaches that the differences between organisms are part of this hierarchical order. Hierarchical classifications of all organisms can be understood as partial summaries or estimates of ontogenetic histories. Phylogeny is the concept that these hierarchical classifications reflect diversification through time of an underlying, or ancestral, ontogenetic pattern. The intertwined concepts of natural order, hierarchy, and transformation have direct empirical ties to the ontogenetic process."

And Rosen simply got more emphatic throughout the rest of the paper (1984:78):

> "In all these investigations ontogeny has had a special status because it can reject hierarchical theories derived from comparative anatomy, but the reverse is not true. Ontogeny can serve as an arbiter of what is noise and what is signal."

Rosen 1982 and 1984 leave little doubt from where this assumption came from.

D. Biogeography
1. *All distribution patterns result from vicariance.*
2. *All distribution patterns result from dispersal.*

I treat these two inhibiting assumptions together because they represent two sides of the same coin. There really are only two general natural explanations for the distribution of life, fragmentation of a continuous biota, or the active movement of biota across space. The specific explanations depend on the taxa, the time, and the space. Before the acceptance of plate tectonic theory and before the English translation of Hennig (1966), Rosen gave typical dispersalist explanations, such as the dispersal scenario for West Indian poeciliids in Rosen and Bailey (1963:146–147):

> "If zoogeographic significance may be deduced from the endemism and distinctiveness of the West Indian poeciliids it is that, although powerfully held by permanent land, their colonization of the islands was probably gradual and via the adventitious means available to waifs."

This is not, by any means, an indictment of Rosen, because such was the standard explanation at the time. It was rare indeed to find pretectonic explanations for the distribution of island biotas that did not include random over water dispersal events (this may seem ridiculous given that Rosen worked on fishes, but there is a larger point). The skeptics (at the turn of the twentieth Century, Thomas Barbour [1914, 1915] was perhaps the most vocal, Pregill and Crother 1999) argued for concordant overland dispersal to explain the West Indian terrestrial biota, and in essence presaged the tectonic model contributions. Croizat was the main skeptic some 50 years later, but Rosen was not yet there.

Rosen (1972) provided a dispersal explanation for the Central American characid genus *Bramocharax*, replete with a figure with directional arrows, but, significantly, the explanation was constrained by phylogeny and this indicated a change in Rosen's thinking. Two years later, there was a revolution in biogeographic explanation with the publication of "Centers of Origin and Related Concepts," by Leon Croizat, Gareth Nelson, and Donn Rosen (1974). It is clearly here that the two inhibiting assumptions were derived. One needs to look no further than the abstract to find:

> "We admit the reality of dispersal and specify how examples of dispersal may be recognized with reference both to sympatry and to generalized tracks, but we suggest that on a global basis the general features of modern biotic distribution have been determined by subdivision of ancestral biotas in response to changing geography."

And it is clear that while dispersal was recognized as a legitimate process, the explanatory power of dispersal was considered weak. Rosen (1975:446) put it succinctly:

> "Of course dispersals occur, we know they do, but they are not at issue."

Or, as Rosen (1985b:656) later wrote:

> "There is a message here for all interested in biogeography and it is *not* that dispersal never occurs, but that theories of dispersal to explain biotic complexity are no more

informative than theories of relationship based on symplesiomorphy, which wide-spread dispersed taxa resemble in biogeography."

3. *Centers of origin can be found.*

Nelson (1983) argued that biogeographic explanation was a continuous battle between creation myth and science because creation (at least specific creation locations like the Garden of Eden, or perhaps Mt. Ararat, the landing spot of Noah's Ark) required a single explanation for the distribution of life: dispersal. If the biota was created or came from a single location, how else could it reach across the globe other than by moving? Thus, centers of origin are intimately tied to the history of dispersal explanations. Early in his writings, Rosen was not immune to this type of explanation; for example, in Rosen and Bailey (1963:144), we find:

> "If we assume that the center of origin of a tribe is in the region where it is represented by the largest concentration of genera, or at least where the genera seem to converge, the picture is as follows."

Rosen changed his view on the matter. There are many places where Rosen disparages centers of origin, but I will present just a single long quote because it clearly explains why believing that *Centers of Origin Can Be Found* is an inhibiting assumption. From the famous "Centers of Origin and Related Concepts" paper by Croizat et al. (1974:273):

> "The relation between centers of origin and the distribution of 'primitive' and 'advanced' taxa may, likewise, lead to conflict ... Some zoogeographers would assume that one species or group, e.g., the species (*G. russus*) in the center of the assemblage, is relatively more primitive (or 'plesiomorphous') than the remaining three, and that it should be assumed to indicate, or to occupy, the center of origin of the group as a whole; these zoogeographers assume that relatively primitive species are generally less apt to disperse than their relatively advanced (or 'apomorphic') relatives. Other zoogeographers would assume that one species or group, e.g., the centrally located species, is advanced and that it indicates, or occupies, the center of origin of the group as a whole; these zoogeographers assume that relatively advanced species are generally less apt to disperse than their primitive relatives. Other zoogeographers would assume that one species or group, especially if it were fossilized and demonstrably older than its relatives (the 'right fossils in the right places' of Darlington) is actually ancestral to the other members of the group, and therefore reveals directly the center of origin. Still other zoogeographers might approach the problem with different sets of apriorisms. The conflict of opinion resulting from different apriorisms raises the question of the applicability of the concept, and even the existence, of a center of origin as envisioned in these discordant approaches. We would point out that, if a center of origin is imaginary, all of its corollaries are equally imaginary; and by an opportune choice of examples, anyone can 'prove' whatever he wishes to 'prove' about it."

4. *Historical geology tests can reject biological theories of area relationships.*

Rosen broached a significant epistemological problem with this inhibiting assumption. In the pursuit of knowledge, we are often confronted with logically dependent data and logically independent data that may bear on testing a hypothesis, or, to step away from Popperian thought, bear on the confirmation of a hypothesis. Whewell (1858) introduced the notion of consilience of inductions as a powerful way to confirm inductive conclusions and it consisted of taking logically independent inductions and seeing if they led to the same conclusion. If so, that consilient result, that confirmation, had more explanatory power than a single induction. The problem Rosen saw was what if the conclusions from the independent data sets were different, which is correct? Can they both be correct and/or can they both be incorrect? The absence of consilience (although Rosen did not use that term), Rosen argued, had no effect on testing hypotheses, specifically, biologically generated hypotheses could not be rejected by geological hypotheses. Why couldn't biology-based hypotheses lead to novel geologic hypotheses? That was a question Rosen asked and that was a direction in which he headed.

Initially, Rosen took the view of reciprocal illumination with regard to the biological and geological data in inferring biogeographic hypotheses (Rosen 1975:433):

> "A particular theory of vicariance may be tested by comparing the number of proposed vicariant events with geological theories of the history of the entire region occupied by the generalized track. Conversely geophysical theories can be tested by comparing hypothesized events of earth history with the vicariance patterns of generalized tracks."

In arguably two of Rosen's most important vicariance papers (1978 and 1985, respectively), "Vicariant Patterns And Historical Explanation In Biogeography" and "Geological Hierarchies and Biogeographic Congruence in the Caribbean," he became quite explicit about his feelings on geological tests of biological hypotheses (Rosen 1978:186):

> "A final, and most necessary, observation is that a geological area-cladogram neither tests nor in any way affects the generality of a biological area-cladogram. A geological area-cladogram differing from the biological pattern does not refute the pattern—it is simply irrelevant to it because it contains no explanatory information regarding the biological pattern."

And (Rosen 1985b:637):

> "Leon Croizat was one of the pioneers who recognized that the biological data tell their own story, which can be at odds with a stabilist geology. Now that stabilist geology has been rejected in favor of a concept of mobilism, some biogeographers accept that biology has an independent story to tell about the history of the world. It is this independence of biological from geological data that makes the comparison of the two so interesting because it is hard to imagine how congruence between the two could be the result of anything but a causal history in which geology acts as the independent variable providing opportunities for change in the dependent biological world."

Rosen (1985b:657) continued:

> "But to accept as true the proposals of one geological model, and to abandon the search for cladistic area-congruence in favor of some a priori notion that all distributions

might be explained by guesswork liberally laced with dispersalist intuition is to ensure
that future generations of biogeographers will regard such proposals lightly."

Rosen (1985b: 658) concluded with a directive to all biogeographers:

> "But, if such corroboration is not forthcoming, as biologists we are bound by the mes-
> sage of biological data in describing a biotic history of those geographic areas regard-
> less of any possible conflicts with geologic theory."

If I may be so bold to speak for Rosen, I would like to extend this inhibiting assumption to a problem Rosen did not encounter, and that is the function of molecular
clocks for testing biogeographic theories. I propose that Rosen would have treated
molecular clock dates the same as geological hypotheses: they do not refute biogeographic (congruent phylogeny/track based) hypotheses. The clocks would merely
function in the context of consilience of inductions, if they say the same things there
is confirmation but if not there is no test but two competing hypotheses. Clocks
have become the modern arbiter of biogeographic hypotheses for many workers,
but are not clocks only hypotheses themselves, built upon several hypotheses and
assumptions (e.g., Hillis et al. 1996; Graur and Martin 2004)? At minimum, the role
of molecular clocks in biogeography should be viewed with skepticism, or as Graur
and Martin (2004:85) put it: "Our advice to the reader is: whenever you see a time
estimate in the evolutionary literature, demand uncertainty!"

5. *Widespread taxa and two taxon systems are informative about the rela-*
tionships of areas.

By way of prologue, there is little to say about the two taxon system: it is the
trivial case without any competing hypotheses. How many ways can A and B be
related? Yet, Rosen spent some time expounding on this issue. The other part of
the inhibiting assumption regards the information content of widespread taxa. If
this confounding problem is looked at through the lens of cladistic biogeography,
there is potential to solve the problem of widespread taxa. Assumption 0 (e.g.,
Wiley 1987) and Assumptions 1 and 2 (e.g., Nelson and Platnick 1981) treat widespread taxa with different levels of information on area relationships: monophyly,
paraphyly, and polyphyly, respectively. Assumption 0 says widespread taxa must
be treated as indicating monophyly of the areas. Assumption 1 says they can
be treated as either monophyletic or paraphyletic and Assumption 2 says with
widespread taxa the areas should be treated with all possible relationships, therefore, monophyletic, paraphyletic, and polyphyletic (see Page 1990). Interestingly,
Rosen did not take this approach to his skepticism of the problem. Nonetheless,
his skepticism and the origins of this inhibiting assumption are on clear display
in the quotes below.
From Rosen (1978:167):

> "What we are seeking is some statement concerning the included taxa of greater gen-
> erality than their simple coincidence. That there exists a disjunction between North
> and Middle American vicariads is one such statement, but this statement (i.e., taxa in

area A are related to taxa in area B), lacks complexity and contains no more information than that originally required to recognize the existence of the general problem. However, if one of the two areas (area A, for example) is subdivided (i.e., a second disjunction is found), a more complex statement becomes possible: A' is more closely related to A" than either is to B. This three-taxon statement implies a historical component not contained in two taxon statements, namely, that a speciation event (second disjunction) affecting the ancestor of A' and A" occurred only after the speciation event (first disjunction) that affected the ancestor of A and B."

And from later in the same paper Rosen (1978:177):

"Some populations, whether recognized as species or not, are informative with respect to a history of geographic isolation because they have differentiated. Others are uninformative either because they haven't differentiated, because they are parts of larger populations that span two or more areas (are insensitive to existing barriers), or because taxonomists have thus far failed to detect the ways in which differentiation has occurred (physiologically, developmentally, behaviorally, etc.)."

7. *A knowledge of geologic processes is necessary for a vicariance analysis of areas of endemism.*

This inhibiting assumption presaged the development of event-based methods (Humphries 2000), which included a radical view taken by Hovenkamp (1997). Hovenkamp argued that areas of endemism probably did not exist (were not real) and biogeographic history was not to be found in the history of these areas, but instead in the history of the events, that is the processes that drove vicariance events. So Hovenkamp essentially flipped Rosen's inhibiting assumption and argued that the most important types of information for vicariance analysis were processes. Since then, Crother and Murray (2011) made the case for areas of endemism as real through ontological arguments, a number of new approaches to the discovery of areas of endemism have been developed (e.g., Szumik and Goloboff 2004; DaSilva et al. 2015), and the continued use of phylogenies to seek general patterns of biogeographic history all point to the conclusion that most workers agree with Rosen's inhibiting assumption, even if unknowingly.

In the following extended quote, Rosen (1978:187) lays out his reasoning behind his use of geological process information and biological information. This is slightly different from his inhibiting assumption that geology cannot refute biological hypotheses. Here, he suggests that perhaps they are reciprocally illuminating:

"Returning now to the problem of the geographic history of *Heterandria* and *Ziphophorus* [sic] in Middle America: the region has been tectonically active since the end of the Mesozoic, and although a great deal has been written on the historical geology of this region, I am unable to obtain enough precise information to produce a cladistic statement that is relevant to the area extending from the Rio Panuco basin southeastward to Nicaragua. This does not mean that I consider a search for such historical explanation fruitless, but only that, as a nongeologist, I am unable to pick and choose among the varied and sometimes conflicting geological interpretations of the Tertiary history of Middle America.

On the other hand, I could abandon the pursuit of a cladistic synthesis of geographic history and simply try to find an event here and there that coincides with some biological disjunction: for example, the fault zones of northern Central America or the east-west oriented Pliocene volcanic zone south of the Rio Panuco basin in Mexico. Or, I could renounce my responsibility altogether and simply assume that the biological patterns have been caused by Quaternary events, as Deevey (1949) said we should assume. Or, I could do something else, which I am inclined to favor. This is to recognize that geology and biogeography are both parts of natural history and, if they represent the independent and dependent variables respectively in a cause and effect relationship, that they can be reciprocally illuminating. But for there to be reciprocal illumination between these two fields there must be a common language. This language must be the language of nested sets, i.e., hierarchical systems of sets and their subsets united by special similarities, i.e., synapomorphies. In short, taxonomists should be encouraged to continue the current salutary trend to organize their data cladistically, and geologists should be encouraged to begin. At least until the geological data pertaining to Middle America are so ordered I am forced to draw the limited conclusion that the observed biological patterns have formed during a period spanning all or part of the last 80 million years."

In the following quote, Rosen (1985b:637, emphasis mine) reemphasizes the independence of biological data and that ultimately it is the cladograms and not inferred processes that tell us more about biogeographic history:

"It is this independence of biological from geological data that makes the comparison of the two so interesting because it is hard to imagine how congruence between the two could be the result of anything but a causal history in which geology acts as the independent variable providing opportunities for change in the dependent biological world. The comparison becomes especially interesting if there is a congruence among geohistories based on different approaches to the geographic problem, and if there is a congruence among cladistic relations of different taxa with respect to the same geographic areas. The specific questions are: 1) do the members of different monophyletic groups of organisms have the same relations to each other with respect to geographic regions in which they are endemic and is their congruence with respect to these areas non-random; and 2) does this non-random congruence of different groups of organisms correspond to a branching diagram that represents part of the history of some geographic region? *The constraint in these comparisons is the branching diagram rather than a process assumed to be of causal importance.*"

8. *Some organisms are better indicators of biotic history than others depending on their ecology and means of dispersal.*

This final inhibiting assumption from Rosen's course list is one that I recall him talking about most vividly and one that I have repeated to many graduate students and biogeography classes. Yet, evidence for a simple explicit statement about this was difficult to find. Perhaps the best is from his classic "A Vicariance Model of Caribbean Biogeography" where he outlined the method and included (Rosen 1975:432):

"Plotting on a map the distributions of many different animal and plant assemblages from a certain region will demonstrate if commonality of distribution pattern occurs."

To me, that is not totally satisfying. A closer look at the aforementioned paper reveals that Rosen used data from a broad diversity of groups of organisms, from various plant groups to flatworms, isopods, and onychophorans to fishes, reptiles, birds, and mammals. The fact that Rosen used such diversity in his vicariance study of the Caribbean suggests that he thought there were no special groups of organisms that held biogeographic secrets in their histories and distributions. To Rosen (1975:432–433), all groups were indicators of biotic history, and those groups that did not conform to general patterns revealed unique histories (or problems) not shared by the congruent groups:

> "An instance of lack of conformity of an individual track with a thoroughly documented generalized track with a thoroughly documented generalized track may signify 1) that the individual track belongs to a different generalized track, 2) that the members of the track have broken away from the parent biota and have dispersed, or 3) that the track is based on a non-monophyletic group."

Like most of the inhibiting assumptions about biogeography Rosen covered, this final one changed the way we viewed biogeographic investigation. Biogeography was no longer about authoritarianism and speculation hidden under the blanket of dogma. It was about looking at multiple groups of organisms and seeking general patterns that could reveal shared histories of groups and areas and reveal unique events for those groups that did not fit the general patterns. It became a science.

4.3 SUMMARY

During the course of the research for this chapter, and during the writing, I learned much about Donn Rosen and appreciated anew his seminal contributions to biogeographic theory and method, his controversial yet forward-thinking views of evolutionary theory, as well as his promotion of the power and utility of cladistic thought. The lesson for all readers, especially for junior members of the scientific community, is to not blindly accept perceived dogma and putative orthodoxy. Question and be critical, and most importantly, do not be afraid to be wrong (General Assumption 6). The hypotheses that came before you, the hypotheses you propose, and the hypotheses that will come after you, are all tentative knowledge propositions. Be bold and accept that your work may very well be falsified later. The key thing is that your hypothesis stimulated someone else to test it and that is how scientific progress is made. Rosen knew that and practiced it.

I also learned much about myself. In many of Rosen's inhibiting assumptions I see reflections of my own views on evolutionary theory, taxonomy, and biogeography. Also in reflection I see my own skepticism. No doubt, the foundation of many of my views originated with Rosen's visit to Miami in 1985. While none of us can revisit Rosen's 1985 course, I recommend revisiting (or reading for the first time!) his works. Yes, I made an effort to do that for the reader in the context of his assumptions, but be assured there is much, much more to be learned from Rosen than what I have presented.

Finally, at the close of Rosen's penultimate paper (because of the chapter in the present volume; Rosen and Patterson 1990; Patterson put together Rosen's last and unpublished work on perciforms), Rosen was still not done. He appeared to pine for still better methods and conceptual thinking and looked to the future (p. 54):

"But perhaps we need to invent a new wheel before some of the larger questions of relationship can be dealt with. The narrative is still unconvincing, adorned as it is with significant bald spots."

ACKNOWLEDGMENTS

C. Beachy shared his knowledge of Dire Straits. R. Mooi, G. Nelson, and L. Parenti shared papers and suggestions. C. Beachy, C. Murray, and M. White read and commented on a draft. All the above are thanked profusely for their help.

REFERENCES

Ashe, A., A. Sapetschnig, E. Weick, et al. 2012. piRNAs can trigger a multigenerational epigenetic memory in the germline of *C. elegans*. *Cell* 150:88–99.
Beatty, J. 1982. Classes and cladists. *Systematic Zoology* 31:25–34.
Brady, R. H. 1979. Natural selection and the criteria by which a theory is judged. *Systematic Zoology* 28:600–621.
Brady, R. H. 1982. Theoretical issues and "pattern cladistics." *Systematic Zoology* 31:286–291.
Carpenter, J. M. 1987. Cladistics of cladists. *Cladistics* 3:363–375.
Croizat, L., G. Nelson, and D. E. Rosen. 1974. Centers of origin and related concepts. *Systematic Zoology* 23:265–287.
Crother, B. I. and C. M. Murray. 2011. Ontology of areas of endemism. *Journal of Biogeography* 38:1009–1015.
DaSilva, M. B., R. Pintoda Rocha, and A. M. DeSouza. 2015. A protocol for the delimitation of areas of endemism and the historical regionalization of the Brazilian Atlantic Rain Forest using harvestmen distribution data. *Cladistics* 31:1–14.
Darwin, C. 1859. *On the origin of the species by natural selection*. London: J. Murray.
Dover, G. A. 1986. Molecular drive in multigene families: How biological novelties arise, spread and are assimilated. *Trends in Genetics* 2:159–165.
Goldschmidt, R. 1940. *The material basis of evolution*. New Haven: Yale University Press.
Gould, S. J. 2002. *The structure of evolutionary theory*. Cambridge: Harvard University Press.
Gould, S. J. and R. C. Lewontin. 1979. The spandrels of San Marco and the Panglossian paradigm: A critique of the adaptationist programme. *Proceedings of the Royal Society, London B: Biological Sciences* 205:581–598.
Graur, D. and W. Martin. 2004. Reading the entrails of chickens: Molecular timescales of evolution and the illusion of precision. *Trends in Genetics* 20:80–86.
Hennig, W. 1966. *Phylogenetic systematics*. Urbana: University of Illinois Press.
Hillis D. M., B. K. Mable, and C. Moritz. 1996. Applications of molecular systematics: The state of the field and a look to the future. In *Molecular systematics*, ed. D. M. Hillis, C. Moritz, and B. K. Mable, 515–543. Sunderland: Sinauer.
Hinegardner, R. and D. E. Rosen. 1972. Cellular DNA content and the evolution of teleostean fishes. *American Naturalist* 106:621–644.
Hovenkamp, P. 1997. Vicariance events, not areas, should be used in biogeographical analyses. *Cladistics* 13:67–79.
Hull, D. L. 1988. *Science as a progress. An evolutionary account of the social and conceptual development of science*. Chicago: University of Chicago Press.

Humphries, C. J. 2000. From, space and time: Which comes first. *Journal of Biogeography* 27:11–15.

Landman, O. E. 1991. The inheritance of acquired characteristics. *Annual Review of Genetics* 25:1–20.

Lewontin, R. C. 1978. Adaptation. *Scientific American*. September:213–230.

Nelson, G. 1983. Vicariance and cladistics: Historical perspectives with implications for the future. In *Evolution, time and space: The emergence of the biosphere*, ed. R. W. Sims et al., 469–492. London & New York: Academic Press.

Nelson, G. and D. E. Rosen. 1981. *Vicariance biogeography: A critique*. New York: Columbia University Press.

Nelson, G. and N. I. Platnick. 1981. *Systematics and biogeography*. New York: Columbia University Press.

Nelson, G., J. W. Atz, K. D. Kallman, and C. L. Smith. 1987. Donn Eric Rosen, 1929–1986. *Copeia* 1987:541–547.

Page, R. D. 1990. Temporal congruence and cladistic analysis of biogeography and cospeciation. *Systematic Biology* 39:205–226.

Patterson, C. 1982. Classes and cladists or individuals and evolution. *Systematic Zoology* 32:284–286.

Patterson, C. 2011. Adventures in the fish trade. *Zootaxa* 2946:118–136.

Patterson, C. and D. E. Rosen. 1977. Review of ichthyodectiform and other Mesozoic teleost fishes, and the theory and practice of classifying fossils. *Bulletin of the American Museum of Natural History* 158:85–172.

Platnick, N. I. 1982. Defining characters and evolutionary groups. *Systematic Zoology* 31:282–284.

Platnick, N. I. and D. E. Rosen. 1987. Popper and evolutionary novelties. *History and Philosophy of the Life Sciences* 9:5–16.

Popper, K. 1959. *The logic of scientific discovery*. London: Hutchinson & Co.

Pregill, G. K. and B. I. Crother. 1999. Ecological and historical biogeography of the Caribbean. In *Caribbean amphibians and reptiles*, ed. B. I. Crother, 335–356. San Diego: Academic Press.

Raff, R. A. 1996. *The shape of life*. Chicago: University of Chicago Press.

Rosen, D. E. 1964. The relationships and taxonomic position of the halfbeaks, killifishes, silversides, and their relatives. *Bulletin of the American Museum of Natural History* 127:217–268.

Rosen, D. E. 1972. Origin of the characid fish genus *Bramocharax* and a description of a second, more primitive, species in Guatemala. *American Museum Novitates* 2500:1–21.

Rosen, D. E. 1974a. Cladism or gradism? A reply to Ernst Mayr. *Systematic Zoology* 23:446–451.

Rosen, D. E. 1974b. Phylogeny and zoogeography of salmoniform fishes and relationships of *Lepidogalaxias salamandroides*. *Bulletin of the American Museum of Natural History* 153:265–326.

Rosen, D. E. 1975. A vicariance model of Caribbean biogeography. *Systematic Zoology* 24:431–464.

Rosen, D. E. 1978a. Darwin's demon. *Systematic Zoology* 27:370–373.

Rosen, D. E. 1978b. Vicariant patterns and historical explanation in biogeography. *Systematic Zoology* 27:159–188.

Rosen, D. E. 1979. Fishes from the uplands and intermontane basins of Guatemala: Revisionary studies and comparative geography. *Bulletin of the American Museum of Natural History* 162:267–376.

Rosen, D. E. 1981. Introduction. In *Systematics and biogeography*, ed. G. Nelson and N. I. Platnick, 1–5. New York: Columbia University Press.

Rosen, D. E. 1982. Teleostean interrelationships, morphological function and evolutionary inference. *American Zoologist* 22:261–273.

Rosen, D. E. 1984. Hierarchies and history. In *Evolutionary theory: Paths into the future*, ed. J. W. Pollard, 77–97. Cambridge: John Wiley & Sons Ltd.

Rosen, D. E. 1985a. An essay on euteleostean classification. *American Museum Novitates* 2827:1–57.

Rosen, D. E. 1985b. Geological hierarchies and biogeographic congruence in the Caribbean. *Annals of the Missouri Botanical Garden* 72:636–659.

Rosen, D. E. 2016. Assumptions that *inhibit* scientific progress in comparative biology. In *Assumptions Inhibiting Progress in Comparative Biology*, ed. B. I. Crother and L. R. Parenti, 1–4. Boca Raton: CRC Press.

Rosen, D. E. and R. M. Bailey. 1963. The poeciliid fishes (Cyprinodontiformes): Their structure, zoogeography, and systematics. *Bulletin of the American Museum of Natural History* 126:1–176.

Rosen, D. R. and D. G. Buth. 1980. Empirical evolutionary research versus neo-Darwinian speculation. *Systematic Zoology* 29:300–308.

Rosen, D. E. and C. Patterson. 1969. The structure and relationships of the paracanthopterygian fishes. *Bulletin of the American Museum of Natural History* 141:357–474.

Rosen, D. E. and C. Patterson. 1990. On Müller's and Cuvier's concepts of pharyngognath and labyrinth fishes and the classification of percomorph fishes: With an atlas of percomorph dorsal gill arches. *American Museum Novitates* 2983:1–57.

Rosen, D. E. and L. R. Parenti, 1981. Relationships of *Oryzias*, and the groups of atherinomorph fishes. *American Museum Novitates* 2719:1–25.

Rosen, D. E. and P. H. Greenwood. 1976. A fourth Neotropical species of synbranchid eel and the phylogeny and systematics of synbranchiform fishes. *Bulletin of the American Museum of Natural History* 157:1–70.

Rosen, D. E., P. L. Forey, B. G. Gardiner, and C. Patterson. 1981. Lungfishes, tetrapods, paleontology, and plesiomorphy. *Bulletin of the American Museum of Natural History* 167:159–276.

Schaeffer, B. and D. E. Rosen. 1961. Major adaptive levels in the evolution of the actinopterygian feeding mechanism. *American Zoologist* 1:187–204.

Szumik, C. A. and P. A. Goloboff. 2004. Areas of endemism: An improved optimality criterion. *Systematic Biology* 53:968–977.

Theißen, G. 2009. Saltational evolution: Hopeful monsters are here to stay. *Theory in Biosciences* 128:43–51.

Waddington, C. H. 1975. *The evolution of an evolutionist*. Cornell: Cornell University Press.

Watkins, J. 1984. *Science and skepticism*. Princeton: Princeton University Press.

Whewell, W. 1858. *Novum organon renovatum*. London: JW Parker and Son.

Wiley, E. O. 1987. Methods in vicariance biogeography. In *Systematics and evolution: A matter of diversity*, ed. P. Hovenkamp, 283–306. Utrecht: Inst. Syst. Bot., Utrech Univ.

Wright, A. M. and D. M. Hillis. 2014. Bayesian analysis using a simple likelihood model outperforms parsimony for estimation of phylogeny from discrete morphological data. *PLoS One*, 9 (10), e109210. http://dx.doi.org/10.1371/journal.pone.0109210.

5 Network Species Model Consociates Process Ecology and Material Object Theory

David Kizirian
American Museum of Natural History

Maureen A. Donnelly
Florida International University

CONTENTS

5.1 INTRODUCTION

Among the issues Donn Rosen thoughtfully considered were the models used to circumscribe units of biodiversity (e.g., Rosen 1978, 1979). In particular, Rosen thought that the predominate models of his time, the Evolutionary Species Concept (ESC) and Biological Species Concept (BSC), failed to identify the fundamental unit for addressing research questions in systematics and evolutionary biology. He also identified an ontological problem with the subspecific taxonomic category: "if a 'subspecies' is, by definition, something less than a species, and yet a 'species' is the smallest cluster of individuals in nature that can be defined, then subspecies are, also by definition, unobservable and undefinable" (Rosen 1979: 227). In response to such issues, Rosen used populations rather than species or subspecies in his bio-geographic studies because he thought they were a better proxy for "the unit of evolutionary significance" (1978: 176). Rosen's criticisms draw attention to the deeper, longstanding problem in biodiversity research: the lack of a sound theoretical model underpinning the term *species*.

Many of the issues raised by Rosen remain problems today such as the ahistorical nature of the BSC and the arbitrary units resulting from implementation of the ESC, yet those models continue to be widely used (see Murray et al. 2016). Herein, we address Rosen's concerns in the light of recent thought about species. Specifically, we will discuss the Network Species Model (NSM; Kizirian and Donnelly 2008), which is similar in implementation to Rosen's pragmatic approach of using populations as operational units in evolutionary studies. Focused on inherent organization and process, the NSM is crafted like models in other scientific disciplines; therefore, we expect to find connections to theory in other fields. Hence, we also explore possible congruence between NMS and two nearly disjunct areas of study: Process Ecology and Material Object Theory.

5.2   NETWORK SPECIES MODEL

The diversity of life is connected through lineage relationships that result from reproductive processes involving organisms. Some organisms are the result of asexual reproduction, such as bacteria and some parthenogenetic vertebrates. In such cases, only one level of organization exists (i.e., organism), hence, no species model (or species taxon) is needed to explain them (Kizirian and Donnelly 2008). Many organisms (e.g., most eukaryotes) exist through processes that combine genomes from two individual organisms, which results in supraorganismal organization (i.e., network of organisms). We could do without recognizing the unique properties of sexual systems and simply name *lineages* on the tree of life (e.g., de Queiroz 1998); however, as we will argue below, there are compelling reasons to distinguish networks of organisms from organisms and other levels of organization. Ignoring such networks would be similar to ignoring patterns of organization described with other models such as cell, organism, and ecosystem.

The term *species* is widely regarded to be associated with a fundamental unit of biological organization (e.g., Russell et al. 2014; see discussion in Kizirian and Donnelly 2008). Ironically, however, most species models are not centered on organization but on contingent properties such as some form of divergence (de Queiroz 1998; Kizirian and Donnelly 2008; Wilkins 2009). Even Rosen (1978, 1979), for example, defined species as "a geographically constrained group of individuals" (1978: 176) or "a population or group of populations *defined by one or more apomorphous features*" (1979: 277; our emphasis). Other species models such as historical individual, replicator continuum, evolutionary lineage, and homeostatic property cluster may be informative general descriptors, but they do not particularize the unique constitution of species; hence, they do not distinguish species from other kinds of entities. Redundancy stemming from the use of terms such as deme, population, Evolutionary Significant Unit, and subspecies (e.g., Du Rietz 1930) does not help matters (Kizirian and Donnelly 2008). The NSM (Kizirian and Donnelly 2008) eschews divergence and all other contingent properties (e.g., longevity, size, fate) and defines species in terms of the (1) unique inherent structure and (2) processes that generate that structure. Specifically, *a species is any network of organisms resulting from reproductive processes (e.g., gametogenesis followed by syngamy).* By focusing on inherent structure and process, the NSM attempts to describe what a species is in an ontological sense, that is, what they are, what they are made of, and how they relate to other things we think are real. Thinking about species in terms of organization also compels us to differentiate between networks and lineages (e.g., de Queiroz 1998). Because lineage relationships exist at many levels of organization (e.g., species; organism; cell; molecule, including hypothesized self-replicating molecules that preceded DNA), they do not identify a unique kind of system in nature (Kizirian and Donnelly 2008). Structurally, a network species (one level of organization) is composed of individual organisms (another level of organization) and is unique in the natural world. The processes that create network species, such as gametogenesis and syngamy, each of which may be composed of numerous other processes (e.g., copulation, acrosome reaction), are also unique to network species. Cases where organisms are produced through conjugation or other forms of horizontal gene exchange may represent species, if a network is manifested.

Some ramifications of viewing units of diversity according to the NSM include the following: (1) Species, subspecies, population, deme, Evolutionary Significant Unit, and related terms (e.g., Du Rietz 1930) cannot be distinguished on the basis of inherent organization and do not reflect a hierarchy of organization. In other words, only one model is needed to explain networks of organisms. Networks may, however, be characterized by subnetworks (e.g., community structure or modularity [Newman 2006, 2012]; as in Figure 5.1). (2) Temporarily isolated networks of organisms possess the same intrinsic organization as permanently isolated ones and, therefore, have the same metaphysical status. (3) Fate is not an intrinsic property of systems and is an inappropriate consideration in species models. (4) Units of diversity are recognized regardless of kind or degree of divergence. (5) Networks are recognized at the moment they become isolated; therefore, resultant classifications reflect the causal events that generate diversity (e.g., vicariance) rather than postvicariance events such as character evolution and, consequently, have greater historical relevance (e.g., biogeography). (6) The NSM will likely result in recognition of smaller and more numerous units of diversity, as in Figures 5.2 and 5.3, than is recognized under other species models (or concepts). (7) Like models for units

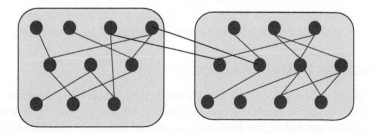

FIGURE 5.1 Conflict between NSM and Process Ecology models. Under strict interpretation of NSM, one species (one network with modularity or community structure) might be recognized. Following Ulanowicz (2009), two species (two networks with leakage) might be recognized if hybridization does not destroy autocatalysis of individual networks.

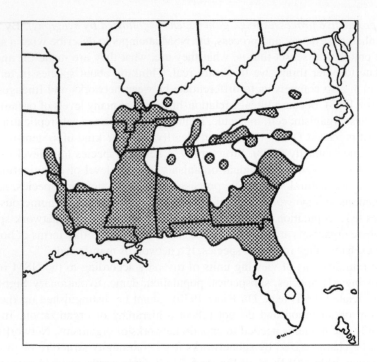

FIGURE 5.2 Distribution map for the mole salamander, *Ambystoma talpoideum*. Under current concepts of species, isolated networks are regarded to be populations of a single species. Under the NSM, at least 10 species would be recognized, which reflects system structure rather than possible future reproductive compatibility, degree of divergence, or fate of isolated networks. Classifications consistent with the proposed species model include (1) "*Ambystoma talpoideum* complex" and (2) unique binominals (appended, or not, with "complex" as appropriate) for each isolated network. (From Conant, R. and J. T. Collins. 1991. *Reptiles and Amphibians of Eastern/Central North America.* Houghton Mifflin Company, Boston.)

(e.g., atom, molecule, cell, organism, ecosystem, solar system) studied in other branches of science, the NSM is explicit about structure and process; consequently, it has the potential to build bridges to models in other disciplines when they are also explicit about structure and/or process. Here, we attempt to reconcile species theory with models in the divergent fields of Process Ecology and Material Object Theory.

5.3 PROCESS IN SPECIES MODELS

Process has been recognized as a key component of species models since ancient times. Aristotle, for example, recognized that, through the process of reproduction, species persist despite the demise of constituent individual organisms (Wilkins 2009: 16). Hohenstaufen (1194–1250) "demonstrated a surprisingly modern view of species not marked by external form ... but by the ability to interbreed" (Wilkins 2005, 2009). Kant stated, "the unity of species is nothing other than the unity of the

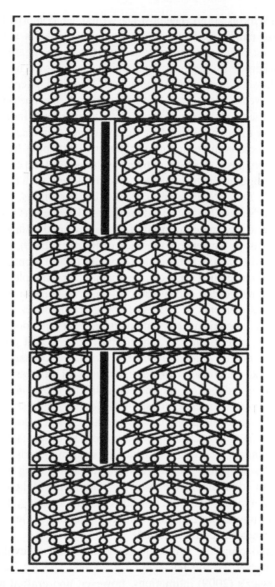

FIGURE 5.3 Diagrammatic representation of seven network species, through time. Circles represent individual organisms, lines represent ancestor–descendant (i.e., epigony, tokogeny) connections, and thick black bars represent vicariance events. Under many species concepts, networks isolated by vicariance are recognized as temporarily isolated populations of the more inclusive species (dashed box). We argue that systems composed of organisms are represented by a single level of organization (e.g., network species, syngamic systems, conjugation networks) that become smaller and more numerous through vicariance and larger and less numerous through hybridization. Species originate at the moment they become isolated and become extinct upon integration with other systems (compare to Hennig 1966: Fig. 6).

generative force" (Wilkins 2009). The post-Darwinian lepidopterist Edward Bagnall Poulton (1904) also emphasized process in his model: species are "societies into which individuals are bound together in space and time by Syngamy and Epigony."† Hence, the idea that process is key to understanding species is not new. More recently, Rieppel (2009) followed Quine (1960) who stated, "Physical objects... are not to be distinguished from events, or processes" and defined species at one point (2009: 42) as "causally integrated relational systems formed by a multitude of individual processes that are held together through species-specific homeostatic mechanisms." Ulanowicz (2009) argued, "process is more important than law in shaping living things" with objects playing a more passive role in complex systems. And, he outlined how viewing systems as "configurations of processes" promises new understanding of living systems not possible under Newtonian or Darwinian paradigms. Lockwood (2012) embraced Ulanowicz's thesis but eschewed the material element as evident here, "species are processes, and they are comprised of processes," and (Ulanowicz 2012) supported him. In contrast to the general approach of Rieppel (2009), Ulanowicz (2009, 2012), and Lockwood (2012), Kizirian and Donnelly (2008) identified specific core configurations of processes that characterize species, for example, the complementary processes of gametogenesis and syngamy, which create network structure.

Briefly, the central elements of Process Ecology (Ulanowicz 2009) include the idea that systems (e.g., organisms, ecosystems, species [Lockwood 2012]) persist in part because they are able to influence themselves in ways that promote their own survival. Persistence entails *autocatalysis* (e.g., links in processes have propensity for positive feedback, stabilization of configurations of processes), *centripetality* (recruitment of matter and energy into the autocatalytic orbit), *mutuality* (cooperatively reinforcing processes such as those that keep an organism alive), *autotomy* (synergistic effects such as the property of being alive), and a complementary relationship between *ascendency* (capacity for a system to order itself) and *overhead* (inefficiency, redundancy), a corollary of which is performance/risk trade-off. In this framework, mutually obligatory *duality* (building up and tearing down) emerges as a common theme. In addition, *chance* events (=contingency; Ulanowicz 2012) may disrupt systems, but may also generate novelties that improve a system's chances for survival. Lastly, configurations of processes have unique histories, "some of which is recorded in their material configurations."

Ulanowicz (2009) proposed a revolutionary new way of explaining complex living systems; however, he did not explicitly address species. Lockwood (2012) recognized Ulanowicz's (2009) intimations, however, and crafted a species model in the new metaphysical framework. For example, he defines a species as "an historically derived, highly stable configuration of processes that resist exogenously imposed change" and "species are processes, and they are comprised of processes." We think process theory has much to offer evolutionary biology; however, the models of Rieppel (2009), Lockwood (2012), and Ulanowicz (2012) do not identify the

---

* Poulton (1904) defined *epigony* as the lineage relationship that exists between a parent and its offspring and seems to be synonymous with *tokogeny* (Hennig 1966). Poulton's species model also includes a network component, namely, *syngamy*, which he regarded to be an organism (or species) level term, in contrast to the cellular level term proposed by Hartog (1913).

core processes associated with species (neither as they define species, nor as we do). That oversight resulted in a misapplication of process theory regarding asexual organisms, which do not form networks. The absence of network structure in asexual lineages precludes the emergence of autocatalysis, centripetality, autotomy, and so on above the organism level, except in an ecological context. In addition, Lockwood (2012) needlessly forgoes the material element, which, we think, renders his model incomplete. Below, we address these issues before recasting species in the Process Ecology framework.

5.4   ASEXUAL SPECIES

Ulanowicz and Lockwood apply process theory wherever systems exist, a sentiment captured by Bickhardt and Campbell (1999), "it's processes all the way down," which is true in one sense, however, such a sweeping generality may lead to confusion in application of the otherwise compelling metaphysical framework. Of particular relevance here, Lockwood (2012) ascribed processes to species on the assumption that asexually reproducing lineages are indeed species, a position evident in his discussion of individuality: he cited Hull (1978) who "contends that the existence of asexual species demonstrates that gene flow can't be the only mechanism that maintains evolutionary unity. There must be some other processes, and Ulanowicz's (2009) self-organization and centripetality would seem to be ideal candidates." In his discussion of Ereshefsky (1999), Lockwood asserts "Asexual species, for which there is no gene flow among organisms, are not individuated by virtue of causal interactions among their parts. … Rather, processes, such as selection, genetic homeostasis, and developmental canalization (all of which are echoed in Ulanowicz's postulates) are what cause organisms to belong to a species."

One problem evident in the claims by Hull, Ereschefsky, and Lockwood is the assumption that asexually reproducing lineages represent a kind of species, which extends from a misunderstanding of the substance and process of organisms versus networks of organisms. As discussed by Kizirian and Donnelly (2008), lineages of asexually reproducing organisms do not form networks; hence, it is impossible for supraorganismic synergy to emerge because there is no supraorganismal whole wherein synergistic properties (e.g., autocatalysis) might be manifested. It is spurious then to invoke more inclusive models of organization to explain asexuals; in other words, "organism" is the highest level of organization evident in asexual lineages. The "unity" exhibited by asexual lineages is better explained as plesiomorphic similarity, rather than evidence of self-organization, centripetality, unity, genetic homeostasis, or developmental canalization. Unless argued in the context of a phylogenetic hypothesis to be convergent, similarities in development or genetics among asexual lineages exist because they were inherited from a common ancestor. It also follows that any change in genetics or development that arises within an asexual lineage cannot have a manifestation outside such lineages because there is no process to export it outside the lineage of origin, as in Figure 5.4. Asexual lineages are generated through a variety of mostly mitotic processes (rare exceptions include paternal apomixis and cytomixis, in which there is no fusion of gametic nuclei). Most eukaryotes, on the other hand, produce new individuals through gametogenesis and

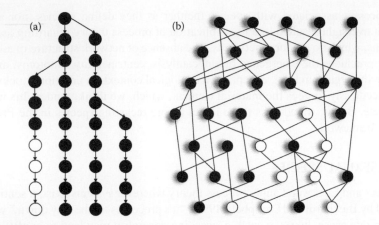

FIGURE 5.4 Propagation of mutation in asexual (a) and sexual (b) lineages. New mutations indicated by white circle, may migrate throughout a network (b) but cannot be exported out of the lineage of origin in asexuals (a).

syngamy (= zygosis [Poulton 1904]), both of which may involve myriad processes (e.g., the former: spermatogenesis, oogenesis, ovulation; the latter: courtship, copulation, acrosome reaction).

Horizontal gene transfer (i.e., transformation, conjugation, transduction) results in a network of organisms; however, such networks are "loose" compared to those in bisexual species. For example, whereas, in bisexual networks, every individual is product of the network and entire genomes are subject to mixis, in unicellular networks, only a portion of the genome is exchanged between organisms and asexual reproduction continues to generate new individuals. Autocatalysis may be manifested in such cases and we would regard such networks to be species, because of their network structure (Kizirian and Donnelly 2008).

While we agree that a lineage of asexually reproducing organisms "persists despite the demise of individual organisms" (a key manifestation of autocatalysis), the same can be said for lineages of cells or molecules (Kizirian and Donnelly 2008). Indeed, lineages of self-replicating molecules may have existed before life did. So, the persistence of lineages is not a valid argument for regarding asexuals as species.

Lockwood's (2012: 251) rationale for lumping asexual and sexual organisms includes supporting examples that we identify as ecological phenomena. He argues, "… species engage in epidemics and outbreaks which exhibit non-linearities that are not explicable in terms of individual actions … and which manifest features that are not reducible to individuals (e.g., forming swarms, swamping predators, and overwhelming host defenses)." We counter that such examples reflect ecological relationships rather than evidence for existence of evolutionary cohesion. Perhaps the point can be made with the following example. Birds frequently form small feeding flocks that include multiple species, including those from divergent lineages. Mixed feeding flocks may increase foraging efficiency (e.g., some birds feed on the ground, while others feed on vegetation) over feeding independently, and individuals of one species may "understand" an alarm call made by an individual of a different species. Despite the interesting dynamics, we do not recognize the birds participating in this process as a single species. In the case of very similar looking, closely related, mutualistic symbionts, such as those that comprise a bacterial mat, it is seemingly rational to refer to them as a species; however, we argue that the autocatalysis, centripetality, autotomy, and so on, exhibited in such cases are ecosystem-level, rather than species-level, phenomena.

Thinking about the material organization of species has helped us see where core processes are at work. Nevertheless, one could reach a similar conclusion about the status of asexuals without leaving the process framework because asexual and sexual lineages differ in the configurations of their processes. In asexual lineages, organism-level processes (e.g., fission) generate new organisms that may subsequently interact with close relatives in an ecological context (e.g., bacterial mat, swarm). Sexual reproduction, by contrast, results in networks (i.e., species) that exist only through unique configurations of core processes (e.g., gametogenesis, syngamy).

5.5 SPECIES AS CONFIGURATIONS OF PROCESSES

With a revised model for species in hand, we revisit the "species as processes" proposal. As Lockwood (2012: 251) states, "... we ought to be able to at least gesture toward properties of species that are unique or not manifested by the lower-order processes of the constituent organisms. Most fundamentally, species (or interbreeding populations or lineages which variously stand in for species) evolve and organisms do not."[†] Here, we pursue the same general goal, however, because we restrict species to networks, we will find a better fit with Process Ecology and we will be more specific about the configurations of processes that characterize species.

Ulanowicz (2009: First Postulate) and Lockwood (2012) state, "the operation of any system is vulnerable to disruption by chance (or contingency; Ulanowicz 2012)" but that "the juxtaposition of propensities (Popper 1990) could serve to counter the ubiquity of chance perturbations." In other words, internal (e.g., deleterious mutation) and external (e.g., climate change) pressures have the potential to disrupt the integrity of a given species, however, configurations of processes are generally robust to some level of disruption. At the species level, chance is at play at various points during the ploidy reduction/restoration cycle (e.g., independent assortment, synapsis, gamete fusion), which results in new combinations of genes, some of which may impact the survivability of the species. An example of this feature might be the many chance deleterious mutations that have arisen in the *Homo sapiens* complex, none of which so far have jeopardized survival of the species. Other chance events are just as likely to involve asexuals or network species; however, in the former, there is no process whereby innovations may be exported among asexual lineages, as seen in Figure 5.4; except horizontal gene transfer.

[†] If evolution is simply change over time, then, the universe and everything in it (at least above the atomic level) evolves, including organisms. Of course, things evolve in unique ways depending on their constitution (i.e., organization and processes).

Ulanowicz (2009: Second Postulate) proposed, "a process, via mediation by other processes, may be capable of influencing itself," which is manifested as autocatalysis, centripetality, autotomy, mutuality, and ascendency/overhead. *Autocatalysis* occurs when "the effect of every consecutive link in a feedback loop is positive." Autocatalysis may explain the evolution of tightly coordinated processes associated with the cycle of ploidy reduction/restoration such as gametogenesis (e.g., meiosis, synapsis) and syngamy (e.g., ovulation, acrosome reaction), if, over time, increases in efficiency at one step were propagated to other steps in the cycle. Other possible examples involving multiple species include mimicry and the Red Queen effect (Van Valen 1973) where an adaptation in one species results in selective pressure on another species, in a positive feedback cycle.

Centripetality is the recruitment of matter and energy into the autocatalytic orbit or "a dynamic that draws a network of processes toward a stable configuration" (Ulanowicz 2009). For network species, this propensity may be interpreted as the biomass of the network of organisms with its stockpile of, for example, unique genomes, gametes, fertilized eggs, fats stores for reproduction, antlers, and so on, accumulated through the stable configurations of processes of gametogenesis, syngamy, and associated processes. In contrast, asexual lineages may produce myriad individual organisms, spores, unfertilized eggs, and accumulate fat bodies; however, although they do evolve, asexual lineages do not exhibit variation resulting from synapsis (except mitotic synapsis), independent assortment, mate choice, and random gamete fusion. Furthermore, in asexuals, there is no supraorganismal entity in which matter and energy may be sequestered. Whereas, asexual lineages invest in individuals and bisexuals invest in the network with its accumulation of genetic variation.

Autotomy is the synergistic effect that emerges in the whole and extends from relationships among its parts. All lineages, including network species and asexuals, exhibit autonomy in that they outlive the constituent organisms (Lockwood 2012; also observed by Aristotle [Wilkins 2009]). At the organism level, the property of being alive emerges from the myriad configurations of processes (e.g., metabolism) operating at lower levels of organization. Network species are unique in that they evolve *as a single entity* in a way not possible without syngamy. In the words of Ghiselin (1974a), "they possess stores of latent genetic variability, and also have a capacity for mobilizing it that would not be possible were they mere collectivities of asexual or automictic individuals." For example, synapsis and random gamete fusion may result in novel combinations of traits, which may diffuse throughout a network and possibly increase survival of the network under different selective regimes. As noted above, such synergy is not possible among lineages of asexuals because they do not comprise a larger entity. Individual asexual lineages evolve, but do so independent of all other lineages and without the benefit of synapsis and so on. Again, groups of asexually reproducing individuals may, for example, respond in like fashion to changes in the environment, but in an ecological context.

In configurations of processes, *mutuality* is manifested when "advantage anywhere in the autocatalytic circuit propagates so as to share that advantage with all other participants" (Ulanowicz 2009). An individual organism, for example, is a "walking 'orgy of mutual benefactions' within itself" (May 1981), referring to the mutual relationships of configurations of processes associated with homeostasis. In network species, mutuality is also evident at various levels. For example, despite their genetically diverse constituency, genetic compatibility is necessary to keep networks intact. Hence, mutuality exists among the core processes that maintain network viability including gametogenesis and syngamy, and the processes that comprise them (e.g., meiosis, synapsis, acrosome reaction). In addition, mutuality may be evident in species-specific mating behaviors, co-parenting of offspring, and so on. Moreover, sexual selection may reinforce mutuality over the long term. Obviously, network integration processes do not exist in strictly asexual lineages, although groups of asexual organisms may respond in like fashion to the environment (e.g., multiple species at a locality respond to a change in oxygen content of atmosphere), in what we classify as an ecological event.

The last component of Ulanowicz's Second Postulate is the complementary relationship between ascendency and overhead. Whereas *ascendency* is "the capacity of a system to order itself and its environment," *overhead* refers to the disorganization, inefficiency, or redundancy found in a system. Network species, characterized by individuals with unique combinations of genes, both products of mutation and core processes (e.g., gametogenesis, syngamy), may have a distinct advantage over asexuals when it comes to persisting in ecosystems, even in the face of great disturbance. Ascendency is counterbalanced by overhead that may derive from deleterious mutations and disruption of co-adapted gene complexes. The degree to which a network species can withstand perturbation, for example, depends partly on its size and organization. For example, the *Homo sapiens* complex is characterized by high ascendency concomitant with large population size, but its closest relatives may exhibit relatively low ascendency in part because their networks have been decimated. A corollary of ascendency/overhead is the trade-off between performance and reliability. At the species level, for example, pandas are highly efficient at making use of bamboo forests, but are more sensitive to perturbation than coyotes, which are generalists in the variety of ecosystems they inhabit. Overhead includes "conditional entropy," (Rutledge et al. 1976), which at the organismal level includes the cost of living, that is, metabolism. In ecosystems, parallel paths of energy transfer are counted as entropy because of the uncertainty associated with redundant pathways (e.g., A $\Rightarrow$ B $\Rightarrow$ C or A $\Rightarrow$ Z $\Rightarrow$ C). In network species, entropy may derive from the chance gene combinations (resulting from independent assortment, synapsis, mate choice, and gamete fusion), some of which may be better adapted than others.

Ulanowicz's (2009) Third Postulate, "systems differ with regard to their histories, which may be evident in their material configurations," is an idea that needs no explanation among evolutionary biologists. Indeed, evolutionary trees are inferred on the basis of material features of organisms (e.g., nucleotide sequences). Ulanowicz (2009) also argues, however, that "the mode of recording doesn't even have to imprint upon a persistent object" and "the trajectory of a system through time conceivably could be used in lieu of a set of existing properties." We see a parallel here with the NSM, under which species are not recognized according to a particular type or degree of divergence; rather, species are recognized where isolated networks of organisms and their processes exist.

The complementary nature of gametogenesis/syngamy, or, ploidy reduction followed by ploidy restoration, is also consistent with the hypothesis that "agonistic transactions" (= "duality") represent a common theme in living systems (Ulanowicz 2009; Lockwood 2012). In asexual lineages, duality exists at the organism level or in an ecological context, but autocatalysis cannot entail multiple clonal lineages as a single evolutionary entity.

In sum, network species, as viewed through the process window, are primarily networks that generate and stockpile genetic variation through reproductive processes (e.g., gametogenesis and syngamy), with which they are able to evolve as a single entity (e.g., respond to environmental pressure). Chance, as manifested in mutation and gene mixing (e.g., synapsis, independent assortment, random gamete fusion), is essential to their existence. Autocatalysis, centripetality, mutuality, and ascendency/overhead are evident especially in the core processes that maintain networks (e.g., gametogenesis/syngamy). Network species are spatiotemporally isolated and some of their history may be evident in their material configurations. Finally, autocatalysis and history determine a network's trajectory from which perturbation is resisted.

5.6 SPECIES AS MATERIAL OBJECTS

We concur with Ulanowicz (2009) regarding the enormous explanatory power of process theory and its relevance to species theory (Lockwood 2012); however, process theory is incomplete without the relevant material components. For example, gametogenesis defined solely in terms of process, in the same way that Lockwood defined species ("species are processes, and they are comprised of processes"), such as, "gametogenesis is a process, and is comprised of processes" or "ploidy reduction" is vague and/or incomplete without the material players and products. In other words, exclusion of *gametes* from a definition of *gametogenesis* is pointlessly minimalistic. If objects are relevant, then models are needed to describe them. The objective of this section is to reconcile species theory with Material Object Theory, which deals with the nature of the material content of the natural world. Here, we follow Moltmann's (2007) interpretation of Varzi (2007).

Varzi (2007) thought that material objects might be characterized by the following: (1) they are physical, concrete objects, accessible to sensory experience, (2) they have properties (e.g., color, shape), (3) they are particulars (or historical individuals), and (4) they have structure or organization. Each of these four components is treated in turn below.

Some scientists (e.g., Kinsey 1930; Kizirian and Donnelly 2008; contra Bessey 1908) have argued that species are real in a material sense in that same way that organisms are real. That is, it is possible to see, touch, and/or interact causally with species, though not with all members of a species simultaneously. For example, it is possible to experience or interact with multiple individual organisms of a given species or witness mating and parturition events. By observing or interacting with individual organisms, one experiences parts of a network species, or participants in the processes that produce a given network, or products of such networks. It is not necessary to experience at a single moment all the individuals of a network in order to have experience with a given species any more than it is necessary for one to witness the embryonic development of a given individual for one to "experience" that individual. So, it may be valid to claim experience with a species through observation of partial networks (or network processes). At the same time, we recognize that organisms have a different kind of continuity not seen in network species. The fact that space and other objects exist between individual organisms in a network does not mean the network cannot act as a single entity. In that respect, they are, organizationally speaking, more like ecosystems than organisms. Nevertheless, we propose that network species are physical, concrete objects, accessible to sensory experience.

Another aspect of Varzi's model is that material objects have qualifying properties (e.g., red chair, heavy locomotive, etc.). Network species have properties, which may include various manifestations of evolutionary divergence including intrinsic reproductive isolation, unique behavior, or unique color pattern. The third component of Varzi's model, that material objects are particulars or historical individuals, has long been recognized as a key aspect of species (Ghiselin 1974b). The last component of the Varzi model is structure, the models for which are fraught with ambiguity. Below, we discuss prevailing models of structure and how they relate to network species.

According to substrata theory, material objects are composed of attributes or properties (e.g., striped, compressed) and the substratum, which underlies the properties and gives them unity (Moltmann 2007: equivalent to the bare particular of Russell [1911] and Bergmann [1967]). In the context of the present discussion, the network represents the bare particular or substrata and Henderson's Lorikeet (*Vini stepheni*) is an example of an individual particular. Moltmann (2007) noted two features of substrata that are consistent with how the NSM might be implemented. First, the object is not the substratum itself, in other words, "particulars are not found 'bare' in nature: they always come around dressed in some set of properties." Hence, Henderson's Lorikeet is a particular species that has its own particular properties (e.g., dark green above shading to golden-yellow tip of tail; red on cheeks and underparts; dark purple central belly; belt across chest green at sides, purple in center; golden-yellow bill and eyes; shrill screech call). A second feature of substrata theory that integrates with evolutionary theory and Process Ecology, is that change over time is accommodated. That is, an individual substratum can evolve without ceasing to exist. It has been argued however that, in order for substratum to play its role, it has to be bare, devoid of properties. If so, the model does not make sense (Sellars 1952), or it is impossible to know (Hume 1888). In that case, substrata theory is inconsistent with the NSM, which proposes unique underlying structure.

According to Bundle Theory, material objects consist of bundles of simultaneously present properties, but lack substratum-like unifying component. This theory does not mesh with the network species for two reasons. First, as per Bundle Theory things do not change, which is in obvious conflict with evolutionary theory. Also, material objects discussed in Bundle Theory lack contingent properties and underlying structure. According to the NSM, species may exhibit any number of contingent properties (e.g., unique color pattern, unique behavior, or shared features in unique combinations) that result in part because of the network structure that underlies species (e.g., mutations appear and spread throughout the network through gene flow, allowing networks to be recognized). Hence, we see little concordance between Bundle Theory and the NSM.

Theory of Tropes, a modification of Bundle Theory, regards the attributes of material objects to be individual objects themselves (Moltmann 2007; Varzi 2007). For example, the redness of a red chair is a trope or a primary entity. Thus, a red chair is not a bundle of properties, but co-localized tropes of different kinds. Trope theory and NSM are antithetical because the latter (and numerous other models such as atom, molecule, cell, organism, ecosystem, and solar system) argues for underlying structure.

5.7 APPLICATION

With a new lens built largely from ideas in Kizirian and Donnelly (2008), Ulanowicz (2009), and Lockwood (2012), one may evaluate alternative species models, including those excoriated by Rosen (1978, 1979). Mayr's 1942 version of the BSC (groups of actually or potentially interbreeding natural populations, which are reproductively isolated from other such groups) incorporates a structural element, population; however, he did not clearly define that notoriously ambiguous term (e.g., Jonckers 1973). The BSC also incorporates process but in the somewhat vague terms used by earlier scientists, namely, interbreeding and reproduction (Wilkins 2009). Mayr (e.g., 1942) also incorporated another problematic element, potentially interbreeding, which refers to a hypothesized future process. Such predictions, however, particularly those involving biological systems may not be justified, as we discuss below. The polytypic species model and use of the subspecific taxon follow from the BSC and present the ontological problem astutely noted by Rosen (1979: 227; quoted above), a problem that exists with regard to the many terms used at the infraspecific level (e.g., deme, population, subspecies, ESU) and extends from their lack of unique inherent organization (Kizirian and Donnelly 2008). According to the NSM, only one model is needed to explain networks of organisms, that is, species, and the capacity for isolated networks to "potentially interbreed" has no bearing on their status.

Rosen (1978, 1979) recognized that implementation of the ESC is arbitrary because there is no objective criterion by which a lineage or group of lineages ("ancestral-descendent sequence of populations evolving separately from others and with its own unitary evolutionary role and tendencies; Simpson 1961) may be recognized as a species. This problem results partly because the structural component of the ESC (and General Lineage Concept [GLC]; de Queiroz 1998), *lineage*, is not unique to a single level of organization, let alone species (Kizirian and Donnelly 2008). Like the BSC, the ESC also includes the perennially problematical term *population* (Jonckers 1973). The only process component of the ESC and GLC is evolution, which, as noted above, is not unique to species or even living systems. The ESC and GLC are silent, however, regarding autocatalytic properties that sustain species (e.g., gametogenesis/syngamy) and the unique ways those systems evolve.

The emphasis of ESC and GLC on lineage and *trajectory* accords with Ulanowicz's Third Postulate, *history*. While trajectory (e.g., fossorial ectotherm, sessile filter-feeder) may be estimated from history and current configuration of networks, predicting the *fate* of such systems is quite another matter. According to some species concepts (e.g., Wiley 1978; Frost and Hillis 1990; Chippendale 2000; see also O'Hara 1993), fate carries decisive weight in evaluating the status of isolated networks. Grant (2002) showed, however, that consideration of fate might result in "problematic" conclusions and Kizirian and Donnelly (2008) impugned fate as a criterion in species models because it is not an intrinsic feature of systems. While (1) the tendency of configurations of processes, including network species, is to resist perturbations and (2) their histories might set species on a trajectory, their fates are not predictable partly because of the "unique combinatorial diversity" exhibited by living systems. In other words, "the universe above the level of atoms is grossly non-repeating" (Elsasser 1969, 1981) and, for that reason, Ulanowicz (2009) argued against the use of probability analysis in biology. If we suspend such concerns for the moment, ignore other complexity (e.g., splitting of networks, network size, distance between networks, ecology), and assume arbitrarily high probability (e.g., 0.99), there are less than even chances any 70 networks or more will reintegrate in the future. Cumulative probabilities decrease precipitously assuming lower initial probabilities of reintegration or greater numbers of coalescent events, as in Figure 5.5. From a number of perspectives then, the inclusion of fate in species models seems to be highly problematic.

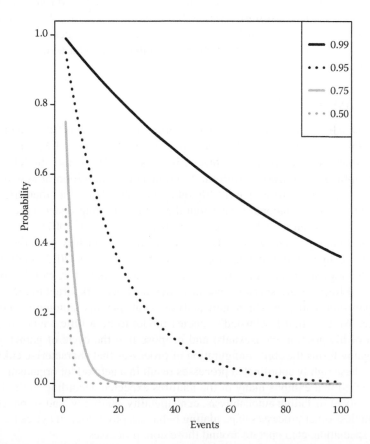

FIGURE 5.5 Cumulative probabilities for N coalescent events. All models assume just two possibilities, that is, no coalescence/total coalescence.

A primary concern for Kizirian and Donnelly (2008) is the status of isolated networks, especially those regarded to be subgroups (e.g., deme, populations, subspecies, ESU) of a more inclusive species. In their view, isolated networks are species regardless of their level of divergence, time of isolation, fate, and so on. Consequently, compared to other species models/concepts, network species tend to be geographically smaller and, in a practical sense, may be too numerous to name. Indeed, according to NSM, most currently recognized binominals are likely species complexes. Less clear, however, is the status of networks among which rare reproduction occurs (e.g., Figure 5.1), where the number of species recognized depends on the model employed. The essence of the NSM is that wherever network structure exists, synergy will be evident; hence, under a strict interpretation of that model, one species (one network characterized by modularity) might be recognized when subnetworks are connected by reproduction. According to Process Ecology, however, hybridization that does not destroy the autocatalytic properties of the involved species may be viewed as a form of overhead, a parallel pathway that brings novel variation into systems (Ulanowicz 2009); hence, multiple species (two networks) could be recognized. Also unclear is the status of loose networks formed through horizontal gene transfer in unicellulars—if autocatalysis does not emerge, then, perhaps no species exists. We have no hard answers for such ambiguous scenarios, rather, we acknowledge that there are limits to the understanding we can achieve with these (or any) models.

5.8　SUMMARY

Our goal here has been to consider species theory in light of ideas from the divergent fields of Process Ecology, which is primarily about process, and Material Object Theory, which is primarily about objects. Because the NSM, like models used in other disciplines, is concerned with the fundamental composition of systems (specific organization and process), it readily integrates with theory in other disciplines, including the transformational ideas about the nature of living systems proposed by Ulanowicz (2009) and models for nonliving objects (e.g., Varzi 2007).

Process Ecology (Ulanowicz 2009) has promising explanatory potential for evolutionary biology and, in principal, we agree with Lockwood's (2012) application of Process Ecology to species theory. We are at variance, however, with his position that asexual lineages are species—because they lack network structure, there is no supraorganismal entity in which autocatalysis, centripetality, autotomy, and so on can arise. We also find Lockwood's species model to be a little vague (in part, a function of his position on asexuals) and propose that the cycle of gametogenesis and syngamy forms the core configuration of processes that characterize eukaryotic species. These tightly coordinated processes result in a network of organisms (= network species) that exhibits many of the features expected according to the Process Ecology model including autocatalysis, centripetality, autotomy, and so on. In addition, countless other processes (e.g., mating behavior, physiological cycles that effect mating, speciation, etc.) operate around those core processes.

Despite their minor role under the process umbrella, material objects comprise essential elements of systems. We also recognize that living systems are fundamentally different from nonliving ones; however, one point of this exercise was to test the idea that the NSM would facilitate the unification of thought among scientific disciplines. If scientists think alike about objects and processes, it should be possible to find common ground among models used in diverse disciplines. We find *substrata* theory (Moltmann 2007; Varzi 2007) to be most consistent with NSM (Kizirian and Donnelly 2008) because of its recognition of underlying structure.

Circling back to the issues that concerned Rosen, the NSM will tend to identify smaller and greater numbers of species than most other species models, in many cases corresponding with what some would identify as populations, the operational unit preferred by Rosen (1978, 1979). In addition, we find Rosen's concerns about the arbitrary nature of species models and the questionable reality of subspecies to be valid. The NSM represents an attempt to rectify those issues by prioritization of core structure and processes, which, consequently, facilitates integration with models in other branches of science.

ACKNOWLEDGMENTS

We thank Frank Burbrink, Marcelo Gehara, Arianna Kuhn, and Brendan Reid for their comments; Arianna Kuhn for creating Figure 5.5; and Paul Sweet for information on mixed feeding flocks in birds.

REFERENCES

Bergmann, G. 1967. *Realism.* University of Wisconsin Press, Madison.
Bessey, C. E. 1908. The taxonomic aspect of the species questions. *American Naturalist* 42:496:218–224.
Bickhardt, M. H. and D. T. Campbell. 1999. Emergence. In Anderson, P. B., Emmeche, C., Finneman, N. O. and Christiansen, P. V. (eds.), *Downward Causation.* Aarhus University Press, Aarhus, pp. 322–348.
Chippendale, P. T. 2000. Species boundaries and species diversity in the central Texas hemidactyliine plethodontid salamanders, Genus Eurycea. In Bruce, R. C., Jaeger, R. G. and Houck, L. D. (eds.), *The Biology of Plethodontid Salamanders.* Plenum Publishers, New York, pp. 149–165.
Conant, R. and J. T. Collins. 1991. *Reptiles and Amphibians of Eastern/Central North America.* Houghton Mifflin Company, Boston.
de Queiroz, K. 1998. The general lineage concept of species, species criteria, and the process of speciation: A conceptual unification and terminological recommendations. In Howard, D. J. and Berlocher, S. H. (eds.), *Endless Forms: Species and Speciation.* Oxford University Press, New York, pp. 57–75.
Du Rietz, G. E. 1930. The fundamental units of botanical taxonomy. *Svensk Botanisk Tidskrift* 24:333–428.
Elsasser, W. M. 1969. A causal phenomena in physics and biology: A case for reconstruction. *American Scientist* 57:502–516.
Elsasser, W. M. 1981. A form of logic suited for biology? In Rosen, R. (ed.), *Progress in Theoretical Biology,* Vol. 6. Academic Press, New York, pp. 23–62.
Ereshefsky, M. 1999. Species and the Linnaean hierarchy. In Wilson R. A. (ed.), *Species: New Interdisciplinary Essays.* MIT Press, Cambridge, pp. 285–305.
Frost, D. R. and D. M. Hillis. 1990. Species in concept and practice: Herpetological applications. *Herpetologica* 46:87–104.

Ghiselin, M. T. 1974a. *The Economy of Nature and the Evolution of Sex.* University of California Press, Berkeley.

Ghiselin, M. T. 1974b. A radical solution to the species problem. *Systematic Zoology* 23:536–544.

Grant, T. 2002. Testing methods: The evaluation of discovery operations in evolutionary biology. *Cladistics* 18:94–111.

Hartog, M. 1913. *Problems of Life and Reproduction.* Putnam, New York.

Hennig, W. 1966. *Phylogenetic Systematics.* University of Illinois Press, Champaign.

Hull, D. L. 1978. A matter of individuality. *Philosophy of Science* 45:335–360.

Hume, D. 1888. *A Treatise of Human Nature.* Oxford, Clarendon Press, London.

Jonckers, L. H. M. 1973. The concept of population in biology. *Acta Biotheoretica* 22:78–108.

Kinsey, A. C. 1930. The gall wasp genus *Cynips.* A study in the origin of species. *Indiana University Studies* 16 (84–86):1–577.

Kizirian, D. A. and Donnelly, M. A. (2008). The Network Species Model. arXiv:0808.1590v1 [q-bio.PE].

Lockwood, J. A. 2012. Species are processes: A solution to the "species problem" via an extension of Ulanowicz's Ecological Metaphysics. *Axiomathes* 22:231–260.

May, R. M. 1981. *Theoretical Ecology: Principles and Applications.* Sinauer, Sunderland, MA.

Mayr, E. 1942. *Systematics and the Origin of Species.* Columbia University Press, New York.

Moltmann, F. 2007. The Structure of Material Objects. Structure in Ontology [lecture notes for 15/11/2007; http://semantics.univ-paris1.fr/pdf/handout%2015-11.pdf].

Newman, M. E. J. 2006. Modularity and community structure in networks. *Proceedings of National Academy of Sciences of the United States of America* 103:8577–8582.

Newman, M. E. J. 2012. Communities, modules and large-scale structure in networks. *Nature Physics* 8:25–31.

O'Hara, R. J. 1993. Systematization, generalization, historical fate, and the species problem. *Systematic Biology* 42:231–246.

Popper, K. R. 1990. *A World of Propensities.* Thoemmes, Bristol.

Poulton, E. B. 1904. What is a species? *Proceedings of the Entomological Society of London* 1903, lxxvii–cxvi.

Quine, W. V. O. 1960. *Word and Object.* MIT Press, Cambridge.

Rieppel, O. 2009. Species as process. *Acta Biotheoretica* 57:33–49.

Rosen, D. 1978. Vicariant patterns and historical explanation in biogeography. *Systematic Zoology* 27:159–188.

Rosen, D. 1979. Fishes from the uplands and intermontane basins of Guatemala: Revisionary studies and comparative geography. *Bulletin of the American Museum of Natural History* 162:267–376.

Russell, B. 1911. On the relation of universals and particulars. *Proceedings of the Aristotelian Society* 12:1–24.

Russell, P. J., P. E. Hertz and B. McMillan. 2014. *Biology: The Dynamic Science.* Third Edition. Cengege Learning, Boston.

Rutledge, R. W., B. L. Basorre and R. J. Mulholland. 1976. Ecological stability: An information theory viewpoint. *Journal of Theoretical Biology* 57:355–371.

Sellars, W. 1952. Particulars. *Philosophy and Phenomenological Research* 13:184–199.

Simpson, G. G. (1961). *Principles of Animal Taxonomy.* Columbia University Press, New York.

Ulanowicz, R. E. 2009. *The Third Window: Natural Life beyond Newton and Darwin.* Templeton Foundation Press, West Conshohocken, Pennsylvania.

Ulanowicz, R. E. 2012. Widening the third window. *Axiomathes* 22:269–289.

Van Valen, L. 1973. A new evolutionary law. *Evolutionary Theory* 1:1–30.

Varzi, A. C. 2007. "La natura e l'identità degli oggetti materiali" [The Nature and Identity of Material Objects]. In A. Coliva (ed.), *Filosofia Analitica. Temi e Problemi.* Carocci, Roma, pp. 17–56.

Wiley, E. O. 1978. The evolutionary species concept reconsidered. *Systematic Zoology* 27: 17–26.

Wilkins, J. S. 2005. A scientific modern amongst medieval species. *University of Queensland Historical Proceedings* 16:1–5.

Wilkins, J. S. 2009. *Defining Species: A Sourcebook from Antiquity to Today.* American University Studies, Series V, Philosophy. Vol. 23. Peter Lang, New York.

6 The Inhibition of Scientific Progress
Perceptions of Biological Units

Christopher M. Murray
Tennessee Technological University

Caleb D. McMahan
The Field Museum

Brian I. Crother
Southeastern Louisiana University

Craig Guyer
Auburn University

CONTENTS

6.1 INTRODUCTION

The species problem has thus involved difficulties in understanding the ontological status of fundamental biological units, and failure to interrelate levels
of integration in the appropriate manner.... In addition we should note that

we often fail to solve our problems because we cannot even identify them. Under such circumstances, conceptual investigations do more than just help. They are the only way out.

Ghiselin (1974, p. 543)

Donn Rosen's (2016) list of "assumptions that *inhibit* scientific progress in comparative biology" intended to notarize the principles that handicap the biological thinker. Rosen's goal to label and display some general, evolutionary, taxonomic, and biogeographic skepticism served as a warning, and helped banish such notions to a quarantine of plaguing ideals. Rosen's efforts to clear the current and possible future roadblocks are applicable to three academic categories: general mindsets (General 1–7), epistemological validity (Evolutionary theory 1, 4, 8, 9–11, Taxonomy 5–7, Biogeography 1–4), and operation (Evolutionary theory 2, 3, 5–7, Taxonomy 1–4, Biogeography 5–8), all fundamental to critical biological progress.

Moreover, Rosen indirectly affirmed the importance of biological units, implying that an understanding of species is fundamental to avoid 13 of his 26 evolutionary, taxonomic, and biogeographic hindering tenets. The importance of fundamental units in biology has been endorsed previously (e.g., Frost and Kluge 1994; de Queiroz 2005). Whether an investigator is using species in an evolutionary investigation or areas of endemism in a biogeographic analysis, these things (species) serve as the fundamental unit with which our comparative deductions are made. Just as all endeavors rely on some unit, in biology, we must rely on natural units.

One such unit, the species, is recognized as critical to our comparative understanding by Rosen, and has been the subject of more decades of debate, relative to its use, than any other. The species debate entertained opinions on the metaphysics of species, aimed at determining their nature of being, their reality or lack thereof, and to clarify their epistemological identity and utility. While discussion of this kind dates back to the 1850s, when species were argued to be notions of a Creator, and thus "real" (Agassiz 1857), the ontological exchange that is "the species debate" peaked between the late 1960s (roughly beginning with Hennig 1966) and early 1990s (seemingly smoldering out with Frost and Kluge 1994, although some have continued the debate, e.g., Rieppel 2005, 2007). Although our goal is not to exhaustively review the species debate, we briefly review the four predominant ontological species categories that maintained academic sponsorship throughout the debate, including noteworthy advocates of each.

It was argued that species are…

Individuals—Real, ostensively defined particulars that lack instances, have spatial and temporal boundaries, and whose parts respond cohesively to change (e.g., the person Donn Rosen) (Hennig 1966; Ghiselin 1974, 1981, 1987; Hull 1976; Wiley 1980; Frost and Kluge 1994; Mayden 2002).

Artificial Classes—Extensionally defined groups of items, sets defined by rules existing only in human perception (e.g., chairs) (Kitcher 1984, 1987, 1989; Bernier 1984; Grene 1989).

Mereological Sums—Cohesive sums of parts (e.g., the liver) (Brogaard 2004).

Natural Kinds—Described by J. S. Mill in the 1870s as like things known to be indefinitely similar, based on infinite induction, that were not causal products, but rather things subject to simultaneous change in property (Magnus 2014, in a work regarding history of the term "natural kind"). Such a category quickly morphed into the designation of Putnam (1975) and Kripke (1971), in which members of a Kind, despite differing properties, are held together by some essence (Magnus 2014). Thus the disparity in the definition of the term "natural kind" is overwhelming, ranging from a definition more similar to that of a mereological sum ["an artificial grouping of like things each with like parts and connection of those parts (Baum 1998)"], to a definition seemingly more artificial than a class ["a thing with an inductive or explanatory role in satisfying the accommodation demands of a disciplinary matrix (Boyd 1999)"].

None of the Above/A Combination of the Above—Rieppel (2007) argues that species do not fit exclusively into any of the above categories, but are perhaps best considered "individuals of a kind" and possessed with criteria of individuals, but are also kinds, "each being one of its kind."

Currently, few attempts are made to continue this debate and discussion. Aside from direct mention of ontology with species as examples (Murray and Crother 2015), discussion of the metaphysics surrounding the species unit has all but fizzled out. Such ignorance of perhaps the most puzzling and fundamental of biological concepts can only be attributed to one of three explanations. (1) The problem is solved and we have come to some consensus, (2) a species is what any competent taxonomist says it is (Regan 1926), or (3) trends in comparative biology have largely shifted away from discussion of units, and been replaced by a general indifference towards metaphysical discussion. A glance at the glossary of any general biology textbook over the past few decades reveals a definition of species revolving around a taxonomic level with phenetic similarity and interbreeding capabilities (Appendix 6.1). Given the growing attention to bioinformatics and a shift to a more analytical approach to biology, coupled with a stagnant epistemology, apathy to species may be a more valid hypothesis. The above hypotheses warrant census of current perceptions of species among comparative biologists. Here, we implement anonymous survey methods to census the current ontological perceptions of species among academic cohorts (e.g., graduate student, faculty, etc.) and field of teaching (e.g., evolution, ecology, zoology, etc.). We hypothesize no familiarity with ontological literature (difference between individuals and classes and the definition of natural kinds) assessed via identification of contradictory responses among questions and evaluate the confusion associated with the variety of definitions for natural kinds.

6.2 METHODS

A survey consisting of six questions was generated to assess contemporary views on the nature of species (Appendix 6.2). All aspects of the distribution of the survey and compilation of responses were performed in accordance with the Institutional

Review Board at Auburn University (Protocol #14–141 EX 1404). Questions centered on the reality of species (and, by extension, subspecies) and their philosophical identity, as well as two questions to evaluate the participant's background that inform his or her views on the nature of species. The survey was built using a host website (Qualtrics Survey Platform, Auburn University).

The survey was posted in four different locations in an effort to attract attention from a broad survey of evolutionary biologists. Although these groups of scientists are certainly not the only ones who may have opinions on the nature of species, this is the group that has historically been most active in the species dialogue. The survey was posted at the Evolution Directory (EvolDir) online listserv, as well as the Facebook pages for the Society for the Study of Evolution, Willi Hennig Society, and the Society of Systematic Biologists.

The survey was available from May 21, 2014 through March 1, 2015 (284 days). At the conclusion of the survey period, a total of 162 responses were received from over 16 different countries. Histograms were generated for selected responses to each question in the survey. G-tests for independence were used to assess differences in the proportion of responses among options, positions, and/or fields of study.

6.3   RESULTS

All academic positions were represented in the responses (Curator [8], Professor [47], Post Doctoral [29], Graduate student [51], Other [24]). Participants represented 65 institutions. Responses to individual questions, among cohorts and fields of study, as well as intentional combination responses aimed at assessing the ontological knowledge behind answers, are expanded below.

6.3.1   ARE SPECIES REAL?

This question received 162 responses. Seventy-eight percent of participants said species are real. This opinion did not differ among positions and no position was in 100% agreement regarding the reality of species. Among fields of study, geneticists and evolutionary biologists were least likely to respond that species are real.

6.3.2   PHILOSOPHICALLY, SPECIES ARE (CATEGORY) ...

The number of responses significantly differed among philosophical categories ($G = 38.7$, DF = 4, $p \leq 0.001$, as in Figure 6.1). Out of 158 responses, 47% of participants said species were natural kinds. The next most popular response was artificial classes (24%). Graduate students made up the majority of participants who claimed species were artificial classes and natural kinds.

In combination with question 1, 16 participants who said species were real also said they were artificial classes. Among cohorts, 20% of graduate students and 25% of postdoctorates provided this combined response. Fifty-five percent of people who responded in favor of the reality of species (70 participants) placed them in the philosophical category natural kinds. That proportion did not vary among cohorts.

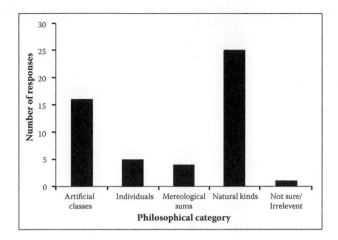

FIGURE 6.1 Histogram showing the number of responses among philosophical category for the species unit ($G = 38.7$, DF = 4, $p = <0.001$).

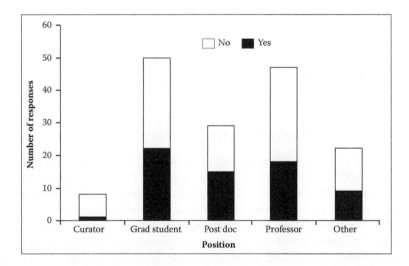

FIGURE 6.2 Histogram showing the proportion of responses concerning the philosophical reality of subspecies among academic cohorts ($G = 10.24$, DF = 4, $p = 0.037$).

6.3.3 ARE SUBSPECIES REAL?

The proportion of responses that claimed subspecies are real differed among participant position ($G = 10.24$, DF = 4, $p = 0.037$, as in Figure 6.2). Roughly half of all positions, besides curator, advocated for the reality of subspecies. In combination, 40% of participants claim species and subspecies are both real. The same proportion advocates for the reality of species but not subspecies. Twenty percent of participants do not advocate for the reality of species or subspecies, and a negligible proportion claim subspecies are real, but not species.

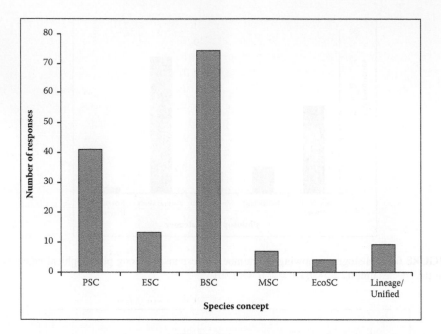

FIGURE 6.3 Histogram showing the proportion of responses among the following species concepts; phylogenetic species concept (PSC), evolutionary species concept (ESC), biological species concept (BSC), morphological species concept (MSC), ecological species concept (EcoSC) and Lineage/Unified species concept ($G = 59.6$, DF = 5, $p \leq 0.001$).

6.3.4 What Species Concept Do You Follow?

The number of responses differed significantly among species concepts ($G = 59.6$, DF = 5, $p = <0.001$ as in Figure 6.3). The top six species concepts reported were phylogenetic, evolutionary, biological, morphological, ecological, and lineage/unified. Forty-six percent of all participants reportedly use the biological species concept followed by 25% who report the phylogenetic species concept. Among fields of study, the biological species concept was the only one reported in all fields (zoology/botany, anatomy/physiology, evolution, ecology/behavior, genetics, introductory biology). Roughly half of participants involved in evolutionary biology report exclusive use of the biological species concept while nearly one-quarter report the use of multiple species concepts. Ecology/behavior and zoology/botany were the two other fields that reported the use of multiple species concepts depending on the empirical endeavor.

6.4 DISCUSSION

Despite decades of effort intended to clarify the philosophy behind the existence of species, results here indicate continued disparity and confusion among comparative biologists. In summary, the majority of participants indicated that species are real things, despite the result that the field potentially most likely to utilize the species unit empirically (evolutionary biologists/geneticists) is least likely to think they are real. Opinions regarding the philosophical category to which species belong remain disparate. Further, the divide in opinions regarding the reality of subspecies could not be more contrasting. Such contrast is maintained among cohorts as well. Lastly, pertaining to direct survey responses, the disparity in species concept is surprisingly small. While many species concepts were offered, the biological species concept was the predominate response, present in all fields of study. Some participants refused to choose a single concept, suggesting that different questions/organisms require different ways to diagnose a species.

6.5 A RESPONSE: INCONGRUENCE AND IMPLICATIONS

Opinions regarding the philosophical nature of species remain disparate despite decades of discussion. Worse, many opinions appear to be contradictory and thus potentially misguided, which indicates not only a dampening in correspondence on such a topic, but a more recent disinterest in arguably one of the most important discussions in comparative biology. Such ontological errors are not position-specific, indicating that the trend is not career or temporally explained.

Certain combined responses indicate either a lack of ontological principles or general ambiguity in ontological terms. For instance, participants who declared both the reality of species and assigned them to artificial classes lack a philosophical understanding of the topic. Perhaps this is a result of our oversimplified use of the word "real" in that some things exist outside of human perception but are defined, not diagnosed. It is not our goal to expand on the gamut of philosophical terminology here, however, within our use, real entities existing outside of human perception cannot be defined solely by human rules, such as a chair. This problematic observation was evident among graduate students, postdoctorates, and professors alike. Perhaps this results from a semantic misconception of the word "real," potentially mistaking the indication of an ontological entity for the utility and presence of the term.

Half of every cohort that supported the reality of species also indicated that species were natural kinds. Therefore, half of every cohort that advocates for the reality of species does not see natural kinds as the best philosophical category for real entities. This indicates the disparity of the definition of natural kinds, a term whose meaning ranges from something potentially even more artificial than classes (Boyd 1999), to something akin to Mill's original definition that resembles real entities in terms of cohesiveness (Magnus 2014).

Our results also indicate that the proportion of people who think species are real and subspecies are not (the philosophical category: Individual) and the proportion who believe both are real is the same. The fact that subspecies, and not individuals, can exist only under the class categories (because if species are individuals they cannot have instances, i.e., subspecies) further suggests a lack of understanding of these categories. If philosophical perceptions supported individuals, results should indicate that species are real and subspecies are not. Rather, if we supported classes, then the reality (diagnosis outside of human perception) of neither should be indicated.

The results surrounding species concepts presented here corroborate the need for Rosen's (2016) assumptions and oppose Rosen's (1979) efforts. Rosen explicitly included the biological species concept as an assumption that inhibits scientific progress. He asserted that reproductive compatibility is a primitive trait retained or altered in a mosaic fashion across an evolutionary landscape and points to hybridization as this concept's most fatal flaw (Rosen 1979). Rosen asserted that "[the biological species concept] will lead to inferences that are in direct conflict with the avowed aims of systematics … (Rosen 1979)." Despite these and many other criticisms, the majority of biologists still think this a useful concept, which we attribute to apathy or inertia.

6.6 CONCLUSIONS

The goal of this manuscript is not to impose an opinion on the reality of species, their philosophical category, or species concepts but rather to illuminate the general apathy for the species debate and species ontology that exists in present-day comparative biology. To participate in biological research without understanding the ontology of the fundamental unit—the species—is like bringing your car to a mechanic who did not care to know what a car was. It doesn't work. The ontology of the unit provides the foundation for its correct use, as in Figure 6.4. This foundation provides a hierarchical diagnosis of philosophical category, criteria of that category, and subsequent utility for any unit. Further, operation can then be assessed via a feedback mechanism that rechecks the use of that unit. A lack of consideration of the nature of being for any unit results in the random and possible misguided use of that unit. Overlaid on this pyramid in Figure 6.4 are epistemological considerations,

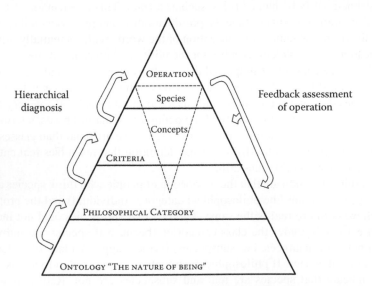

FIGURE 6.4 Schematic illustrating the hierarchical relationship between levels of thinking in comparative biology and the foundational role of ontology. Overlaid are the proportions of the pyramid incorporating species concept thinking.

Magnus, P. D. 2014. No grist for Mill on natural kinds. *Journal for the History of Analytical Philosophy* 2:1–15.

Mayden, R. L. 2002. On biological species, species concepts, and individuation in the natural world. *Fish and Fisheries* 3:171–196.

Murray, C. M. and Crother, B. I. 2015. Entities on a temporal scale. *Acta Biotheoretica* 64:1–10. DOI 10.1007/s10441–015–9269–5.

Putnam, H. 1975. Mind, language and reality. *Philosophical Papers*. Volume 2. Cambridge: Cambridge University Press.

de Queiroz, K. 2005. A unified concept of species and its consequences for the future of taxonomy. *Proceedings of the California Academy of Sciences* 56:196–215.

Regan, C. T. 1926. Organic evolutions. *Rep. 93rd Meeting of the British Association for the Advancement of Science 1925*, pp. 75–86.

Rieppel, O. 2005. Monophyly, paraphyly, and natural kinds. *Biology and Philosophy* 20:465–487.

Rieppel, O. 2007. Species: Kinds of individuals or individuals of a kind. *Cladistics* 23:373–384.

Rosen, D. E. 1979. Fishes from the uplands and intermontane basins of Guatemala: Revisionary studies and comparative geography. *Bulletin of the AMNH* 162: article 5.

Rosen, D. E. 2016. Assumptions that inhibit scientific progress in comparative biology. In: Crother, B. I. and Parenti, L. R. (eds.), *Assumptions Inhibiting Progress in Comparative Biology*. Boca Raton: CRC Press, pp. 1–4.

Wiley, E. O. 1980. Is the evolutionary species fiction?—A consideration of classes, individuals, and historical entities. *Systematic Zoology* 29:76–80.

7 Neo-Darwinism, Hopeful Monsters, and Evo-Devo's Much Expanded Evolutionary Synthesis

Mary E. White
Southeastern Louisiana University

CONTENTS

7.1 INTRODUCTION

In 1985, when Donn Rosen proposed his list of assumptions that inhibit progress in comparative biology (Rosen 2016), the resurgence of interest in evolutionary developmental biology was, if not in its infancy, certainly not much more than a toddler. The first usage of the term evolutionary developmental biology was apparently only two years earlier in 1983 (according to Hall 2012), and the homeobox had just been discovered (McGinnis et al. 1984a,b; Scott and Weiner 1984). It is therefore that much more impressive that Rosen's Evolutionary Theory list included three items that have been reinvigorated by the subsequent decades of research in "evo-devo." Two of these inhibiting assumptions will be the main focus of this essay: Number 5 read, "Random mutation, natural selection and microevolution combine as a progressive research program to explain the hierarchy of organisms," and number 10 stated, "Goldschmidt's ideas must be wrong because they conflict with neo-Darwinism." Number 10 on his list was followed by number 11, "Ditto for neolamarckism," a related topic that will be dealt with only briefly herein.

7.2 DARWINISM, NEO-DARWINISM, AND THE MODERN SYNTHESIS

In 1858, Darwin and Wallace published separately their hypothesis of evolution by natural selection, followed by Darwin's *On the Origin of Species* in 1859, and the world of evolutionary biology has never been the same. Others quickly weighed in to support, refute or tinker with their hypothesis. Historians of science often consider the neo-Darwinian period to have begun with Weismann's work in the late 1800s to expand the Darwinian hypothesis, emphasizing the importance of natural selection and excluding Lamarckian ideas of inheritance of acquired characteristics (for review, see Gould 2002, p. 198; Kutschera and Niklas 2004). The 1920s saw the beginning of the "modern synthesis," sometimes called the neo-Darwinian synthesis, which describes a view of evolution where natural selection, acting on genetic variation, produces gradual change over time. This synthesis, named by Huxley (1942), but generally associated with Fisher, Wright, Haldane, and Dobzhansky among others, used population genetics to reconcile Mendelian inheritance with the external selective forces espoused by Darwin and Wallace. The modern synthesis posited natural selection as the major, if not sole, driver of evolutionary change. Dobzhansky (1937) is credited with defining evolution as "a change in the frequency of an allele in a gene pool."

In his introduction to the 100th anniversary edition of Darwin's *Origin of Species* (1958), Huxley wrote, "Neo-Darwinism, as we may call the modern theory of gradual transformation operated by natural selection acting on a Mendelian genetic outfit of self-reproducing and self-varying genes, is fully accepted by the great majority of students of evolution. Darwin would have rejoiced to see how it (and it alone) can account for the varied and often puzzling facts of evolution ..." and later added, "Today, a century after the publication of the *Origin*, Darwin's great discovery, the universal principle of natural selection, is firmly and finally established as the sole agency of major evolutionary change." (Darwin and Huxley 1958).

7.3 CRITICISMS OF NEO-DARWINISM AND THE MODERN SYNTHESIS

Let me say this—I am a fan of Charles Darwin. We all grew up with examples of natural selection and change over time. It is easy to imagine how selection acting on natural variation can lead to gradual changes such as different color morphs of moths or the varied finches of the Galapagos Islands, the latter of which might rightly be called macroevolution. However, it is a little harder to reconcile these types of changes with big-picture evolution of such disparate organisms as rotifers and jellyfish and oak trees and turtles and zebras. This quote by Theißen (2006) sums up my own queasiness about the idea that natural selection on different alleles might, over very long periods of time, lead to gradual but very profound changes in form: "Why did bacteria not just give rise to more and more optimized and better and better adapted bacteria forever, but to mushrooms, monkeyflowers and man?"

As early as 1889, Darwin's cousin Galton wrote, "The theory of Natural Selection might dispense with a restriction for which it is difficult to see either the need or the justification, namely, that the course of evolution always proceeds by steps that are severally minute, and that become effective only through accumulation. That the steps may be small and that they must be small are very different views; it is only to the latter that I object, and only when the indefinite word "small" is used in the sense of "barely discernable ...". He added that his objective was, "... to argue that Evolution need not proceed by small steps only." He later noted that while very small steps may be most common, the large steps may be more important in evolution (Galton 1889).

Though Galton did not agree with requiring evolutionary change to proceed gradually, he did not apparently object to the idea of natural selection as a (the?) major driver of evolution. However, some contemporaries of Darwin argued that natural selection alone could not explain evolution. In the introduction to his book, *The Genesis of Species*, Mivart (1871) gave a list of difficulties that natural selection could not explain. Included in that list: "That 'Natural Selection' is incompetent to account for the incipient stages of useful structures," "That it does not harmonize with the co-existence of closely similar structures of diverse origin," and "That there are grounds for thinking that specific differences may be developed suddenly instead of gradually." Subsequent chapters in his book then discussed these and other criticisms in great detail. For example, expanding on the first point, in chapter 2 (p. 26) Mivart added, "But Natural Selection utterly fails to account for the conservation and development of the minute and rudimentary beginnings, the slight and insignificant commencements of structures, however useful those structures may afterwards become." He discussed such varied structures as the head (and particularly eye position) of flatfish, the baleen of whales, the tube feet of sea urchins, and the imaginal disks of fruit flies (Mivart 1871, chapter 2).

Even as the new synthesis was in its infancy, some argued that natural selection was not a creative force, but perhaps a restrictive one—removing unfit organisms rather than originating fit ones. One of Darwin's most serious critics was Haldane (1932), who in suggesting that Darwin's ideas did not explain all facets of evolution, wrote, "In the first place, we have every reason to believe that new species may arise quite suddenly, sometimes by hybridization, sometimes by other means. Such species do not arise, as Darwin thought, by natural selection. When they have arisen they may justify their existence before the tribunal of natural selection, but that is a different matter ..." A particularly striking example of this, unknown in Haldane's time, are the unisexual parthenogenetic lizard species that have arisen, many through hybridization (for review, see Cole 1975). While some might not consider these to be distinct species, Cole (1985) argued strongly that at least some of these clonal unisexual lines deserved species recognition. Mallet (2007) reviewed significant speciation by hybridization in plants, and also gave examples of reproductively isolated hybrid species in fruit flies, butterflies, and fishes. According to Mallet (2007), "... they {interspecific hybrids} are also 'hopeful monsters', with hefty differences from each parent, no adaptive history to any ecological niche, and little apparent scope for survival."

In his criticisms of natural selection as a generative force, Mivart (1871) repeatedly invoked development. Darwin himself recognized the importance of embryology, though perhaps more as evidence for rather than a mechanism of evolution.

However, throughout the first half of the twentieth century, the potential contribution of embryology was all but ignored in favor of population genetics models and the new synthesis. Thus enter Richard Goldschmidt.

7.4 RICHARD GOLDSCHMIDT—SYSTEMIC MUTATIONS AND MACROMUTATIONS

Richard Goldschmidt's life and his work have been analyzed extensively by Dietrich, a historian of biology at Dartmouth, and I have borrowed freely from his work (Dietrich 1995, 1996, 2000, 2003). As an experimental scientist, Goldschmidt amassed volumes of data particularly in the field of physiological genetics. He studied sex determination in gypsy moths and found that when different geographical morphs of the sexually dimorphic moths were mated, the offspring were often intersexes rather than sexually dimorphic. As noted below, he later worked on homeotic mutations in *Drosophila*. However, Goldschmidt is most remembered for his controversial views rather than his experimental science, particularly his idea of the "hopeful monster."

Goldschmidt's first published usage of the term "hopeful monster" was in 1933, when he wrote, "I further emphasized the importance of rare but extremely consequential mutations affecting rates of decisive embryonic processes which might give rise to what one might term hopeful monsters, monsters which would start a new evolutionary line if fitting into some empty environmental niche." According to Goldschmidt (1933), "The dachshund and the bulldog are monsters. But the first reptiles with rudimentary legs or fish species with bulldog-heads were also monsters." Note that at the time, amphibians were considered "reptiles" and he was clearly referring to the transition of amphibians onto land. Goldschmidt's concept of the "hopeful monster" was certainly influenced by the discovery of homeotic mutations in *Drosophila* by Bridges and Morgan among others in the early part of the twentieth century. Single mutations such as bithorax, which transforms the third thoracic segment with its halteres into an additional second thoracic segment with a second pair of wings, and antennapedia, in which the distal portion of the antenna is transformed somewhat improbably into the distal portion of a leg, provide the raw material from which monsters are imagined.

In 1940, Goldschmidt published his book *The Material Basis for Evolution* that he hoped would help span what he called the "bridgeless gap" between microevolution and macroevolution. Keep in mind, when this book was written, Avery, MacLeod, and McCarty, and Hershey and Chase had not yet performed their seminal experiments that convinced most scientists that DNA, not protein, was the genetic material (Avery et al. 1944; Hershey and Chase 1952). In his book, Goldschmidt proposed two very different possible mechanisms for macroevolution: systemic mutations and the previously mentioned macromutations.

Goldschmidt's idea of systemic mutations suggested that chromosomal rearrangements rather than changes in specific genes were responsible for big scale evolutionary changes. This was supported to some extent by his own work on gypsy moths and by the work of prominent geneticists working on *Drosophila* at the time. He was particularly influenced by Muller's experiments that showed the importance of position effect on scute formation in *Drosophila* (discussed in Dietrich 2003). Interestingly, and perhaps ironically, conversations with Dobzhansky about position effects led in part to Goldschmidt's conclusion (Dietrich 2000).

The idea of systemic mutations was highly controversial, and rightly so, because it denied the importance of the Morgan model of particulate genes. In apparently characteristic Goldschmidt fashion, he took valid data from a few studies and formulated an overarching theory that fit some available data but is certainly untenable now. We do recognize some examples of the importance of position effects, including the effects of translocation and inversions. For example, when genes are translocated into regions in or near heterochromatin, gene expression can be eliminated or diminished. A number of models have been developed that associate chromosome rearrangements with speciation (for review, see Rieseberg 2001) and empirical evidence for this was presented in *Drosophila* (Brown et al. 2004), among other species. Parris (2011) reviewed work indicating that pericentric inversions may be important in speciation, including the speciation that led to the divergence of chimps and humans. Despite these observations, it does not seem that position effects are the dominant driver of macroevolution.

While he may have been reviled in his day for the concept of systemic mutations, Goldschmidt is far better remembered today for his alternate idea of macroevolution by macromutation, or the "hopeful monster." This idea was less controversial at the time and even found some support from scientists such as Simpson and Wright, among others. Goldschmidt (1940) wrote, "A monstrosity appearing in a single genetic step might permit the occupation of a new environmental niche and thus produce a new type in one step." Goldschmidt (1933) recognized that most mutations that led to large phenotypic changes would produce "monsters" and most of these monsters would not be viable. The "hopeful monster" was therefore the admittedly rare macromutation that could survive and leave behind progeny. As he stated, "I can not see any objection to the belief that occasionally, though extremely rarely, such a mutation may act on one of the few open avenues of differentiation and actually start a new evolutionary line."

7.5   HOMEOTIC GENES AND HOPEFUL MONSTERS

One obvious place to start looking for potential hopeful monsters was in the aforementioned homeotic mutations, which began to be isolated and characterized in 1915 (reviewed in Bridges 1944). Mutations such as bithorax and antennapedia might not make viable hopeful monsters, but they at least showed that large scale changes could occur quickly, in a single genetic step. Goldschmidt himself turned his experimental efforts to homeotic mutations in *Drosophila* such as podoptera, which transforms wings to legs, and tetraltera, which transforms wings to halteres. His results were eventually incorporated into Wright's shifting balance theory (Dietrich 2000).

Lewis's careful analysis of the bithorax complex beginning in the 1950s (reviewed in Lewis 1978, 1992) provided much information on the regulation of segment identity in *Drosophila*, for which he was awarded the 1995 Nobel Prize in Physiology or Medicine along with Nüsslein-Volhard and Wieschaus. With the characterization of control of axial patterning by *Hox* complexes in fruit flies (e.g., see McGinnis and Krumlauf 1992) and then in many other organisms (Burke et al. 1995; Carroll 1995; Krumlauf 1994) a whole new world of embryonic regulation began to open up. Changes in *Hox* gene expression can lead to what almost sounds like Rudyard Kipling "Just So" stories, such as "How the Snake Lost its Legs" in which elegant work by Cohn and Tickle (1999) suggested that the pattern of *Hoxc6* and *Hoxc8* expression in snakes compared to chickens could explain the absence of forelimbs. Or "How the Goose Got a Long Neck" in which patterns of expression of *Hoxc6* are correlated with neck length, or more correctly number of cervical vertebrae, with very few in the mouse, for example, and many in the goose (Gaunt 1994; Burke et al. 1995). Notice that these examples do not require mutations in the protein-coding regions of the genes themselves, but rather in the expression patterns. Expression of *Hox* genes was also associated with changes in body plans of arthropods (e.g., Hughes and Kaufman 2002).

It is not just the *Hox* genes that can make big changes. The Just So story of "How the Chicken Lost its Teeth" may very well involve the *talpid*^{2} gene (Harris et al. 2006). A gradualist idea might suggest that as the beak evolved, the need for teeth became less, and thus the genes that produced components such as the enamel and the dentin would degenerate and the teeth might become smaller and then disappear. However, it turns out that formation of teeth requires communication between two different tissues in the embryonic mouth, dorsal ectoderm and the ectomesenchyme ventral to it. During evolution of the avian beak, the signaling center in the epithelium became disjoint from the mesenchyme tissue, preventing formation of the normal integumentary appendages (Harris et al. 2006). A homozygous mutant of *talpid*^{2} repositions the two tissues to the plesiomorphic position seen in crocodilians, and produces chicken teeth very similar to alligator teeth (Harris et al. 2006). This is not a hopeful monster, as it is an embryonic lethal mutation, but again it demonstrates a large change in a single step due to embryonic signaling.

Developmental biologists have manipulated many aspects of vertebrate limb development, producing extra limbs, limbs with extra bones, mirror image duplications of digits and other mutations simply by altering expression of signaling molecules such as sonic hedgehog (SHH) and fibroblast growth factors (FGFs) (for review, see Bénazet and Zeller 2009; Diaz and Trainor 2015). While these may be "monsters," they demonstrate yet again that big changes can occur in rapid fashion. It seems fairly easy, for example, to turn a chicken foot into a duck paddle, by manipulating the regulatory signals involved in interdigital apoptosis (Zou and Niswander 1996). More bizarre even than the limb manipulations produced in the laboratory by scientists are the cleft hands and feet of chameleons. Diaz and Trainor (2015) detailed changes in interdigital apoptosis that led in part to the evolution and development of these fascinating appendages.

Interestingly, while no tetrapods alive today have more than five digits per appendage (Tabin 1992), several early tetrapod fossils such as *Acanthostega*, with eight digits on its front limbs, and *Ichthyostega*, with seven digits on its hindlimbs, exceeded this number (Coates and Clack 1990).

7.6 EVOLUTION OF DEVELOPMENT AND THE IMPORTANCE OF REGULATORY GENES

While the "modern" or "Neo-Darwinian" synthesis focused on the interaction of natural selection and population genetics, the relatively new science of evo-devo has turned its scrutiny to the importance of development for evolution. Many early studies in the molecular era concentrated on changes in protein coding regions, often due to gene duplication and formation of gene families. To the surprise of most of us, important developmental genes are astonishingly conserved, and it is the regulation of these genes that seems important for evolutionary change. As far back as 1971, Britten and Davidson (1971) wrote, "Evolutionary changes in the developmental process could certainly come about by alterations of individual genes expressed at given stages of development. It is clear, however, that alterations in the genes which determine the regulative programs could cause enormous changes in the developmental process and this would be a much more potent source of evolutionary change." King and Wilson (1975) noted how little genetic difference there is to explain the morphological differences between humans and chimpanzees and wrote that "evolutionary changes in anatomy and way of life are more often based on changes in the mechanisms controlling the expression of genes than on sequence changes in proteins." These and similar statements far pre-dated the large body of recent work that has implicated changes in gene regulatory networks (GRNs) due to *cis*-regulatory elements (CREs) as a major cause of evolution of body form.

Perhaps one of the most striking findings of evo-devo in the past two decades has been the shocking conservation of developmental regulatory genes that have come to be known as the "toolkit" for development of body form. These genes encode proteins that serve as transcription regulators and as signaling molecules. Carroll (2005, 2008) reviewed the functional equivalence of toolkit genes over essentially the evolution of metazoan life. Carroll (2008) detailed eight principles for extending the evolutionary synthesis using developmental biology. Among these were (1) deep homologies, (2) heterotopy, and (3) the importance and modularity of CREs.

Numerous studies revealed the deep homologies (and functional equivalence) of toolkit genes such *Hox* genes, which control axial patterning across animal life. *Pax-6* is a regulator of eye development in insects, fishes, mammals, and even ribbon worms (reviewed in Harris 1997). Homologues of the *tinman* gene (*Nkx2-5*) from *Drosophila* are involved in heart development in insects, frogs and mice (reviewed in Patterson et al. 1998). Although the structures themselves may be convergent, the genes that control them are not.

Heterotopy, or changes in patterns of spatial expression of genes, is associated with great morphological change in animals. The loss of limbs in snakes and the length of vertebrate necks (number of cervical vertebrae) mentioned earlier are both the result of heterotopies in expression of *Hox* genes, as is segment number in crustaceans (Averoff and Patel 1997). The genes themselves do not necessarily change, but their expression patterns do.

Davidson and Erwin (2006) detailed the importance of changes in GRNs in the evolution of body form. Carroll (2008) proposes evolution of CREs, as they participate in GRNs, to be the dominant moderator of evolution of morphology. CREs are DNA sequences to which transcription factors can bind to alter gene expression. The level of expression and spatial and temporal pattern of expression of a particular gene can be regulated by transcription factors binding to the CREs. One gene can have many CREs, and each CRE can interact with multiple transcription factors. Changes in both CREs and available transcription factors can lead to changes in gene expression. And the interaction of multiple transcription factors and multiple CREs affords new combinations for the control of gene expression. Changes in gene expression can include changes in expression of signaling molecules such as SHH or transcription factors such as the *Hox* proteins. One signaling molecule can lead to cascades of other changes, and one transcription factor may interact with hundreds of downstream genes, so small changes in CREs that result in changes in spatial or temporal expression of these regulatory proteins can have large impacts.

Because transcription factors recognize short DNA sequences, there is ample opportunity for evolution of new CREs (Carroll 2008). Carroll reviews mechanisms for the evolution of CREs, including co-option of old CREs by mutation to produce new ones, loss of existing CREs to mutation, and formation of new CREs by transposable elements (Carroll 2008). The presence of many transposable elements near developmentally regulated genes was documented (Lowe et al. 2007). This type of information, the product of studies in the evolution of development, provided the basis for understanding how regulatory elements change, and how changes in regulatory elements can lead to morphological evolution.

7.7 CAN CHANGES IN DEVELOPMENT LEAD TO HOPEFUL MONSTERS?

Is there evidence that some of this morphological evolution may represent hopeful monsters? Olivier Rieppel (2001) made an argument for the ancestor to turtles as a hopeful monster based on morphological research by Gilbert et al. (2001). Their work on the growth and development of turtle shells, along with examination of the fossil record, led Rieppel to write, "The evolution of the highly derived adult anatomy of turtles is a prime example of a macroevolutionary event triggered by changes in early embryonic development. Early ontogenetic deviation may cause patterns of morphological change that are not compatible with scenarios of gradualistic, stepwise transformation." Subsequent work has only strengthened this argument. Cebra-Thomas et al. (2005) showed evidence that FGF signaling maintained the carapacial ridge (CR) and directed the ribs to grow into the dermis. The ribs then secreted bone morphogenetic proteins (BMPs) that signaled ossification in surrounding dermal cells. Inhibiting FGF caused the CR to degenerate and the ribs to grow outward as in other vertebrates, obliterating many features of the carapace (Cebra-Thomas et al. 2005). They suggested that this coordinated means of carapace formation allowed evolution of the carapace without intermediate forms. Interestingly, Mivart (1871, p. 150) in his rejection of intermediate forms, wrote, "The singular order Chelonia, including the tortoises, turtles and terrapins (or fresh-water tortoises), is another instance of an extreme form without any, as yet known, transitional stages." Although recent fossil discoveries of early turtle relatives *Odontochelys* (Li et al. 2008) and *Eunotosaurus* (Lyson et al. 2010) have made this no longer strictly correct, Lyson et al. (2013) used paleontological evidence in light of developmental signaling discoveries to modify a previous hypothesis (Kuratani et al. 2011) and develop essentially a four-step model of turtle shell evolution with significant morphological steps, such as the formation of the carapace, occurring in concert.

My work on germ cell determination with Johnson, Crother and others led to our hypothesis that the evolution of predetermined germ cells may promote macro-evolutionary change (allow hopeful monsters?). While many animals use inductive signals to specify primordial germ cells, this requires specific tissues to be in specific places at specific times during development, and thus may serve to constrain body shape (Johnson et al. 2003; Crother et al. 2007). Multiple instances of evolution of a predetermined germ line (Extavour and Akam 2003; Johnson et al. 2003) including in fruit flies, frogs, and zebrafish, in which germ cell determinants are localized in specific regions of the oocyte or early embryo, may serve to release the developmental constraints imposed by inductive signaling and allow macro-evolutionary changes in body plans as shown in frogs and teleost fish (Johnson et al. 2003; Crother et al. 2007). In fish, the anteriorized body plan, as indicated by the forward position of the pelvic fin, is associated with predetermined germ cells. Interestingly, Mivart (1871, p. 44) commented on the position of pelvic fins in teleosts: "Yet we find in many fishes the pair of fins, which correspond to the hinder limbs of other animals, placed so far forwards as to be either on the same level with, or actually in front of, the normally anterior pair of limbs; …" In comparing this to the relatively fixed position of limbs in other vertebrates he added, "… if then such a change can have taken place in the comparatively short time occupied by the evolution of these special fish forms, we might certainly expect that other and far more bizarre structures would (did not some law forbid) have been developed from other rugosities, in the manifold exigencies of the multitudinous organisms which must (on the Darwinian hypothesis) have been gradually evolved during the enormous period intervening between the first appearance of vertebrate life and the present day. Yet with these exceptions, the position of the limbs is constant from the lower fishes up to man, there always being an anterior pectoral pair placed in front of a posterior or pelvic pair when both are present, and in no single instance are there more than these two pairs."

In his 2005 book, *Endless Forms Most Beautiful*, Carroll wrote, "For a half century since the Modern Synthesis, the specter of a 'hopeful monster' has lingered. The facts of Evo Devo squash this." He later added, "The continuity of the tool kit and the continuity of structures throughout this vast time illustrate that we need not invoke very rare or special mechanism to explain large-scale change. The extrapolation from small-scale variation to large-scale evolution is well justified. In evolutionary parlance, Evo Devo reveals that macroevolution is the product of microevolution writ large." (Carroll 2005). And yet this strikes me as a (surely unintentional) misrepresentation of Goldschmidt's hopeful monster. Recall that in his initial description Goldschmidt (1933) said, "rare but extremely consequential mutations affecting rates of decisive embryonic processes which might give rise to what one might term hopeful monsters…" Although he certainly did in other writings,

Goldschmidt did not invoke any mechanisms that deny the importance of Mendelian genetics, but rather stressed that not all evolutionary change requires gradual accumulation of small phenotypic variations.

7.8   NON-MENDELIAN INHERITANCE AND MACROEVOLUTION

Are non-Mendelian mechanisms involved in macroevolution? I think it would be hard to argue otherwise. Surely two of the major macroevolutionary events in the history of life include the endosymbiotic origins of mitochondria and chloroplasts in the evolution of eukaryotes. This theory, first articulated in the early part of the twentieth century, was re-discovered primarily by Margulis (for review, see Margulis 2010) and has gained almost universal acceptance. Far from gradual evolution, endosymbiosis was evolution in one big gulp. While there is little doubt that natural selection has tinkered with these organelles, the initial events are hard to reconcile with gradual changes in allele frequencies. Kutschera and Niklas (2008) reviewed numerous cases of secondary endosymbiosis that they referred to as "neo-Goldschmidtian hopeful monsters."

Studies in development have even re-invigorated long discredited Lamarckian ideas of inheritance of acquired characteristics. McGrath and Solter (1984) showed that maternal and paternal pronuclei in mice were not equivalent, and that both were necessary to support complete embryonic development. These observations led eventually to the discovery of parent-specific genomic imprinting by epigenetic modifications (for review, see Barlow and Bartolomei 2014). Cytosine methylation and histone modification are two types of acquired (epigenetic) changes that impact gene expression leading to phenotypic differences not coded in the DNA. These epigenetic modifications can persist through many cell and even organismal generations. For example, Heijmans et al. (2008) showed that prenatal famine impacted gene methylation patterns that persisted decades later into adulthood. Vargas (2009) reexamined the infamous "The case of the midwife toad" (Koestler 1971), in which Paul Kammerer was discredited after he suggested that midwife toads inherited environmentally acquired traits. Kammerer committed suicide soon after an article suggesting his results were fraudulent was published. In his work, Kammerer noted differences depending on the parent of origin, a phenomenon that was not easily explained in the 1920s. This led Vargas to suggest that Kammerer, rather than committing fraud, was actually the first to discover genomic imprinting and epigenetic inheritance (Vargas 2009).

Describing all of evolution as changing allele frequencies in gene pools does not encompass endosymbiosis, hybridization and other mechanisms of change such as lateral gene transfer, polyploidy in plants, and the alterations of body plans that can occur due to changes in GRNs. Yes, once a transposable element has positioned a new CRE, it may be inherited in a Mendelian fashion, but the initial event is far beyond what most biologists (with the possible exception of Barbara McClintock) could have imagined as the modern synthesis was being formulated.

Goldschmidt wrote in 1940, "... a single mutational step, affecting the right process at the right moment can accomplish everything, providing it is able to set in motion the ever-present potentialities of embryonic regulation." That is a statement that I find little to disagree with 76 years later. One might imagine, for example, a transposable element adding a CRE to change the expression of a developmental regulatory gene. This single mutation might indeed accomplish great changes. And while it is true that some of these changes due to development may not happen in a single step, as per Goldschmidt, they also may not conform to the very gradual accumulation of small mutations that the modern synthesis posits.

7.9   DONN ROSEN'S ASSUMPTIONS AND GOLDSCHMIDT

I would like to go back to the assumptions that inhibit scientific progress (Rosen 2016). Number 6 under "General" states, "Discovering you were wrong is bad in some sense (never publish until you're convinced you've found the truth)." As Gould (2002, p. 463) pointed out, Goldschmidt's concept of the hopeful monster was presented in 1933, well before the development of his controversial concepts of systemic mutations. One can only surmise that Goldschmidt, unlike many of his contemporaries, realized that the current state of knowledge in genetics and development did not fully explain the evolution of the diversity of life. Who could have imagined the amazing discoveries of evo-devo that were decades in the future? While his concept of systemic mutations is generally untenable today, we should respect his realization that there must be something more. Goldschmidt's ideas were incorrect in many ways, but as he himself wrote in a letter in 1940, he would rather be wrong than a "terribly cautious agnostic" (as quoted in Dietrich 1996). Here we are today still talking about his anything but cautious ideas as we develop a much-expanded synthesis of the evolution of animal form.

ACKNOWLEDGMENTS

Thank you to Brian Crother and Lynne Parenti for organizing the Rosen symposium and this symposium volume. Thanks also to Brian for your assistance and your patience. Thank you to Raul Diaz for excellent suggestions, editing assistance, and for introducing me to Vargas's midwife toad paper and more. Thank you to an anonymous reviewer whose editorial suggestions improved this manuscript.

REFERENCES

Averoff, M. and N. H. Patel 1997. Crustacean appendage evolution associated with changes in *Hox* gene expression. *Nature* 388:682–686.

Avery, O. T., C. M. MacLeod and M. McCarty 1944. Studies on the chemical nature of the substance inducing transformation of pneumococcal types: Induction of transformation by a deoxyribonucleic acid fraction isolated from pneumococcus type III. *Journal of Experimental Medicine* 79:137–158.

Barlow, D. P and M. S. Bartolomei 2014. Genomic imprinting in mammals. *Cold Spring Harbor Perspectives in Biology* 6(2):1–20.

Bénazet, J.-D. and R. Zeller 2009. Vertebrate limb development: Moving from classical morphogen gradients to an integrated 4-dimensional patterning system. *Cold Spring Harbor Perspectives in Biology* 1(4):a001339.

Bridges, C. B. 1944. *The Mutants of Drosophila melanogaster.* Carnegie Institute of Washington Publication 552. D. S. Brehme, Ed. The Lord Baltimore Press. Baltimore, MD. 1–257.

Britten, R. J. and E. H. Davidson 1971. Repetitive and non-repetitive DNA sequences and a speculation on the origin of evolutionary novelty. *The Quarterly Review of Biology* 46:111–138.

Brown, K. M., L. M. Burk, L. M. Henagan and M. A. F. Noor 2004. A test of the chromosomal rearrangement model of speciation in *Drosophila pseudoobscura. Evolution* 58(8):1856–1860.

Burke, A. C., C. E. Nelson, B. A. Morgan and C. Tabin 1995. Hox genes and the evolution of vertebrate axial morphology. *Development* 121:333–346.

Carroll, S. B. 1995. Homeotic genes and the evolution of arthropods and chordates. *Nature* 376:479–485.

Carroll, S. B. 2005. *Endless Forms Most Beautiful: The New Science of Evo-Devo.* W.H. Norton and Co. New York, NY.

Carroll, S. B. 2008. Evo-Devo and an expanding evolutionary synthesis: A genetic theory of morphological evolution. *Cell* 134:25–36.

Cebra-Thomas, J., F. Tan, S. Sistla, E. Estes, G. Bender, C. Kim, P. Riccio and S. F. Gilbert 2005. How the turtle forms its shell: A paracrine hypothesis of carapace formation. *Journal of Experimental Zoology, Part B Molecular and Developmental Evolution* 304:558–569.

Coates, M. I. and J. A. Clack 1990. Polydactyly in the earliest known tetrapod limbs. *Nature* 347:66–69.

Cohn, M. J. and C. Tickle 1999. Developmental basis of limblessness and axial patterning in snakes. *Nature* 399:474–479.

Cole, C. J. 1975. Evolution of parthenogenetic species of reptiles. In: *Intersexuality in the Animal Kingdom.* R. Reinboth, Ed. Springer-Verlag. Berlin.

Cole, C. J. 1985. Taxonomy of parthenogenetic species of hybrid origin. *Systematic Zoology* 34(3):359–363.

Crother, B. I., M. E. White and A. D. Johnson 2007. Inferring developmental constraint and constraint release: Primordial germ cell determination mechanisms as examples. *Journal of Theoretical Biology* 248:322–330.

Darwin, C. and J. Huxley 1958. *The origin of species/Charles Darwin introduction by Sir Julian Huxley.* The New American Library. Mentor Books. New York, NY.

Davidson, E. H. and D. H. Erwin 2006. Gene regulatory networks and the evolution of animal body plans. *Science* 311(5762):796–800.

Diaz Jr., R. E. and P. Trainor 2015. Hand/foot splitting and the "re-evolution" of mesopodial skeletal elements during the evolution and radiation of chameleons. *BMC Evolutionary Biology* 15:184. DOI: 10.1186/s12862-015-0464-4.

Dietrich, M. R. 1995. Richard Goldschmidt's "Heresies" and the evolutionary synthesis. *Journal of the History of Biology* 28:431–461.

Dietrich, M. R. 1996. On the mutability of genes and geneticists: The Americanization of Richard Goldschmidt and Victor Jollos. *Perspectives on Science* 4:321–345.

Dietrich, M. R. 2000. From hopeful monsters to homeotic effects: Richard Goldschmidt's integration of development, evolution and genetics. *American Zoologist* 40(5):738–747.

Dietrich, M. R. 2003. Richard Goldschmidt: Hopeful monsters and other "Heresies". *Nature Reviews Genetics* 4:68–74.

Dobzhansky, T. 1937. *Genetics and the Origin of Species.* Columbia University Biological Series (vol. 11). Columbia University Press. New York.

Extavour, C. G. and M. Akam. 2003. Mechanisms of germ cell formation across the metazoans: Epigenesis and preformation. *Development* 130:5869–5884.

Galton, F. 1889. *Natural Inheritance.* Macmillan. London.

Gaunt, S. J. 1994. Conservation in the Hox code during morphological evolution. *International Journal of Developmental Biology* 38:549–552.

Gilbert, S. F., G. A. Loredo, A. Brukman and A. C. Burke 2001. Morphogenesis of the turtle shell: The development of a novel structure in tetrapod evolution. *Evolution & Development* 3:47–58.

Goldschmidt, R. 1933. Some aspects of evolution. *Science* 78:539–547.

Goldschmidt, R. 1940. *The Material Basis of Evolution*. Yale University Press. New Haven, CT.

Gould, S. J. 2002. *The Structure of Evolutionary Theory*. Belknap Press of Harvard University Press. Cambridge, MA.

Haldane, J. B. S. 1932. *The Causes of Evolution*. Longmans, Green and Co. London, New York and Toronto.

Hall, B. K. 2012. Evolutionary biology (Evo-Devo): Past, present, and future. *Evolution, Education and Outreach* 5:184–193. DOI: 10.1007/s12052-012-0418-x.

Harris, M. P., S. M. Hasso, M. W. J. Ferguson and J. F. Fallon 2006. The development of archosaurian first-generation teeth in a chicken mutant. *Current Biology* 16:371–377.

Harris, W. A. 1997. *Pax-6*: Where to be conserved is not conservative. *Proceedings of the National Academy of Sciences USA* 94:2098–2100.

Heijmans, B. T., E. W. Tobi, A. D. Stein, H. Putter, G. J. Blauw, E. S. Susser, P. E. Slagboom and L. H. Lumey 2008. Persistent epigenetic differences associated with prenatal exposure to famine in humans. *Proceedings of the National Academy of Sciences USA* 105(44):17046–17049.

Hershey, A. and M. Chase 1952. Independent functions of viral protein and nucleic acid in growth of bacteriophage. *Journal of General Physiology* 36:39–56.

Hughes, C. L. and T. C. Kaufman 2002. Hox genes and the evolution of the arthropod body plan. *Evolution & Development* 4:459–499.

Huxley, J. 1942. *Evolution: The Modern Synthesis*. Allen and Unwin Publishers. London.

Johnson, A. D., M. Drum, R. F. Bachvarova, T. Masi, M. E. White and B. I. Crother 2003. Evolution of predetermined germ cells in vertebrate embryos: Implications for Macroevolution. *Evolution and Development* 5:414–431.

King, M.-C. and A. C. Wilson 1975. Evolution at two levels in humans and chimpanzees. *Science* 188:107–116.

Koestler, A. 1971. *The Case of the Midwife Toad*. Hutchinson and Co. London.

Krumlauf, R. 1994. Hox genes in vertebrate development. *Cell* 78:191–201.

Kuratani, S., S. Kuraku and H. Nagashima. 2011. Evolutionary developmental perspective for the origin of turtles: The folding theory for the shell based on the nature of the carapacial ridge. *Evolution and Development* 13:1–14.

Kutschera, U. and K. J. Niklas 2004. The modern theory of biological evolution: An expanded synthesis. *Naturwissenschaften* 91:255–276.

Kutschera, U. and K. J. Niklas 2008. Macroevolution via secondary endosymbiosis: A Neo-Goldschmidtian view of unicellular hopeful monsters and Darwin's primordial intermediate form. *Theory in Biosciences* 127:277–289.

Lewis, E. B. 1978. A gene complex controlling segmentation in *Drosophila*. *Nature* 276:565–570.

Lewis, E. B. 1992. Clusters of master control genes regulate the development of higher organisms. *Journal of the American Medical Association* 267:1524–1531.

Li, C., X.-C. Wu, O. Rieppel, L.-T. Wang and L.-J. Zhao 2008. An ancestral turtle from the Late Triassic of southwestern China. *Nature* 456:497–501.

Lowe, C. B., G. Bejerano and D. Haussler 2007. Thousands of human mobile element fragments undergo strong purifying selection near developmental genes. *Proceedings of the National Academy of Sciences USA* 104:8005–8010.

Lyson, T. R., G. S. Bever, B.-A. S. Bhullar, W. G. Joyce and J. A. Gauthier 2010. Transitional fossils and the origin of turtles. *Biology Letters* 6(6):830–833.

Lyson, T. R., G. S. Bever, T. M. Scheyer, A. Y. Hsiang and J. A. Gauthier 2013. Evolutionary origin of the turtle shell. *Current Biology* 23:1113–1119.

Mallet, J. 2007. Hybrid speciation. *Nature* 446:279–283. DOI: 10.1038/nature05706.

Margulis, L. 2010. Symbiogenesis. A new principle of evolution: Rediscovery of Boris Mikhaylovich Kozo-Polyansky (1890–1957). *Paleontological Journal* 44(12):1525–1539.

McGinnis, W. and R. Krumlauf 1992. Homeobox genes and axial patterning. *Cell* 68:283–302.

McGinnis, W., R. L. Garber, J. Wirz, A. Kuroiwa and W. J. Gehring 1984a. A homologous protein-coding sequence in Drosophila homeotic genes and its conservation in other metazoans. *Cell* 37(2):403–408.

McGinnis, W., M. S. Levine, E. Hafen, A. Kuroiwa and W. J. Gehring 1984b. A conserved DNA sequence in homoeotic genes of the *Drosophila Antennapedia* and *bithorax* complexes. *Nature* 308(5958):428–433. DOI: 10.1038/308428a0.

McGrath, J. and D. Solter 1984. Completion of mouse embryogenesis requires both the maternal and paternal genomes. *Cell* 37:179–183.

Mivart, St.-G. J. 1871. *On the Genesis of Species*. Macmillan and Co. (Reissued by Cambridge University Press, 2009).

Parris, G. E. 2011. The hopeful monster finds a mate and founds a new species. *Hypotheses in the Life Science* 1:32–37.

Patterson, K. D., O. Cleaver, W. V. Gerber, M. W. Grow, C. S. Newman and P. A. Krieg 1998. Homeobox genes in cardiovascular development. *Current Topics in Developmental Biology* 40:1–44.

Rieppel, O. 2001. Turtles as hopeful monsters. *Bioessays* 23:987–991.

Rieseberg, L. H. 2001. Chromosomal rearrangements and speciation. *Trends in Ecology and Evolution* 16(7):351–358.

Rosen, D. E. 2016. Assumptions that *inhibit* scientific progress in comparative biology. In: *Assumptions Inhibiting Progress in Comparative Biology.* B. I. Crother, and L. R. Parenti, Eds. CRC Press. Boca Raton, 1–4.

Scott, M. P. and A. J. Weiner 1984. Structural relationships among genes that control development: Sequence homology between the *Antennapedia, Ultrabithorax,* and *fushi tarazu* loci of *Drosophila. Proceedings of the National Academy of Sciences USA* 81(13): 4115–4119. DOI: 10.1073/pnas.81.13.4115.

Tabin, C. J. 1992. Why we have (only) five fingers per hand: Hox genes and the evolution of paired limbs. *Development* 116:289–296.

Theißen, G. 2006. The proper place of hopeful monsters in evolutionary biology. *Theory in Biosciences* 124:349–369.

Vargas, A. O. 2009. Did Paul Kammerer discover epigenetic inheritance? A modern look at the controversial midwife toad experiments. *Journal of Experimental Zoology, Part B Molecular and Developmental Evolution* 312:1–12.

Zou, H. and L. Niswander 1996. Requirement for BMP signaling in interdigital apoptosis and scale formation. *Science* 272:738–741.

8 Does Competition Generate Biodiversity? An Essay in Honor of Donn Eric Rosen

Maureen A. Donnelly
Florida International University

CONTENTS

The patterns we observe in nature may therefore be influenced or explained by processes other than competition alone.

Wiens (1977:592)

Positive interactions are diverse and have a well-documented influence on every ecosystem on earth.

Bruno et al. (2003:124)

Certainly we were never justified in thinking that the ecological world was so simple as to be largely explainable on the basis of a single interaction. New discoveries are continually refining our understanding of the domain of competition, and we are well on the way to developing a multifaceted theory to match what is clearly a highly diverse natural world.

Schoener (1982:594)

Based on our analyses, current competition theory inadequately predicts the nature of interactions occurring in herbivorous insects; virtually every fundamental paradigm that we tested was violated to some degree, suggesting a poor concordance between theory and empirical patterns.

Kaplan and Denno (2007:990)

8.1   AN ESSAY

Competition is one of several biotic interactions and occurs when two or more individuals attempt to utilize the same *limited* resource. Competition is generally considered to be a negative type of interaction for both participants, but as Wiens (1977) stated in the quote above, other processes are important for generation of the patterns we see in nature (i.e., biodiversity), and Bruno et al. (2003) suggested that positive interactions may be important for generation of natural patterns. Bronstein (2009) noted that the study of mutualisms has lagged behind studies of antagonistic interactions. Although Schoener (1982) was certain that a "theory" would be developed that would explain the multifaceted wonders of the natural world, the mathematics of competition do not fare well in the nonequilibrium conditions that characterize nature (Wiens 1977; Brown 1981; Simberloff 1982). Volterra (1926b, 1928) was the first to describe competition mathematically, and two types of competition (interference and exploitative) are recognized by most ecologists, but Schoener (1983) described six forms of competition. Much of organismal biology is focused on understanding how resource utilization affects populations, and how interacting populations function together in ecosystems (Stiling 2012; Ricklefs and Relyea 2014). Current studies of populations of wild organisms are particularly urgent as a variety of taxa face unparalleled extinction threats in our rapidly changing world (Wake and Vredenburg 2008). The "biodiversity crisis" that began in the late 1970s means that we are losing biodiversity faster than we can name it, but the academic community is still not clear on the processes that generate biodiversity.

Hutchinson (1959) asked the fundamental question regarding why there are so many species while giving a nod to Santa Rosalia when I was five years old. We are still trying to answer the fundamental question about the process or processes that generate biological diversity. While scientists agree that "evolution" is responsible for the generation of diversity, we have not yet determined the mechanism or suite of mechanisms that are responsible for speciation. Recently, facilitation (i.e., positive interactions that benefit one participant and harm neither participant; see Box 1 in Bruno et al. 2003) has re-emerged as an explanation for global patterns of diversity (McIntire and Fajardo 2014), and as Schoener said in 1982, seeking a single mechanism is probably not going to yield a satisfactory answer to the most elusive of biological processes. In most of the studies in the edited volume, *Evolution in Action* (Glaubrecht 2010), multiple mechanisms were invoked to explain speciation for the focal groups under study.

Valiente-Banuet and Verdú (2007:1034–1035) eloquently summarized how competition theory asserted its dominance in ecology and evolution: "The origin of this competitive paradigm is based on Malthusian population theory and attributed to Darwin, who thought competition was logically connected to the assumed universal density-dependent mechanism of natural selection (Den Boer 1986). Thus competition was given pre-eminence as the driver of numerical processes in both populations and communities. More recent research has found the role of facilitation is ubiquitous in community structure, and often as important as competition (Callaway 1995, Bruno et al. 2003)." Indeed Rosen's views on the importance of competition evolved over time and while he felt it was not important for evolution (see Crother 2016) he clearly felt it could inhibit progress in comparative biology because competition theory made his list.

Competition was first modeled by Volterra (1926b, 1928*): "§2. Biological Association of two Species which contend for the same Food" (Volterra 1928:7). Lotka (1920) applied logistic equations to an herbivory example, and only referred to competition between humans in his later landmark publication (Lotka 1925). He did say in his Summary of Chapters (Lotka 1925:30) that "…competition is principally within the species." Competition theory predicts that intraspecific competition should be stronger than interspecific competition because members of the same species share the same fundamental niche and could compete for resources if they were limiting. The Lotka–Volterra equations (Lotka 1920, 1925; Volterra 1926a,b, 1928) are staples of all modern ecology courses (Stiling 2012; Ricklefs and Relyea 2014) and form the foundational bases for competition models. The modeling of competition started in the 1960s in the "evolutionary ecology" school (Brown 1981). Ecologists in this school adopted competition mathematics to explain community structure (Wiens 1977) as well as evolutionary change as a result of natural selection. The equilibrium conditions required by the Lotka and Volterra equations allowed for development of a variety of mathematical models to explain competition, and Wiens (1977) argued that the development of the mathematics of competition helped the theory attain dominance in ecology. Because equilibrium conditions are rare to nonexistent in nature, Wiens (1977) pointed out that "Support for the theory, however, seems largely based upon intuition and indirect evidence." While both Schoener (1983) and Connell (1983) reviewed the literature dealing with data from field experiments, they found varying levels of support for competition from those data. Connell et al. (2004) found that competition among corals varied as a function of local environmental conditions. The meta-analysis by Kaplan and Denno (2007) should have put competition theory to rest given that *every* assumption of the theory was violated in their analysis. Herbivorous insects are a diverse group that includes specialist and generalist feeding species across a variety of phylogenetic lineages and they do not appear to follow *any* predictions of competition theory. If the theory fails for this group of organisms, when can it explain patterns in nature? Although the equilibrium conditions required by the models are rarely seen in the field, the development of equilibrium mathematics has continued unabated. Most environmental effects on populations are not mathematically tractable (Wiens 1977) making predictions derived from equilibrium models unrealistic and untestable (Brown 1981; Simberloff 1982).

Interactions among organisms are affected by biotic (e.g., disease and parasitism) and abiotic (e.g., climatic conditions, geological changes) factors. Biotic interactions

---

* The 1928 paper is a translation of Volterra (1926b) published in Italian.

(e.g., mutualism, facilitation), in addition to competition, have been identified as potent evolutionary forces. Recently, Robert Ricklefs suggested that pathogens may have as much or more to do with species diversity than does competition when he was awarded the Grinnell Medal in 2015 (Charles Crumly, pers. comm.). When Ricklefs was awarded the Ramon Margalef Prize in Ecology in 2015 the jury noted that "Specifically, he challenged the belief that local interactions control the diversity of species, proposing instead a key role of large-scale and historical processes in shaping current patterns of diversity, from tropical to temperate latitudes. Its unifying research has led him to investigate beyond the traditional boundaries of ecology to focus on the aging process through genetic analysis and duration of the evolving life." (http://ub.edu/web/ub/en/menu_eines/noticies/2015/10/066.html).

To answer the question posed in the title of this chapter, I gathered three types of data: I sampled the literature over a 50-year timespan to discover if "Competition Theory is well supported and informative" by identifying the driver of speciation to see if competition and/or "natural selection" were responsible for speciation. After I finished this data-gathering phase, I used Google Scholar to search for papers that had Competition and Speciation in the title because this should link the phenomena if there is a link, and I reviewed the edited volume *Evolution in Action* (Glaubrecht 2010) as my third data source. Donn Rosen directly challenged my thinking about the role natural selection plays in the generation of diversity so I selected my symposium topic and book contribution to determine if there was a linkage between competition and speciation. My overall findings support an observation made by Hood et al. (2012): "No empirical study has yet to directly tie shifts in resource utilization due to intraspecific competition to the evolution of reproductive isolation and speciation."

Donn challenged me to think critically about the role played by natural selection in the generation of diversity. He was the first person, besides creationists, who questioned the paradigm of evolutionary change as the result of survival of the fittest. Donn's questions prompted me to become a more critical scientist than I was before I met him. He taught me to examine evidence carefully and ensure that the methods used by the investigators could answer the questions being asked. Before I describe the results of my literature survey, I want to describe how Donn interacted with students because his example as a mentor plays a role in how I try to interact with my young colleagues. I hope his example for me lives on through the students I have interacted with during my career so they can pay it forward to their future colleagues.

I encountered Donn Rosen at my first national scientific meeting in 1978. We did not actually meet during that meeting, but by the end of the first day of the meeting, I knew who he was. The Arizona State University meeting in Tempe, Arizona, was the first joint meeting of the three U.S. herpetological societies: the American Society of Ichthyologists and Herpetologists (ASIH), the Herpetologists' League (HL), and the Society for the Study of Amphibian and Reptiles (SSAR). I was a first year Master's student at California State University, Fullerton, and was coauthor on a project that examined sleep in the Western Toad. I stayed in the plenary room on the first day of the meeting to attend an ichthyology contributed paper session. I wanted to hear a presentation by Jack Briggs on the biogeography of fishes. Robert Rush Miller was moderating the session, and after Dr. Briggs completed his talk, Donn Rosen stood up and began to ask Jack pointed and difficult questions that left Dr. Briggs speechless. When Donn was done, Jay Savage stood up and asked additional challenging questions. When Jay was done, Donn popped up again and Dr. Miller quickly took control of the session and urged the three of them to find a spot to continue their spirited discussion at a later time. I sat there and wondered if I really wanted to be part of a career track that included such intense public interactions among colleagues. From my youthful perspective, the interaction on the stage seemed harsh and confrontational. I was not deterred from herpetology and started the PhD program at the University of Southern California (USC) with Jay Savage in January 1979.

I actually met Donn and interacted with him when he came to USC in the early 1980s as a visiting professor. I was back from a summer of fieldwork in the tropics and full of bug bites, wonder, and amazement. Donn engaged with all of the students in Savage's laboratory group, and he then sought us out for individual discussions. His kindness and interest in our projects was encouraging for me as a young doctoral student. During his USC visit, he asked me why I was so interested in tropical biology. I told him that the diversity of the tropics captivated me—I liked the messiness and the intensity of my tropical experience—smells, sights, sounds, sensations, and encountering a variety of species of many lineages. He then asked me what factors were responsible for generation of diversity in the tropics. I replied that evolution was responsible for the generation of diversity. Donn asked me to explain how that occurred, and I told him that diversification as a result of natural selection was the mechanism that generated diversity. He asked for an example of this mechanism, and I recounted the *Biston betularia* case study about moths changing coloration to match the environment. Predation pressures changed the population through time as their environment changed. He then pointed out that the species changed through time but at the end it was still *Biston betularia*. He then asked me what any of that example had to do with why we have lions, tigers, and bears. I was stunned.

Before Donn died, he came to the University of Miami in 1985 as a visiting professor, and he taught a course on biogeography. Donn continued to be a warm, nurturing mentor until the end of his life. He still interacted with the group, and still peeled us off individually to see how we were faring as young colleagues. His encouragement and his kindness made his critical style of questioning a welcome break from faculty who were less nurturing in their mentoring styles. I am grateful that I got to know Donn during my career, that he pushed my limits, made me question assumptions, and critically examine evidence presented in literature or during presentations.

Donn's gentle questioning changed my life as a scientist, and from that point onward I became critical of the papers and books I read, lectures I heard, and became open to other possible explanations for generation of biological diversity. Competition as an explanation for everything (e.g., community structure and evolutionary change) came under challenge from ecologists in the 1970s (Peters 1976, 1991; Schoener 1983) who studied other types of interactions and showed convincingly that predation, disease dynamics, mutualisms, and sexual selection were also important interactions, and at least as important as competition. As I started work on the literature survey for my symposium contribution in 2015, a tweet about a review paper by McIntire and Fajardo (2014) appeared and I obtained a reprint. Facilitation is another important driver of biodiversity and interest in this topic has increased over time since it was demonstrated experimentally in 1914 (Callaway 2007). Like the study of mutualisms, the study of facilitation has lagged behind studies of competition and predation (Bronstein 2009).

I selected the first assumption thought by Donn to inhibit progress in Evolutionary Theory: "Competition Theory is important and well formulated" because I had questioned the role competition played in structuring ecological communities as a doctoral candidate. Connell (1980, 1983), Menge and Sutherland (1976), Simberloff (1982), Strong et al. (1979), Wiens (1977), and other scientists challenged the competition paradigm, as did many botanists who argued that facilitation was more important ecologically and evolutionarily than competition (Hacker and Gaines 1997; Stachowicz 2001; Bruno et al. 2003). Many ecological studies of competition fail to identify the limiting resource (Brown 1981), fail to attain the equilibrium population sizes required by the mathematical models (Wiens 1977; Brown 1981; Simberloff 1983), and while some systems are not ideal for experimentation, many systems can be manipulated to determine if competition occurs providing strong evidence for or against competition.

To address the assumption that competition is important for the generation of diversity, I searched Google Scholar (an open source academic search engine) for all papers on "Speciation" because the generation of diversity occurs when a taxon produces daughter species. When I started data collection I almost withdrew from the symposium because Google Scholar indicated that over half a million papers existed on the topic. I knew I would never be able to complete data collection by the summer symposium. In part the "large numbers" reflected the fact that speciation is a chemical process that has sparked considerable academic research across a variety of fields. Instead of a survey of the literature, I sampled the literature every five years beginning in 1965 (1965, 1970, 1975, 1980, 1985, 1990, 1995, 2000, 2005, and 2010; ten time steps in total). I only included refereed journal articles that actually addressed the process of lineage divergence written in English for inclusion in the data set in Appendix 8.1. I recorded year, authors, taxon, title, citation, and driver of speciation (if indicated) and found 504 articles from over 8,000 articles that mentioned biological speciation, in Table 8.1. After I completed the first 10 searches, I searched Google Scholar for journal titles in English that included "speciation" and "competition" to directly determine if competition was a driver of biological diversity without time restriction on the search (37 titles). I also sampled an edited volume (*Evolution in Action*) to determine how many papers on evolution identified competition as a driver of diversity (25 papers, Glaubrecht 2010). The third part of the sample includes papers from laboratories that received funding from the Deutsche Forschungsgemeinschaft (DFG) from 2002 to 2008 (Glaubrecht 2010).

Table 8.1 summarizes the sampling of the literature. The number of "hits" is the number the Scholar search estimates are in the overall search result, the papers that deal with biological diversification (as opposed to chemical speciation) are in the second column of the table, and the actual papers are those that actually address speciation through modeling, discussion of theory, or presentation of results. Many papers mentioned speciation, but a small fraction in any year actually examined the phenomenon. While the number of hits shows a steady increase over time, the search engine yielded over 900 papers from 1980 onwards. There was a lull in studies of

TABLE 8.1

The Number of Articles Obtained by Searching on Speciation with Google Scholar for a Given Year (No Patents or Citations)

Year	Number of Hits in Google Scholar	Number of Papers That Included Biological Speciation Hits	Number of Papers on Speciation That Included in Dataset
1965	377	377	35
1970	686	686	40
1975	884	884	40
1980	1,670	938	37
1985	2,810	943	47
1990	4,420	932	11
1995	6,560	949	14
2000	12,400	965	65
2005	21,500	983	69
2010	29,800	984	146

The number of hits is the number provided by Google Scholar for the search, the number of actual papers is the number of papers that were biological and purported to examine speciation, and the number of papers on speciation is the number of papers that included a discussion of the process or data obtained from research into the process.

speciation in the 1990s, but those numbers have rebounded in the new century, as in Figure 8.1. Perhaps the emergence of phylogeography in the 1990s resulted in a reduced focus on speciation during that decade.

Chromosomal changes and other types of genetic changes (e.g., founder effects, hybridization) were merged into a single category "genetics" for the summarization of research from the three sources. Plants and insects have been model systems for studies of speciation (Table 8.2), genetic changes are frequently linked to speciation events (Table 8.3), and the number of research foci, types of speciation, and number of journals describing speciation have increased over time (Figure 8.1). Most of the studies of speciation over the 50-year time period focus on plants, insects, theory and/or models; fish are the most studied vertebrates (Table 8.2). Natural Selection and Competition represent a small fraction of the studies (2.5% of the 504 papers included in the sample of the speciation literature, 12% of the 17 papers that had Competition and Speciation in the title, and none of the papers in *Evolution in Action*). Thirty-four of the 66 modes of speciation were only described once (Table 8.3). The journals that have published studies of speciation have also increased, more dramatically than the number of study systems or modes of speciation (Figure 8.1, Appendices 8.1 and 8.4).

The results obtained from the three sources (papers that included "specia-tion" in the paper, titles that linked speciation and competition, and the edited 2010 volume) seem to refute the assumption that Competition Theory is important

TABLE 8.2

The Focus of the 504 Studies of Speciation from 1965 to 2010 Obtained by Google Scholar Searching for Speciation by Year

	1965	1970	1975	1980	1985	1990	1995	2000	2005	2010	Total	Titles	Evolution in Action
Plants	8	12	12	4	11	3	3	7	9	23	92	1	8
Insects	6	4	6	9	9	2	3	11	14	24	88	1	2
Theory/Models	5	4	7	7	9	0	0	12	9	18	71	10	0
Fishes	4	3	3	3	2	2	0	8	11	11	47	3	1
Birds	0	3	3	3	4	1	0	6	4	12	36	0	1
Mammals	1	3	1	3	2	0	2	6	2	7	27	0	0
Gastropods	0	0	1	3	1	2	0	2	2	6	17	0	6
Amphibians	1	0	1	2	1	0	2	1	2	4	14	0	1
Fungi	0	1	1	1	0	0	1	1	2	7	14	0	0
Reptiles	0	2	1	0	0	0	1	1	1	8	14	0	0
Bacteria	6	2	0	0	0	0	0	1	2	2	13	0	0
Coevolution	0	1	2	0	0	0	0	0	2	5	10	1	5
Crustaceans	2	0	0	0	1	0	0	3	1	2	9	1	1
Parasites	0	0	1	0	1	0	2	1	1	1	7	0	0
Protists	1	1	1	0	0	0	0	0	0	3	6	0	0
Arachnids	1	1	0	0	1	0	0	2	0	0	6	0	0
Algae	0	0	0	0	1	0	0	0	1	1	4	0	0
Flatworms	0	0	0	0	1	0	0	1	0	2	4	0	0
Animals	0	1	0	0	2	0	0	0	0	0	3	0	0
Life/Organisms	0	1	0	0	0	0	0	0	0	1	2	0	0
Nematodes	0	0	0	1	0	0	0	0	1	0	2	0	0
Echinoderms	0	0	0	0	0	0	0	1	0	1	2	0	0

(Continued)

TABLE 8.2 (Continued)

The Focus of the 504 Studies of Speciation from 1965 to 2010 Obtained by Google Scholar Searching for Speciation by Year

	1965	1970	1975	1980	1985	1990	1995	2000	2005	2010	Total	Titles	Evolution in Action
Virus	0	0	0	0	0	0	0	0	1	1	2	0	0
Rotifers	0	0	0	0	0	0	0	0	2	0	2	0	0
Vertebrates	0	0	0	0	0	0	0	0	0	2	2	0	0
Ecosystem	0	1	1	0	0	0	0	0	0	0	1	0	0
Tunicates	0	0	0	0	0	0	0	0	0	0	1	0	0
Conservation Biology	0	0	0	0	1	0	0	0	0	0	1	0	0
Hawaiian Biota	0	0	0	0	0	0	0	1	0	0	1	0	0
Jellyfishes	0	0	0	0	0	0	0	0	1	0	1	0	0
Porifera	0	0	0	0	0	0	0	0	0	1	1	0	0
Arrow Worms	0	0	0	0	0	0	0	0	0	1	1	0	0
Annelids	0	0	0	0	0	0	0	0	0	1	1	0	0
Onychophora	0	0	0	0	0	0	0	0	0	1	1	0	0
Marine Fauna	0	0	0	0	0	0	0	0	0	1	1	0	0
Totals	35	40	40	37	47	11	14	65	69	146	504	17	25

Note: The foci of the additional data sets (see text) are also provided in this table.

TABLE 8.3

The Mode of Speciation in 504 Studies from 1965 to 2010 Obtained by Google Scholar Searching for Speciation by Year

	1965	1970	1975	1980	1985	1990	1995	2000	2005	2010	Totals	Titles	Evolution in Action
No Mechanism	15	13	6	5	6	0	1	2	7	8	63	6	3
Multiple Types	0	1	4	2	4	1	0	6	6	9	33	2	10
Models/Theory	2	0	1	0	2	0	0	1	1	1	8	1	0
Genetics	10	16	16	10	11	5	6	8	9	35	126	0	3
Coevolution	1	2	2	4	4	0	1	2	4	15	35	0	0
Climate/Geology	1	1	3	1	0	1	1	6	4	13	31	0	1
Allopatric Speciation	1	0	0	6	2	2	2	5	4	8	30	0	1
Sympatric Speciation	0	1	0	2	1	1	0	7	5	4	21	1	0
Sexual Selection	0	0	1	0	1	1	0	7	4	3	17	1	0
Ecological Speciation	0	0	0	0	2	0	0	1	3	11	17	0	3
Cryptic Speciation	0	0	0	0	0	0	0	2	7	8	17	0	0
Natural Selection	2	1	0	2	0	0	1	0	0	1	7	0	0
Adaptive Differentiation/Speciation	0	0	2	0	1	0	0	2	2	0	7	0	0
Geographic Speciation	0	0	0	0	1	0	0	2	1	3	7	0	2
Character Displacement	0	2	1	1	0	0	0	0	0	2	6	0	0
Vicariance	0	0	2	0	1	0	0	2	0	1	6	0	0
Competition	0	0	0	0	1	0	0	0	0	5	6	2	0
Incipient Speciation	1	0	1	0	0	0	0	1	1	0	4	1	0
Parapatric Speciation	0	0	0	1	0	0	0	1	1	1	4	0	0
Fragmentation	0	0	0	0	1	0	0	0	0	2	3	0	0
Habitat Specialization/Selection	0	0	0	0	1	0	0	0	1	1	3	0	0
Reproductive Isolation	0	0	0	0	0	0	0	1	0	2	3	0	0
Adaptive Radiation	0	0	0	0	0	0	0	0	2	1	3	0	2

(Continued)

TABLE 8.3 *(Continued)*

The Mode of Speciation in 504 Studies from 1965 to 2010 Obtained by Google Scholar Searching for Speciation by Year

	1965	1970	1975	1980	1985	1990	1995	2000	2005	2010	Totals	Titles	Evolution in Action
Sensory Drive	0	0	0	0	0	0	0	0	0	3	3	0	0
Dispersal	0	1	0	0	1	0	0	0	0	0	2	0	0
Phyletic Change	0	1	0	0	0	0	1	0	0	0	2	0	0
Species Evolution	0	0	1	0	0	0	0	0	0	1	2	0	0
Punctuated Events	0	0	0	0	2	0	0	0	0	0	2	0	0
Predation	0	0	0	0	0	0	0	0	0	2	2	0	0
Allochronic Speciation	0	0	0	0	0	0	0	1	0	0	1	0	0
Alloxenic Speciation	0	0	0	0	0	0	0	1	0	0	1	0	0
Autochthonous Speciation	0	0	0	0	0	0	0	1	0	0	1	0	0
Centrifugal Speciation	0	0	0	0	0	0	0	1	0	0	1	0	0
Determinism	0	0	0	0	0	0	0	1	0	0	1	0	0
Ecological Adjustment	0	1	0	0	0	0	0	0	0	0	1	0	0
Elevational Speciation	0	0	0	0	0	0	0	0	1	0	1	0	0
Environmental Selection	0	0	0	0	0	0	0	1	0	0	1	0	0
Force of Distance	0	0	0	0	0	0	0	0	1	0	1	0	0
Frequency Dependent Selection	0	0	0	0	0	0	0	0	1	0	1	0	0
Heteropatric Speciation	0	0	0	0	0	0	0	0	0	1	1	0	0
Historic Factors	0	0	0	0	0	0	0	0	1	0	1	0	0
Horizontal Gene Transfer	0	0	0	0	0	0	0	1	0	0	1	0	0
Humans as Selective Agents	0	0	0	0	0	0	0	0	1	0	1	0	0
Infectious Speciation	0	0	0	0	0	0	0	0	0	1	1	0	0
Insular Speciation	0	0	0	0	1	0	0	0	0	0	1	0	0
Invasion	0	0	0	1	0	0	0	0	0	0	1	0	0

(Continued)

TABLE 8.3 (Continued)

The Mode of Speciation in 504 Studies from 1965 to 2010 Obtained by Google Scholar Searching for Speciation by Year

	1965	1970	1975	1980	1985	1990	1995	2000	2005	2010	Totals	Titles	Evolution in Action
Isolation	0	0	0	1	0	0	0	0	0	0	1	0	0
Life Cycle Change	0	0	0	0	0	0	0	1	0	0	1	0	0
Morphic Speciation	0	0	0	0	0	0	0	0	0	1	1	0	0
Organization Theory	0	0	0	0	1	0	0	0	0	0	1	0	0
Parasitism	0	0	0	1	0	0	0	0	0	0	1	0	0
Phenotypic Plasticity	0	0	0	0	0	0	0	0	0	1	1	0	0
Phylogenetic Differentiation	0	0	0	0	0	0	0	0	0	1	1	0	0
Premating Isolation	1	0	0	0	0	0	0	0	0	0	1	0	0
Random Morphological Change	0	0	0	0	1	0	0	0	0	0	1	0	0
Rapid Speciation	0	0	0	0	0	0	1	0	0	0	1	0	0
Reinforcement	0	0	0	0	0	0	0	0	1	0	1	0	0
Replacement Pattern Speciation	0	0	0	0	1	0	0	0	0	0	1	0	0
Sequential Speciation	0	0	0	0	0	0	0	0	0	1	1	0	0
Speciation in situ	0	0	0	0	0	0	0	1	0	0	1	0	0
Stasipatric Speciation	0	0	0	0	1	0	0	0	0	0	1	0	0
Symbiogenesis	0	0	0	0	0	0	0	0	1	0	1	0	0
Transformation	1	0	0	0	0	0	0	0	0	0	1	0	0
Assortative Mating	0	0	0	0	0	0	0	0	0	0	0	1	0
Biological Degeneracy	0	0	0	0	0	0	0	0	0	0	0	1	0
Biotic Constraints	0	0	0	0	0	0	0	0	0	0	0	1	0
Totals	35	40	40	37	47	11	14	65	69	146	504	17	25

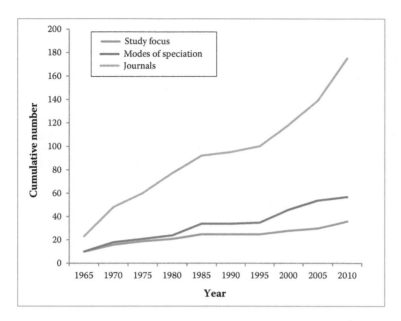

FIGURE 8.1 Accumulation of the number of focal systems (e.g., taxa, models, and theory), speciation modes, and the journals publishing papers on speciation.

and well formulated for evolution (Tables 8.2 and 8.3, Appendix 8.1). Adherence to Competition Theory has clearly not stifled research on speciation (Figure 8.1, Appendices 8.2–8.4). The linkage between competition and speciation is weak, and other factors may be strong evolutionary forces that generate diversity (Tables 8.2 and 8.3). In 2016, we are still paying homage to Santa Rosalia, and the search for the answer to the question of "what generates diversity" is still an active field of inquiry. It does not appear that "competition theory" is well founded in evolution nor has it thwarted active research in the field.

ACKNOWLEDGMENTS

I thank Brian Crother and Lynne Parenti for inviting me to participate in the Rosen Symposium held during the 2015 Joint Meeting of Ichthyologists and Herpetologists in Reno, Nevada. I thank Jon Kastendiek for helping me shed my allegiance to evolutionary ecology during my years at USC, to the late Robert F. Denno for exposing flaws in a theory he held dear during his career, and to Donn Eric Rosen for helping me be a better scientist.

APPENDIX 8.1 FULL CITATIONS FOR THE PAPERS INCLUDED IN THE PRESENT SURVEY OF SPECIATION

The papers separated by years were obtained by searching Google Scholar for Speciation. The other groups of papers were obtained by a different search and from an edited volume on evolution (see text for explanation).

1965

Bock, WJ. 1965. The role of adaptive mechanisms in the origin of higher levels of organization. *Systematic Zoology* 14:272–287.

Bose, S. 1965. Polyploidy in the genus *Crinum*. *Cytologia* 30:349–353.

Bruce-Chwatt, LJ. 1965. Paleogenesis and paleo-epidemiology of primate malaria. *Bulletin of the World Health Organization* 32:363–387.

Bullock, GL, SF Snieszko & CE Dunbar. 1965. Characteristics and identification of oxidative pseudomonads isolated from diseased fish. *Journal of General Microbiology* 38:1–7.

Colwell, RR, RV Citarella & I Ryman. 1965. Deoxyribonucleic acid base composition and Adansonian analysis of heterotrophic aerobic pseudomonads. *Journal of Bacteriology* 90:1148–1149.

Dos Passos, CF. 1965. Review of the Nearctic species of *Pieris "napi"* as classified by androconial scales and description of a new seasonal form (Lepidoptera: Pieridae). *Journal of the New York Entomological Society* 73:135–137.

Falkow, S. 1965. Nucleic acids, genetic exchange and bacterial speciation. *The American Journal of Medicine* 39:753–765.

Frost, WE. 1965. Breeding habits of Windermere charr, *Salvelinus willughbii* (Gunther), and their bearing on speciation of these fish. *Proceedings of the Royal Society of London. Series B* 163:232–284.

Fryer, G. 1965. Predation and its effects on migration and speciation in African fishes: A comment. *Journal of Zoology* 144:301–310.

Greenwood, PH. 1965. A further comment. *Journal of Zoology* 144:310–313.

Hecht, MD. 1965. The role of natural selection and evolutionary rates in the origin of higher levels of organization. *Systematic Zoology* 14:301–317.

Hennig, W. 1965. Phylogenetic systematics. *Annual Review of Entomology* 10:97–116.

Hoogstraal, H. 1965. *Haemaphysalis tibetensis* sp. n., and its significance in elucidating phylogenetic patterns in the genus (Ixodoidea, Ixodidae). *The Journal of Parasitology* 51:452–459.

Hubby, JL & LH Throckmorton. 1965. Protein differences in Drosophila. II. Comparative species genetics and evolutionary problems. *Genetics* 52:203–215.

Jackson, PBN. 1965. Reply to G. Fryer and P.H. Greenwood. *Journal of Zoology* 144:313–321.

John, B & KR Lewis. 1965. Genetic speciation in the grasshopper *Eyprepocnemis plorans*. *Chromosoma* 16:308–344.

Littlejohn, M. 1965. Premating isolation in the *Hyla ewingi* complex (Anura: Hylidae). *Evolution* 19:234–243.

MacPherson, AH. 1965. The origin of diversity in mammals of the Canadian arctic tundra. *Systematic Zoology* 14:153–173.

Malik, CP. 1965. Cytology of some Indian species of Rosaceae. *Caryologia* 18:13–149.

Moll, RH, JH Lonnquist, JV Fortuno & EC Johnson. 1965. The relationship of heterosis and genetic divergence in maize. *Genetics* 52:139–144.

Mordukhai-Boltovskoi, PD. 1965. Polyphemidae of the Pontocaspian basin. *Hydrobiologia* 25:212–219.

Pridham, TG & AJ Lyons. 1965. Further taxonomic studies on straight to flexuous streptomycetes. *Journal of Bacteriology* 89:331–342.

Rana, RS. 1965. Somatic reduction in an intervarietal hybrid of chrysanthemum. *The Japanese Journal of Genetics* 40:199–201.

Sen, M & SP Sen. 1965. Interspecific transformation in *Azotobacter*. *Journal of General Microbiology* 41:1–6.

Smith, SG. 1965. Cytological species-separation in Asiatic *Exochomus* (Coleoptera: Coccinellidae). *Canadian Journal of Genetics and Cytology* 7:363–373.

Stebbins, GL & J Major. 1965. Endemism and speciation in California plants. *Ecological Monographs* 35:1–35.

Stolp, H, MP Starr & NL Baigent. 1965. Problems in speciation of phytopathogenic pseudomonads and xanthomonads. *Annual Review of Phytopathology* 3:231–264.

Thomas, H & ML Jones. 1965. Chromosomal differentiation in diploid species of *Avena*. *Canadian Journal of Genetics and Cytology* 7:108–111.

Venkatesh, CS & S Kedharnath. 1965. Genetic improvement of *Eucalyptus* in India. *Silvae Genetica* 14:155–159 (cited incorrectly in a number of places).

Vorzimmer, P. 1965. Darwin's ecology and its influence upon his theory. *Isis* 56:148–155.

Warburg, MR. 1965. The evolutionary significance of the ecological niche. *Oikos* 16:205–213.

White, MJD. 1965. Chiasmatic and achiasmatic meiosis in African eumastacid grasshoppers. *Chromosoma* 16:271–307.

Wilson, EO. 1965. A consistency test for phylogenies based on contemporaneous species. *Systematic Zoology* 14:214–220.

Wright, S. 1965. Factor interaction and linkage in evolution. *Proceedings of the Royal Society B* 162 (986): 80–104.

Zhukovsky, PM. 1965. Main gene centres of cultivated plants and their wild relatives within the territory of the USSR. *Euphytica* 14:177–188.

1970

Aldrich, JW & KP Baer. 1970. Status and speciation in the Mexican duck (*Anas diazi*). *The Wilson Bulletin* 82:63–73.

Antonovics, J. 1970. The evolution of heavy metal tolerance in plants: An example of sympatric speciation. *Transactions and Proceedings of the Botanical Society of Edinburgh* 630pp.

Aubreville, A. 1970. Speciation in wet tropical forests: Monotypic or small genera. *Adansonia* 10:301–307.

Avery, DF & WW Tanner. 1970. Speciation in the Fijian and Tongan iguana *Brachylophus* (Sauria, Iguanidae) with the description of a new species. *Great Basin Naturalist* 30:166–172.

Ayala, SC. 1970. Lizard malaria in California: Description of a strain of *Plasmodium mexicanum*, and biogeography of lizard malaria in western North America. *Journal of Parasitology* 56:417–425.

Baker, HG. 1970. Evolution in the tropics. *Biotropica* 2:101–111.

Barber, HN. 1970. Hybridization and the evolution of plants. *Taxon* 19:154–160.

Bock, WJ. 1970. Microevolutionary sequences as a fundamental concept in macroevolutionary models. *Evolution* 24:704–722.

Brink, FH. 1970. Distribution and speciation of some carnivores. *Mammal Review* 1:67–79.

Carson, HL. 1970. Chromosomal tracers of founder events. *Biotropica* 2:3–6.

Cook, CDK. 1970. Hybridization in the evolution of *Batrachium*. *Taxon* 19:161–166.

Darlington, PJ, Jr. 1970. A practical criticism of Hennig-Brundin "Phylogenetic Systematics" and Antarctic biogeography. *Systematic Zoology* 19:1–18.

Darlington, PJ, Jr. 1970. Carabidae on tropical islands, especially the West Indies. *Biotropica* 2:7–15.

Darnell, RM. 1970. Evolution and the ecosystem. *American Zoologist* 10:9–15.

De Wet, JMJ & JR Harlan. 1970. Apomixis, polyploidy, and speciation in *Dichanthium*. *Evolution* 24:270–277.

Dillon, LS. 1970. Speciation and changing environment. *American Zoologist* 10:27–39.

Hagen, DW & JD McPhail. 1970. The species problem within *Gasterosteus aculeatus* on the Pacific coast of North America. *Journal of the Fishery Research Board of Canada* 27:147–155.

Helle, W, J Gutierrez & HR Bolland. 1970. A study on sex-determination and karyotypic evolution in Tetranychidae. *Genetica* 41:21–32.

Jones, D & PH Sneath. 1970. Genetic transfer and bacterial taxonomy. *Bacteriological Review* 34:40–81.

Kogan, M & EF Legner. 1970. A biosystematic revision of the genus *Muscidifurax* (Hymenoptera: Pteromalidae) with descriptions of four new species. *The Canadian Entomologist* 102:1268–1290.

Levin, SA. 1970. Community equilibria and stability, and an extension of the competitive exclusion principle. *American Naturalist* 104:413–423.

Navasin, MS & NA Cusksanova. 1970. Chromosome number and evolution. *Genetika* 6:71–83.

Olorode, O. 1970. The evolutionary implications of interspecific hybridization among four species of *Zinnia* sect. Mendezia (Compositae). *Brittonia* 22:207–216.

Pitt, JI & MW Miller. 1970. Speciation in the yeast genus *Metschnikowia*. *Antonie van Leeuwenhoek* 36:357–381.

Ramírez, W. 1970. Host specificity of fig wasps (Agaonidae). *Evolution* 24:680–691.

Rand, AL. 1970. Species formation in the blue monarch flycatchers genus *Hypothymis*. *Natural History Bulletin of the Siam Society* 23:353–365.

Rasch, EM, LM Prehn & RW Rasch. 1970. Cytogenetic studies of *Poecilia* (Pisces). *Chromosoma* 31:18–40.

Roberts, FL. 1970. Atlantic salmon (*Salmo salar*) chromosomes and speciation. *Transactions of the American Fisheries Society* 99:105–111.

Rupert, EA, B Dehgan & GL Webster. 1970. Experimental studies of relationships in the Genus *Jatropha*. I. *J. curcas* x *integerrima*. *Bulletin of the Torrey Botanical Club* 97:321–325.

Ruston, AR. 1970. Cytotaxonomy and chromosomal evolution of the bats. *Annals de Génétique et de Sélection Animale* 2:457–469.

Short, LL. 1970. A reply to Uzzell and Ashmole. *Systematic Zoology* 19:199–202.

Sinha, BMB & AK Sinha. 1970. Translocation heterozygosity in *Pimpinella bracteata* Haines. *Cytologia* 35:593–600.

Stebbins, GL. 1970. Adaptive radiation of reproductive characteristics in angiosperms, I: Pollination mechanisms. *Annual Review of Ecology & Systematics* 1:307–326.

Todd, NB. 1970. Karyotypic fissioning and canid phylogeny. *Journal of Theoretical Biology* 26:445–480.

Trejo, WH. 1970. Section of microbiology: An evaluation of some concepts and criteria used in the speciation of *Streptomycetes**. *Transactions of the New York Academy of Sciences* 32:989–997.

Tryon, R. 1970. Development and evolution of fern floras of oceanic islands. *Biotropica* 2:76–84.

Uzzell, T & NP Ashmore. 1970. Suture-zones: An alternative view. *Systematic Zoology* 19:197–199.

Vanzolini, PE & EE Williams. 1970. South American anoles: The geographic differentiation and evolution of the *Anolis chrysolepis* species group (Sauria, Iguanidae). *Arquivos de Zoologia* 19:1–124.

Wagner, WH, Jr. 1970. Biosystematics and evolutionary noise. *Taxon* 19:146–151.

White, MJD. 1970. Cytogenetics of speciation. *Australian Journal of Entomology* 9:1–6.

1975

Avise, JC & FJ Ayala. 1975. Genetic change and rates of cladogenesis. *Genetics* 81:757–773.

Avise, JC, JJ Smith & FJ Ayala. 1975. Adaptive differentiation with little genic change between two native California minnows. *Evolution* 29:411–426.

Axelrod, DI. 1975. Evolution and biogeography of Madrean-Tethyan sclerophyll vegetation. *Annals of the Missouri Botanical Garden* 62:280–334.

Benson, WW, KS Brown, Jr. & LE Gilbert. 1975. Coevolution of plants and herbivores: Passion flower butterflies. *Evolution* 29:659–680.

Burkart, A. 1975. Evolution of grasses and grasslands in South America. *Taxon* 24:53–66.

Bush, GL. 1975. Modes of animal speciation. *Annual Review of Ecology & Systematics* 6:339–364.

Candela, M & J-R Lacadena. 1975. Chromosomal polymorphism by reciprocal translocations in a natural population of cultivated rye, *Secale cereale* L.: A preliminary note. *Anales del Instituto Botánico A. J. Cavanilles* 32:649–657.

Carson, HL. 1975. The genetics of speciation at the diploid level. *American Naturalist* 109:83–92.

Carson, HL, WE Johnson, PS Nair & FM Sene. 1975. Allozymic and chromosomal similarity in two *Drosophila* species. *Proceedings of the National Academy of Sciences USA* 72:4521–4525.

Cracraft, J. 1975. Historical biogeography and earth history: Perspectives for a future synthesis. *Annals of the Missouri Botanical Garden* 62:227–250.

Delgadillo, C. 1975. Taxonomic revision of *Aloina, Aloinella* and *Crossidium* (Musci). *The Bryologist* 78:245–303.

Dobzhansky, T. 1975. Analysis of incipient reproductive isolation within a species of *Drosophila*. *Proceedings of the National Academy of Sciences USA* 72:3638–3641.

Fouquette, MJ. 1975. Speciation in chorus frogs. I. Reproductive character displacement in the *Pseudacris nigrita* complex. *Systematic Zoology* 24:16–23.

Harlan, JR & JM DeWit. 1975. On Ö. Winge and a prayer: The origins of polyploidy. *The Botanical Review* 41:361–390.

Heller, J. 1975. The taxonomy, distribution and faunal succession of *Buliminus* (Pulmonata: Enidae) in Israel. *Zoological Journal of the Linnean Society* 57:1–57.

Hunziker, JH. 1975. On the geographical origin of *Larrea divaricata* (Zygophyllaceae). *Annals of the Missouri Botanical Garden* 62:497–500.

Irwin, MR. 1975. Phylogenetic evolution in Columbidae as revealed by interrelationships of genetic systems of antigenic characters. *Evolution* 29:72–86.

Jotterand, M. 1975. The African *Mus* (Pigmy-Mice): The role of chromosomal polymorphism in speciation. *Caryologica* 23:335–344.

Kellogg, DE. 1975. Character displacement in the radiolarian genus, *Eucyrtidium*. *Evolution* 29:736–749.

Kornfield, IL & RK Koehn. 1975. Genetic variation and speciation in New World cichlids. *Evolution* 29:427–437.

Levin, DA. 1975. Interspecific hybridization, heterozygosity and gene exchange in *Phlox*. *Evolution* 29:37–51.

Lowther, PE. 1975. Geographic and ecological variation in the family Icteridae. *The Wilson Bulletin* 87:481–495.

Mabberley, DJ. 1975. The giant lobelias: Pachycauly, biogeography, ornithophily and continental drift. *New Phytologist* 74:365–374.

Montanucci, RR, RW Axtell & HC Dessauer. 1975. Evolutionary divergence among collared lizards (*Crotaphytus*), with comments on the status of *Gambelia*. *Herpetologica* 31:336–347.

Moore, DM. 1975. The alpine flora of Tierra del Fuego. *Anales del Instituto Botánico A. J. Cavanilles* 32:419–440.

Nei, M, T Maryuama & R Chakraborty. 1975. The bottleneck effect and genetic variability in populations. *Evolution* 29:1–10.

Nijalingappa, BHM. 1975. Cytological studies in *Fimbristylis* (Cyperaceae). *Cytologia* 40:177–183.

Ownbey, GB, PH Raven & DW Kyhos. 1975. Chromosome numbers in some North American species of the genus *Cirsium*. III. Western United States, Mexico, and Guatemala. *Brittonia* 27:297–304.

Papavero, N. 1975. Studies of Asilidae (Diptera) systematics and evolution: IV. Tribe Megapodini Carrera (Dasypogoninae), with a review of the neotropical species. *Arquivos de Zoologia* 27:191–315.

Richardson, RH & PE Smouse. 1975. Ecological specialization of Hawaiian *Drosophila*. *Oecologia* 22:1–13.

Rieux, R. 1975. Host specificity in the genus *Matsucoccus* (Homoptera: Margarodidae) with special reference to the host plants of *M. pini*. Classification of *Matsucoccus* species according to their hosts. *Annales des Sciences Forestieres* 32:157–168.

Rosen, DE. 1975. A vicariance model of Caribbean biogeography. *Systematic Zoology* 24:431–464.

Sage, RD & RK Selander. 1975. Trophic radiation through polymorphism in cichlid fishes. *Proceedings of the National Academy of Sciences USA* 72:4669–4673.

Sneath, PHA. 1975. Cladistic representation of reticulate evolution. *Systematic Zoology* 24:360–368.

Snelling, RE & JJ Hunt. 1975. The ants of Chile (Hymenoptera: Formicidae). *Revista Chilena de Entomología* 9:63–129.

Stanley, SM. 1975. A theory of evolution above the species level. *Proceedings of the National Academy of Sciences USA* 72:646–650.

Stebbins, GL. 1975. The role of polyploid complexes in the evolution of North American grasslands. *Taxon* 24:91–106.

Toussoun, TA & PE Nelson. 1975. Variation and speciation in the Fusaria. *Annual Review of Phytopathology* 13:71–82.

Van Soest, RVM. 1975. Zoogeography and speciation in the Salpidae (Tunicata, Thaliacea). *Beaufortia* 23:181–215.

Yoon, JS, K Resch, MR Wheeler & RH Richardson. 1975. Evolution in Hawaiian Drosophilidae: Chromosomal phylogeny of the *Drosophila crassifemur* complex. *Evolution* 29:249–256.

1980

Brooks, DR. 1980. Allopatric speciation and non-interactive parasite community structure. *Systematic Zoology* 29:192–203.

Bryan, G. 1980. The British species of *Achrysocharoides* (Hymenoptera, Eulophidae). *Systematic Entomology* 5:245–262.

Conant, DS & G Cooper-Driver. 1980. Autogamous allohomoploidy in *Alsophila* and *Nephelea* (Cyatheaceae): A new hypothesis for speciation in homoploid homosporous ferns. *American Journal of Botany* 67:1269–1288.

Early, JW. 1980. The Diapriidae (Hymenoptera) of the southern islands of New Zealand. *Journal of the Royal Society of New Zealand* 10:153–171.

Fischer, EA. 1980. Speciation in the hamlets (*Hypoplectrus*: Serranidae): A continuing enigma. *Copeia* 1980:649–659.

Fry, CH. 1980. The origin of Afrotropical kingfishers. *Ibis* 22:57–74.

Futuyma, DJ & GC Mayer. 1980. Non-allopatric speciation in animals. *Systematic Zoology* 29:254–271.

Gabor Miklos, GL, DA Willcocks, PR Baverstock. 1980. Restriction endonuclease and molecular analyses of three rat genomes with special reference to chromosome rearrangement and speciation problems. *Chromosoma* 76:339–363.

Glazier, DS. 1980. Ecological shifts and the evolution of geographically restricted species of North American *Peromyscus* (mice). *Journal of Biogeography* 7:63–83.

Grant, PR. 1980. Colonization of Atlantic islands by chaffinches. *Bonner Zoologische Beiträge* 31:311–317.

Kaneshiro, KY. 1980. Sexual isolation, speciation and the direction of evolution. *Evolution* 34:437–444.

Khonglam, A & A Singh. 1980. Cytogenetic studies on the weed species of *Eupatorium* found in Meghalaya, India. *Proceedings of the Indian Academy of Science* 89:237–241.

Kilias, G, SN Alahiotis & M Pelecanos. 1980. A multifactorial genetic investigation of speciation theory using *Drosophila melanogaster*. *Evolution* 34:730–737.

Lande, R. 1980. Genetic variation and phenotypic evolution during allopatric speciation. *American Naturalist* 116:463–479.

Layzer, D. 1980. Genetic variation and progressive evolution. *American Naturalist* 115:809–826.

Lee, DW & JB Lowry. 1980. Plant speciation on tropical mountains: *Leptospermum* (Myrtaceae) on Mount Kinabalu, Borneo. *Botanical Journal of the Linnean Society* 80:223–242.

Levinton, JS & MC Simon. 1980. A critique of the punctuated equilibria model and implications for the detection of speciation in the fossil record. *Systematic Zoology* 29:130–142.

Lewis, RW. 1980. Evolution: A system of theories. *Perspectives in Biology and Medicine* 23:551–572.

Martin, J, C Kuvangkadilok, DH Peart & BTO Lee. 1980. Multiple sex determining regions in a group of related *Chironomus* species (Diptera: Chironomidae). *Heredity* 44:367–382.

McCracken, GF & RK Selander. 1980. Self-fertilization and monogenic strains in natural populations of terrestrial slugs. *Proceedings of the National Academy of Sciences USA* 77:684–688.

McKaye, KR. 1980. Seasonality in habitat selection by the gold color morph of *Cichlasoma citrinellum* and its relevance to sympatric speciation in the family Cichlidae. *Environmental Biology of Fishes* 5:75–78.

Mitton, JB, KB Sruncsox & ML Davis. 1980. Genetic differentiation in ponderosa pine along a steep elevational transect. *Silvae Genetica* 29:100–103.

Morescalchi, A. 1980. Evolution and karyology of the amphibians. *Bolletino di Zoologia* 22:57–74.

Müller, FP. 1980. *Ovatus crataegarius* (Walker, 1850) and *O. insitus* (Walker, 1849) as a model for sympatric speciation without use of ecological niches. *Deutsche Entomologische Zeitschrift* 27:199–217.

Murray, J & B Clarke. 1980. The genus *Partula* on Moorea: Speciation in progress. *Proceedings of the Royal Society of London. Series B* 21:83–117.

Nelson, G & NI Platnick. 1980. A vicariance approach to historical biogeography. *Bioscience* 30:339–343.

Ratti, JT. 1980. The classification of avian species and subspecies. *American Birds* 1980:860–866.

Reid, GM. 1980. 'Explosive Speciation' of carps in Lake Lanao (Philippines)—fact or fancy? *Systematic Zoology* 29:314–316.

Starmer, WT, HW Kircher & HJ Phaff. 1980. Evolution and speciation of host plant specific yeasts. *Evolution* 34:137–146.

Tempelton, A. 1980. Modes of speciation and inferences based on genetic distances. *Evolution* 34:719–729.

Tempelton, A. 1980. The theory of speciation via the founder principle. *Genetics* 94:1011–1038.

Triantaphyllou, AC & H Hirschmann. 1980. Cytogenetics and morphology in relation to evolution and speciation of plant-parasitic nematodes. *Annual Review of Phytopathology* 18:333–359.

Wake, DB, SY Yang & TJ Papenfuss. 1980. Natural hybridization and its evolutionary implications in Guatemalan plethodontid salamanders of the genus *Bolitoglossa*. *Herpetologica* 36:335–345.

Ward, PS. 1980. Genetic variation and population differentiation in the *Rhytidoponera impressa* group, a species complex of ponerine ants (Hymenoptera: Formicidae). *Evolution* 34:1060–1076.

Wheatley, BP. 1980. Malaria as a possible selective factor in the speciation of macaques. *Journal of Mammalogy* 61:307–311.

White, MJD. 1980. Modes of speciation in orthopteroid insects. *Italian Journal of Zoology* 47:suppl 83–94.

Wood, TK. 1980. Divergence in the *Enchenopa binotata* Say complex (Homoptera: Membracidae) effected by host plant adaptation. *Evolution* 34:147–160.

1985

Altukhov, YP. 1985. Two kinds of genetic variability and the problem of speciation. *Evolution* 39:223–226.

Barr, TC & JR Holsinger. 1985. Speciation in cave faunas. *Annual Review of Ecology & Systematics* 16:313–337.

Brothers, DJ. 1985. Species concepts, speciation, and higher taxa. *Transvaal Museum Monograph* No. 4, pp 35–42.

Bullini, L. 1985. Speciation by hybridization in animals. *Bolletino di Zoologia* 52:1–2, 121–137.

Burgman, MA. 1985. Cladistics, phenetics and biogeography of populations of *Boronia inornata* Turcz. (Rutaceae) and the *Eucalyptus* diptera Andrews (Myrtaceae) species complex in Western Australia. *Australian Journal of Botany* 33(4). DOI: 10.1071/BT9850419.

Capanna, E, M Corti & G Nascetti. 1985. Role of contact areas in chromosomal speciation of the European long-tailed house mouse (*Mus musculus domesticus*). *Bolletino di Zoologia* 52:1–2, 97–119. DOI: 10.1080/11250008509440345.

Carson, HL. 1985. Unification of speciation theory in plants and animals. *Systematic Botany* 10:380–390.

Coluzzi, M, V Petrarca & MA di Deco. 1985. Chromosomal inversion intergradation and incipient speciation in *Anopheles gambiae*. *Bolletino di Zoologia* 52:1–2, 45–63. DOI: 10.1080/11250008509440343.

Cracraft, J. 1985. Species selection, macroevolutionary analysis, and the "hierarchical theory" of evolution. *Systematic Zoology* 34:222–229.

Davis, JI & AJ Gilmartin. 1985. Morphological variation and speciation. *Systematic Botany* 10:417–425.

de Vries, EJ. 1985. The biogeography of the genus *Dugesia* (Turbellaria, Tricladida, Paludicola) in the Mediterranean region. *Journal of Biogeography* 12:509–518.

Dodd, DMB & JR Powell. 1985. Founder-flush speciation: An update of experimental results with *Drosophila*. *Evolution* 39:1388–1392.

Engen, S & S Bernt-Erik. 1985. The evolutionary significance of sexual selection. *Journal of Theoretical Biology* 117:277–289.

Foote, BA. 1985. Biology of *Rivellia pallida* (Diptera: Platystomatidae, a consumer of the Nitrogen-fixing root nodules of *Amphicarpa brachteata* (Leguminosinae). *Journal of the Kansas Entomological Society* 58:27–35.

Fryer, G & PH Greenwood. 1985. The demonstration of speciation in fossil molluscs and living fishes. *Biological Journal of the Linnean Society* 26:325–336.

Gibby, M. 1985. Hybridization and speciation in the genus *Dryopteris* (Pteridophyta: Dryopteridaceae) on Pico Island in the Azores. *Plant Systematics and Evolution* 149:241–251.

Graves, G. 1985. Elevational correlates of speciation and intraspecific geographic variation in plumage in Andean forest birds. *The Auk* 102:556–579.

Hoberg, EP. 1985. *Reticulotaenia* n. gen. for *Lateriporus australis* Jones and Williams, 1967, and *Lateriporus mawsoni* Prudhoe, 1969 (Cestoda: Dilepididae), from Sheathbills, *Chionis* spp., in Antarctica, with a consideration of infraspecific variation and speciation. *The Journal of Parasitology* 71:319–326.

Ichinose, T & C Horiuchi. 1985. Allopatric speciation of Yaeyama insular isolates in the *Papilio bianor* complex of Japan (Lepidoptera, Papilionidae). *Japanese Journal of Entomology* 53:233–244.

Jermy, JT. 1985. Is there competition between phytophagous insects? *Journal of Zoological Systematics and Evolutionary Research* 23:275–285.

Johnson, NK & CB Johnson. 1985. Speciation in sapsuckers (*Sphyrapicus*): II. Sympatry, hybridization, and mate preference in *S. ruber daggetti* and *S. nuchalis*. *The Auk* 102:1–15.

Kolding, S. 1985. Genetic adaptation to local habitats and speciation processes within the genus *Gammarus*. *Marine Biology* 89:249–255.

Koshy, JK & PM Mathew. 1985. Cytology of the genus *Cleome* Linn. *Cytologia* 50:283–287.

Lande, R. 1985. The fixation of chromosomal rearrangements in a subdivided population with local extinction and colonization. *Heredity* 54:323–332. DOI: 10.1038/hdy.1985.43.

Maurer, BA. 1985. On the ecological and evolutionary roles of interspecific competition. *Oikos* 45:300–302.

Müller, FP. 1985. Biotype formation and sympatric speciation in aphids (Homoptera: Aphidinea). *Entomologia Generalis* 10:161–181.

Nevo, E. 1985. Speciation in action and adaptation in subterranean mole rats: Patterns and theory. *Bolletino di Zoologia* 52:1–2, 65–95. DOI: 10.1080/11250008509440344.

Parris, BS. 1985. Ecological aspects of distribution and speciation in Old World tropical ferns. *Proceedings of the Royal Society of Edinburgh, Section B: Biological Sciences* 86:341–346.

Pohja, P & J Lehtonen. 1985. Simulation of morphogenesis and speciation in the unicellular alga *Micrasterias* (Chlorophyta, Conjugatophyceae). *Journal of Theoretical Biology* 115:401–414.

Rejón, MR, L Pascual, CR Rejón & B Valdés. 1985. A new species of Muscari subgenus *Leopoldia* from the Iberian Peninsula. *Biochemical Systematics and Ecology* 13:239–250.

Ringo, J, D Wood, R Rockwell & H Dowse. 1985. An experiment testing two hypotheses of speciation. *American Naturalist* 126:642–661.

Roberts, JD & LR Maxson. 1985. Tertiary speciation models in Australian anurans: Molecular data challenge Pleistocene scenario. *Evolution* 39:325–334.

Rodin, SN, VA Berdnikov & AA Zharkikh. 1985. A model of genome size dynamics during speciation. *Biometrical Journal* 27:807–821.

Shaw, DD, DJ Coates, ML Arnold & P Wilkinson. 1985. Temporal variation in the chromosomal structure of a hybrid zone and its relationship to karyotypic repatterning. *Heredity* 55:293–306. DOI: 10.1038/hdy.1985.111.

Sick, H. 1985. Observations on the Andean-Patagonian component of southeastern Brazil's avifauna. *Ornithological Monographs 36 Neotropical Ornithology*, pp 233–237.

Sleep, A. 1985. Speciation in relation to edaphic factors in the *Asplenium adiantum-nigrum* group. *Proceedings of the Royal Society of Edinburgh, Section B: Biological Sciences* 86:325–334.

Soulé, ME. 1985. What is conservation biology? A new synthetic discipline addresses the dynamics and problems of perturbed species, communities, and ecosystems. *Bioscience* 35:727–734.

Subramanian, D. 1985. Cytotaxonomical studies in South Indian Ranunculaceae. *Cytologia* 50:759–768.

Taylor, WC, NT Luebke & NB Smith. 1985. Speciation and hybridisation in North American quillworts. *Proceedings of the Royal Society of Edinburgh, Section B: Biological Sciences* 86:259–263.

Tryon, R. 1985. Fern speciation and biogeography. *Proceedings of the Royal Society of Edinburgh, Section B: Biological Sciences* 86:353–360.

Tsurusaki, N. 1985. Geographic variation of chromosomes and external morphology in the *montanum*-subgroup of the *Leiobunum curvipalpe*-group (Arachnida, Opiliones, Phalangiidae) with special reference to its presumable process of raciation (Taxonomy). *Zoological Science* 2:767–783.

Vuilleumier, F. 1985. Forest birds of Patagonia: Ecological geography, speciation, endemism, and faunal history. *Ornithological Monographs* 36:255–304.

Warwick, SI & LD Gottlieb. 1985. Genetic divergence and geographic speciation in *Layia* (Compositae). *Evolution* 39:1236–1241.

Wicken, JS. 1985. Thermodynamics and the conceptual structure of evolutionary theory. *Journal of Theoretical Biology* 117:363–383.

Wiley, EO & RL Mayden. 1985. Species and speciation in phylogenetic systematics, with examples from the North American fish fauna. *Annals of the Missouri Botanical Garden* 72:596–635.

Williamson, PG. 1985. Punctuated equilibrium, morphological stasis and the palaeontological documentation of speciation: A reply to Fryer, Greenwood and Peake's critique of the Turkana Basin mollusc sequence. *Biological Journal of the Linnean Society* 26:307–324.

Wood, TK & SI Guttman. 1985. A new member of the *Enchenopa binotata* Say complex on tulip tree (*Liriodendron tulipifera*). *Proceedings of the Entomological Society of Washington* 87:171–175.

1990

Arnold, ML, JL Hamrick & BD Bennett. 1990. Allozyme variation in Louisiana irises: A test for introgression and hybrid speciation. *Heredity* 65:297–306.

Bullini, L & G Nascetti. 1990. Speciation by hybridization in phasmids and other insects. *Canadian Journal of Zoology* 68:1747–1760.

Chown, SL. 1990. Speciation in the sub-Antarctic weevil genus *Dusmoecetes* Jeannel (Coleoptera: Curculionidae). *Systematic Entomology* 15:283–296.

Cook, LM, RAD Cameron & LA Lace. 1990. Land snails of eastern Madeira: Speciation, persistence and colonization. *Proceedings of the Royal Society of London. Series B* 239. DOI: 10.1098/rspb.1990.0008.

Goto, A & T Andoh. 1990. Genetic divergence between the sibling species of river-sculpin, *Cottus amblystomopsis* and *C. nozawae*, with special reference to speciation. *Environmental Biology of Fishes* 28:257–266.

McDonald, MA & MH Smith. 1990. Speciation, heterochrony, and genetic variation in Hispaniolan Palm-Tanagers. *The Auk* 107:707–717.

McElroy, DM & I Kornfield. 1990. Sexual selection, reproductive behavior, and speciation in the mbuna species flock of Lake Malawi (Pisces: Cichlidae). *Environmental Biology of Fishes* 28:273–284.

Reid, DG. 1990. Trans-Arctic migration and speciation induced by climatic change: The biogeography of *Littorina* (Mollusca: Gastropoda). *Bulletin of Marine Science* 47:35–49.

Rieseberg, LH, R Carter & S Zona. 1990. Molecular tests of the hypothesized hybrid origin of two diploid *Helianthus* species (Asteraceae). *Evolution* 44:1498–1511.

Singh, HV & BR Chaudhary. 1990. Karyological investigations on the genus *Oedogonium* Link (Oedogoniales, Chlorophyceae). *Caryologia* 43:169–181.

Varadarajan, GS. 1990. Patterns of geographic distribution and their implications on the phylogeny of *Puya* (Bromeliaceae). *Journal of the Arnold Arboretum* 71:527–552.

1995

Brasier, CM. 1995. Episodic selection as a force in fungal microevolution, with special reference to clonal speciation and hybrid introgression. *Canadian Journal of Botany* 73(S1):1213–1221.

Charlesworth, D. 1995. Hybrid speciation: Evolution under the microscope. *Current Biology* 5:835–836.

Claridge, MF. 1995. Species concepts and speciation in insect herbivores: Planthopper case studies. *Bolletino di Zoologia* 62:53–58.

Corti, M & M Aguilera. 1995. Allometry and chromosomal speciation of the casiraguas *Proechimys* (Mammalia, Rodentia). *Journal of Zoological Systematics and Evolutionary Research* 33:109–115.

Gusev, AV. 1995. Some pathways and factors of monogenean microevolution. *Canadian Journal of Fisheries and Aquatic Sciences* 52(S1):52–56.

Highton, R. 1995. Speciation in eastern North American salamanders of the genus *Plethodon*. *Annual Review of Ecology & Systematics* 26:579–600.

Hoberg, EP. 1995. Historical biogeography and modes of speciation across high-latitude seas of the Holarctic: Concepts for host–parasite coevolution among the Phocini (Phocidae) and Tetrabothriidae (Eucestoda). *Canadian Journal of Zoology* 73:45–57.

Miura, I, M Nishioka, LJ Borkin & Z Wu. 1995. The origin of the brown frogs with $2n = 24$ chromosomes. *Experientia* 51:179–188.

Moya, A, A Galiana & FJ Ayala. 1995. Founder-effect speciation theory: Failure of experimental corroboration. *Proceedings of the National Academy of Sciences USA* 92:3983–3986.

Noor, MA. 1995. Speciation driven by natural selection in *Drosophila*. *Nature* 375:674–675.

Ortells, MO. 1995. Phylogenetic analysis of G-banded karyotypes among the South American subterranean rodents of the genus *Ctenomys* (Caviomorpha: Octodontidae), with special reference to chromosomal evolution and speciation. *Biological Journal of the Linnean Society* 54:43–70.

Raxworthy, CJ & RA Nussbaum. 1995. Systematics, speciation and biogeography of the dwarf chameleons (*Brookesia*; Reptilia, Squamata, Chamaeleontidae) of northern Madagascar. *Journal of Zoology* 235:525–558.

Rieseberg, LH, C Van Fossen & AM Desrochers. 1995. Hybrid speciation accompanied by genomic reorganization in wild sunflowers. *Nature* 375:313–316.

Wendel, JF, A Schnabel & T Seelanan. 1995. Bidirectional interlocus concerted evolution following allopolyploid speciation in cotton (*Gossypium*). *Proceedings of the National Academy of Sciences USA* 92:280–284.

2000

Arnqvist, G, M Edvardsson, U Friberg & T Nilsson. 2000. Sexual conflict promotes speciation in insects. *Proceedings of the National Academy of Sciences USA* 97:10460–10464.

Barraclough, TG & AP Vogler. 2000. Detecting the geographical pattern of speciation from species-level phylogenies. *The American Naturalist* 155. DOI: 10.1086/303332.

Beheregaray, LB, JA Levy & JR Gold. 2000. Population genetics of the silverside *Odontesthes argentinensis* (Teleostei, Atherinopsidae): Evidence for speciation in an estuary of southern Brazil. *Copeia* 2000:441–447.

Boake, CRB. 2000. Flying apart: Mating behavior and speciation. *Bioscience* 50:501–508.

Briggs, JC. 2000. Centrifugal speciation and centres of origin. *Journal of Biogeography* 27:1183–1188.

Brochmann, C, L Borgen & OE Stabbetorp. 2000. Multiple diploid hybrid speciation of the Canary Island endemic *Argyranthemum sundingii* (Asteraceae). *Plant Systematics and Evolution* 220:77–92.

Buerkle, CA, RJ Morris, MA Asmussen & LH Rieseberg. 2000. The likelihood of homoploid hybrid speciation. *Heredity* 84:441–451.

Chesser, RT. 2000. Evolution in the high Andes: The phylogenetics of *Muscisaxicola* ground-tyrants. *Molecular Phylogenetics and Evolution* 15:369–380.

Collins, AC & JM Dubach. 2000. Biogeographic and ecological forces responsible for speciation in *Ateles*. *International Journal of Primatology* 21:421–444.

Combes, C & A Théron. 2000. Metazoan parasites and resource heterogeneity: Constraints and benefits. *International Journal of Parasitology* 30:299–304.

Craddock, EM. 2000. Speciation processes in the adaptive radiation of Hawaiian plants and animals. *Evolutionary Biology* 31:1–53.

Craze, PG & LA Lace. 2000. Spatial ecology, habitat and speciation in the Porto Santan land snail genus *Heterostoma*. *Biological Journal of the Linnean Society* 71:665–676.

de la Cruz, F & J Davies. 2000. Horizontal gene transfer and the origin of species: Lessons from bacteria. *Trends in Microbiology* 8:128–133.

Deltsher, C. 2000. The endemic spiders (Araneae) of the Balkan Peninsula. *Ekológia* 19:59–65.

Doebeli, M & U Dieckmann. 2000. Evolutionary branching and sympatric speciation caused by different types of ecological interactions. *The American Naturalist* 156:S77–S101.

Esseghir, S, PD Ready & R Ben-Ismail. 2000. Speciation of *Phlebotomus* sandflies of the subgenus *Larroussius* coincided with the late Miocene-Pliocene aridification of the Mediterranean subregion. *Biological Journal of the Linnean Society* 70:189–219.

Gavrilets, S. 2000. Waiting time to parapatric speciation. *Proceedings of the Royal Society of London. Series B* 267(1461). DOI: 10.1098/rspb.2000.1309.

Gavrilets, S, R Acton & J Gravner. 2000. Dynamics of speciation and diversification in a meta-population. *Evolution* 54:1493–1501.

Gill, C & J Martinell. 2000. Phylogeny, speciation and species turnover. The case of the Mediterranean gastropods of genus *Cyclope* Risso, 1826. *Lethaia* 33:236–250.

Gómez-Zurita, J, E Petitpierre & C Juan. 2000. Nested cladistic analysis, phylogeography and speciation in the *Timarcha goettingensis* complex (Coleoptera, Chrysomelidae). *Molecular Ecology* 9:557–570.

Gorur, G. 2000. The role of phenotypic plasticity in host race formation and sympatric speciation in phytophagous insects, particularly in aphids. *Turkish Journal of Zoology* 24:63–68.

Grant, PR, BR Grant & K Petren. 2000. The allopatric phase of speciation: The sharp-beaked ground finch (*Geospiza difficilis*) on the Galápagos islands. *Biological Journal of the Linnean Society* 69:287–317.

Gray, DA & WH Cade. 2000. Sexual selection and speciation in field crickets. *Proceedings of the National Academy of Sciences USA* 97:14449–14454.

Haufler, CH, EA Hooper & JP Therrien. 2000. Modes and mechanisms of speciation in pteridophytes: Implications of contrasting patterns in ferns representing temperate and tropical habitats. *Plant Species Biology* 15:223–236.

Häunfling, B & R Brandl. 2000. Phylogenetics of European cyprinids: Insights from allozymes. *Journal of Fish Biology* 57:265–276.

Jarman, SN, NG Elliott, S Nicol & A McMinn. 2000. Molecular phylogenetics of circumglobal *Euphausia* species (Euphausiacea: Crustacea). *Canadian Journal of Fisheries and Aquatic Sciences* 57(S3):51–58.

Jousson, O, P Bartoli & J Pawlowski. 2000. Cryptic speciation among intestinal parasites (Trematoda: Digenea) infecting sympatric host fishes (Sparidae). *Journal of Evolutionary Biology* 13:778–785.

Kaneko, K & T Yoo. 2000. Sympatric speciation: Compliance with phenotype diversification from a single genotype. *Proceedings of the Royal Society of London. Series B* 267. DOI: 10.1098/rspb.2000.1293.

Kawata, M & J Yoshimura. 2000. Speciation by sexual selection in hybridizing populations without viability selection. *Evolutionary Ecology Research* 2:897–909.

Kirkpatrick, M. 2000. Speciation: Fish found in flagrante delicto. *Nature* 408:298–299.

Knowles, LL. 2000. Tests of Pleistocene speciation in montane grasshoppers (genus *Melanoplus*) from the sky islands of western North America. *Evolution* 54:1337–1348.

Lougheed, SC, JR Freeland, P Handford & PT Boag. 2000. A molecular phylogeny of warbling-finches (*Poospiza*): Paraphyly in a Neotropical emberizid genus. *Molecular Phylogenetics & Evolution* 17:367–378.

Marshall, DC & JR Cooley. 2000. Reproductive character displacement and speciation in periodical cicadas, with description of a new species, 13-year *Magicicada neotredecim*. *Evolution* 54:1313–1325.

Marshall, DJ & L Coetzee. 2000. Historical biogeography and ecology of a continental Antarctic mite genus, *Maudheimia* (Acari, Oribatida): Evidence for a Gondwanan origin and Pliocene-Pleistocene speciation. *Zoological Journal of the Linnean Society* 129:111–128.

Mattern, MY & DA McLennan. 2000. Phylogeny and speciation of felids. *Cladistics* 16:232–253.

McCartney, MA, G Keller & HA Lessios. 2000. Dispersal barriers in tropical oceans and speciation in Atlantic and eastern Pacific sea urchins of the genus *Echinometra*. *Molecular Ecology* 9:1391–1400.

Mulcahy, DG & JR Mendelson, III. 2000. Phylogeography and speciation of the morphologically variable, widespread species *Bufo valliceps*, based on molecular evidence from mtDNA. *Molecular Phylogenetics & Evolution* 17:173–179.

Nevo, E, E Ivanitskaya, MG Filippucci & A Beiles. 2000. Speciation and adaptive radiation of subterranean mole rats, *Spalax ehrenbergi* superspecies, in Jordan. *Biological Journal of the Linnean Society* 69:263–281.

O'Donnell, K, HC Kistler, BK Tacke & HH Casper. 2000. Gene genealogies reveal global phylogeographic structure and reproductive isolation among lineages of *Fusarium graminearum*, the fungus causing wheat scab. *Proceedings of the National Academy of Sciences USA* 97:7905–7910.

Park, LE & KF Downing. 2000. Implications of phylogeny reconstruction for ostracod speciation modes in Lake Tanganyika. *Advances in Ecological Research* 31:303–330.

Pfrender, ME, K Spitze & N Lehman. 2000. Multi-locus genetic evidence for rapid ecologically based speciation in *Daphnia*. *Molecular Ecology* 9:1717–1732.

Qian, H & RE Ricklefs. 2000. Large-scale processes and the Asian bias in species diversity of temperate plants. *Nature* 407:180–182.

Randi, E, V Lucchini, T Armijo-Prewitt, RT Kimball, EL Braun & JD Ligon. 2000. Mitochondrial DNA phylogeny and speciation in the tragopans. *The Auk* 117:1003–1015.

Riddle, BR. 2000. Phylogeography and systematics of the *Peromyscus eremicus* species group and the historical biogeography of North American warm regional deserts. *Molecular Phylogenetics & Evolution* 17:145–160.

Rodríguez-Robles, JA & JM De Jesús-Escobar. 2000. Molecular systematics of New World gopher, bull, and pinesnakes (*Pituophis*: Colubridae), a transcontinental species complex. *Molecular Phylogenetics & Evolution* 14:35–50.

Ruiz, A, AM Cansian, GCS Kuhn, MAR Alves & FM Sene. 2000. The *Drosophila serido* speciation puzzle: Putting new pieces together. *Genetica* 108:217–227.

Rumpler, Y. 2000. What cytogenetic studies may tell us about species diversity and speciation of lemurs. *International Journal of Primatology* 21:865–881.

Rundle, HD, L Nagel, JW Boughman & D Schluter. 2000. Natural selection and parallel speciation in sympatric sticklebacks. *Science* 287:306–308.

Scheffer, SJ & BM Wiegmann. 2000. Molecular phylogenetics of the holly leafminers (Diptera: Agromyzidae: *Phytomyza*): Species limits, speciation, and dietary specialization. *Molecular Phylogenetics & Evolution* 17:244–255.

Schilthuien, M. 2000. Dualism and conflicts in understanding speciation. *BioEssays* 22:1134–1141.

Seehausen, O. 2000. Explosive speciation rates and unusual species richness in haplochromine cichlid fishes: Effects of sexual selection. *Advances in Ecological Research* 31:237–274.

Shaw, AJ. 2000. Molecular phylogeography and cryptic speciation in the mosses, *Mielichhoferia elongata* and *M. mielichhoferiana* (Bryaceae). *Molecular Ecology* 9:595–608.

Shaw AJ & B Allen. 2000. Phylogenetic relationships, morphological incongruence, and geographic speciation in the Fontinalaceae (Bryophyta). *Molecular Phylogenetics & Evolution* 16:225–237.

Shaw, PW, GF Turner, M Rizman Idid, RL Robinson & GR Carvalho. 2000. Genetic population structure indicates sympatric speciation of Lake Malawi pelagic cichlids. *Proceedings of the Royal Society of London. Series B* 267. DOI: 10.1098/rspb.2000.1279.

Simon, C, J Tang, S Dalwadi, G Staley, J Deniega & TR Unnasch. 2000. Genetic evidence for assortative mating between 13-year cicadas and sympatric "17-year cicadas with 13-year life cycles" provides support for allochronic speciation. *Evolution* 54:1326–1336.

Singh, RS & RJ Kulathinal. 2000. Sex gene pool evolution and speciation. A new paradigm. *Genes & Genetic Systems* 75:119–130.

Soto-Adames, FN. 2000. Phylogeny of neotropical *Lepidocyrtus* (Collembola: Entomobryidae): First assessment of patterns of speciation in Puerto Rico and phylogenetic relevance of some subgeneric diagnostic characters. *Systematic Entomology* 25:485–502.

Takimoto, G, M Higashi & N Yamamura. 2000. A deterministic genetic model for sympatric speciation by sexual selection. *Evolution* 54:1870–1881.

Taylor, EB & JD McPhail. 2000. Historical contingency and ecological determinism interact to prime speciation in sticklebacks, *Gasterosteus*. *Proceedings of the Royal Society of London. Series B* 267. DOI: 10.1098/rspb.2000.1294.

Templeton, AR, SD Maskas & MB Cuzan. 2000. Gene trees: A powerful tool for exploring the evolutionary biology of species and speciation. *Plant Species Biology* 15:211–222.

Tregenza, T, RK Butlin & N Wedell. 2000. Evolutionary biology: Sexual conflict and speciation. *Nature* 407:149–150.

Uy, JAC & G Borgia. 2000. Sexual selection drives rapid divergence in bowerbird display traits. *Evolution* 54:273–278.

Wilson, AB, K Noach-Kunnmann & A Meyer. 2000. Incipient speciation in sympatric Nicaraguan crater lake cichlid fishes: Sexual selection versus ecological diversification. *Proceedings of the Royal Society of London. Series B* 267:2133–2141.

Winker, K. 2000. Evolution: Migration and speciation. *Nature* 404:36.

Zeh, DW & JA Zeh. 2000. Reproductive mode and speciation: The viviparity-driven conflict hypothesis. *BioEssays* 22:938–946.

2005

Abbott, RJ, HE Ireland, L Joseph, MS Davies & HJ Rogers. 2005. Recent plant speciation in Britain and Ireland: Origins, establishment and evolution of four new hybrid species. *Biology and Environment: Proceedings of the Royal Irish Academy* 105B:173–183.

Adams, RP, JA Morris, RN Pandey & AE Schwarzbach. 2005. Cryptic speciation between *Juniperus deltoides* and *Juniperus oxycedrus* (Cupressaceae) in the Mediterranean. *Biochemical Systematics and Ecology* 33:771–787.

Barnosky, AD. 2005. Effects of quaternary climatic change on speciation in mammals. *Journal of Mammalian Evolution* 12:247–264.

Birky, CW, Jr., C Wolf, H Maughan, L Herbertson & E Henry. 2005. Speciation and selection without sex. *Hydrobiologia* 546:29–45.

Blair, CP, WG Abrahamson, JA Jackman & L Tyrrell. 2005. Cryptic speciation and host-race formation in a purportedly generalist tumbling flower beetle. *Evolution* 59:304–316.

Boake, CRB. 2005. Sexual selection and speciation in Hawaiian *Drosophila*. *Behavior Genetics* 35:297–303.

Brändle, M, S Knoll, S Eber, J Stadler & R Brandl. 2005. Flies on thistles: Support for synchronous speciation? *Biological Journal of the Linnean Society* 84:775–783.

Braun, MJ, ML Isler, PR Isler, JM Bates & MB Robbins. 2005. Avian speciation in the Pantepui: The case of the Roraiman Antbird (*Percnostola* [*Schistocichla*] "leucostigma" *saturata*). *The Auk* 107:327–341.

Briggs, JC. 2005. The marine East Indies: Diversity and speciation. *Journal of Biogeography* 32:1517–1522.

Cadena, CD, RE Ricklefs, I Jiménez & E Bermingham. 2005. Ecology: Is speciation driven by species diversity? *Nature* 438:E1–E2. DOI: 10.1038/nature04308.

Carleton, KL, JWL Parry, JK Bowmaker, DM Hunt & O Seehausen. 2005. Colour vision and speciation in Lake Victoria cichlids of the genus *Pundamilia*. *Molecular Ecology* 14:4341–4353.

Chain, PSG, DJ Comerci, ME Tolmasky, FW Larimer, SA Malfatti, LM Vergez, F Aguero, ML Land, RA Ugalde & E Garcia. 2005. Whole-genome analyses of speciation events in pathogenic Brucellae. *Infection and Immunity* 73:8353–8361.

Church, SA & DR Taylor. 2005. Speciation and hybridization among *Houstonia* (Rubiaceae) species: The influence of polyploidy on reticulate evolution. *American Journal of Botany* 92:1372–1380.

Dawson, MN. 2005. Incipient speciation of *Catostylus mosaicus* (Scyphozoa, Rhizostomeae, Catostylidae), comparative phylogeography and biogeography in south-east Australia. *Journal of Biogeography* 32:515–533.

Derycke, S, T Remerie, A Vierstraete, T Backeljau, J Vanfleteren, M Vincx & T Moens. 2005. Mitochondrial DNA variation and cryptic speciation within the free-living marine nematode *Pellioditis marina*. *Marine Ecology Progress Series* 300:91–103.

Doebeli, M. 2005. Adaptive speciation when assortative mating is based on female preference for male marker traits. *Journal of Evolutionary Biology* 18:1587–1600.

Doebeli, M, U Dieckmann, JAJ Metz & D Tautz. 2005. What we have also learned: Adaptive speciation is theoretically plausible. *Evolution* 59:691–695.

Forbes, AA, J Fisher & JL Feder. 2005. Habitat avoidance: Overlooking an important aspect of host-specific mating and sympatric speciation? *Evolution* 59:1552–1559.

Gavrilets, S & R Harrison. 2005. "Adaptive speciation"—it is not that easy: A reply to Doebeli et al. *Evolution* 59:696–699.

Gavrilets, S & TI Hayashi. 2005. Speciation and sexual conflict. *Evolutionary Ecology* 19:167–198.

Genner, MJ & GF Turner. 2005. The mbuna cichlids of Lake Malawi: A model for rapid speciation and adaptive radiation. *Fish and Fisheries* 6:1–34.

Gillespie, RG. 2005. Geographical context of speciation in a radiation of Hawaiian Tetragnatha spiders (Araneae, Tetragnathidae). *Journal of Arachnology* 33:313–322.

Glor, RE, JB Losos & A Larson. 2005. Out of Cuba: Overwater dispersal and speciation among lizards in the *Anolis carolinensis* subgroup. *Molecular Ecology* 14:2419–2432.

Gross, BL & LH Rieseberg. 2005. The ecological genetics of homoploid hybrid speciation. *Journal of Heredity* 96:241–252.

Haesler, MP & O Seehausen. 2005. Inheritance of female mating preference in a sympatric sibling species pair of Lake Victoria cichlids: Implications for speciation. *Proceedings of the Royal Society of London. Series B* 272. DOI: 10.1098/rspb.2004.2946.

Hall, JPW. 2005. Montane speciation patterns in *Ithomiola* butterflies (Lepidoptera: Riodinidae): Are they consistently moving up in the world? *Proceedings of the Royal Society of London. Series B* 272:2457–2466.

Hegarty, MJ & SJ Hiscock. 2005. Hybrid speciation in plants: New insights from molecular studies. *New Phytologist* 165:411–423.

Held, C & JW Wägele. 2005. Cryptic speciation in the giant Antarctic isopod *Glyptonotus antarcticus* (Isopoda: Valvifera: Chaetiliidae). *Scientia Marina* 69(Suppl. 2):175–181.

Hillig, KW. 2005. Genetic evidence for speciation in *Cannabis* (Cannabaceae). *Genetic Resources and Crop Evolution* 52:161–180.

Hoskin, CJ, M Higgie, KR McDonald & C Moritz. 2005. Reinforcement drives rapid allopatric speciation. *Nature* 437:1353–1356.

Huyse, T, R Poulin & A Theron. 2005. Speciation in parasites: A population genetics approach. *Trends in Parasitology* 21:469–475.

Irwin, DE, S Bensch, JH Irwin & TD Price. 2005. Speciation by distance in a ring species. *Science* 307:414–416.

Katongo, C, S Koblmüller, N Duftner & L Makasa. 2005. Phylogeography and speciation in the *Pseudocrenilabrus philander* species complex in Zambian Rivers. *Hydrobiologica* 542:221–233.

Kay, KM, PA Reeves, RG Olmstead & DW Schemske. 2005. Rapid speciation and the evolution of hummingbird pollination in neotropical *Costus* subgenus *Costus* (Costaceae): Evidence from nrDNA ITS and ETS sequences. *American Journal of Botany* 92:1899–1910.

Kocher, TD. 2005. Evolutionary biology: Ghost of speciation past. *Nature* 435:29–30.

Kohn, LM. 2005. Mechanism of fungal speciation. *Annual Review of Phytopathology* 43:279–308.

Kutschera, U & KJ Niklas. 2005. Endosymbiosis, cell evolution, and speciation. *Theory in Biosciences* 124:1–24.

Latimer, AM, JA Silander & RM Cowling. 2005. Neutral ecological theory reveals isolation and rapid speciation in a biodiversity hot spot. *Science* 309:1722–1725.

Luchetti, A, M Marini & B Mantovani. 2005. Mitochondrial evolutionary rate and speciation in termites: Data on European *Reticulitermes* taxa (Isoptera, Rhinotermitidae). *Insectes Sociaux* 52:218–221.

Lukhtanov, VA, NP Kandul, JB Plotkin, AV Dantchenko, D Haig & NE Pierce. 2005. Reinforcement of pre-zygotic isolation and karyotype evolution in *Agrodiaetus* butterflies. *Nature* 436:385–389.

Mendelson, TC & KL Shaw. 2005. Sexual behaviour: Rapid speciation in an arthropod. *Nature* 433:375–376.

Morse, GE & BD Farrell. 2005. Interspecific phylogeography of the *Stator limbatus* species complex: The geographic context of speciation and specialization. *Molecular Phylogenetics & Evolution* 36:201–213.

Nies, G & TBH Reusch. 2005. Evolutionary divergence and possible incipient speciation in post-glacial populations of a cosmopolitan aquatic plant. *Journal of Evolutionary Biology* 18:19–26.

Osada, N & CI Wu. 2005. Inferring the mode of speciation from genomic data. *Genetics* 169:259–264.

Pala, I & MM Coelho. 2005. Contrasting views over a hybrid complex: Between speciation and evolutionary "dead-end". *Gene* 347:283–294.

Papakostas, S, A Triantafyllidis, I Kappas & TJ Abatzopoulos. 2005. The utility of the 16S gene in investigating cryptic speciation within the *Brachionus plicatilis* species complex. *Marine Biology* 147:1129–1139.

Polechová, J, NH Barton & S Gavrilefs. 2005. Speciation through competition: A critical review. *Evolution* 59:1194–1210.

Pringle, A, DM Baker, JL Platt, JP Wares, JP Latgé & JW Taylor. 2005. Cryptic speciation in the cosmopolitan and clonal human pathogenic fungus *Aspergillus fumigatus*. *Evolution* 59:1886–1899.

Rice, WR, JE Linder, U Friberg, TA Lew, EH Morrow & AD Stewart. 2005. Inter-locus antagonistic coevolution as an engine of speciation: Assessment with hemiclonal analysis. *Proceedings of the National Academy of Sciences USA* 102(suppl. 1):6527–6534.

Ritchie, MG & CM Garcia. 2005. Evolution of species: Explosive speciation in a cricket. *Heredity* 95. DOI: 10.1038/sjhdy.6899690.

Ritchie, MG, SA Webb, JA Graves, AE Magurran & C Macias Garcia. 2005. Patterns of speciation in endemic Mexican Goodeid fish: Sexual conflict or early radiation? *Journal of Evolutionary Biology* 18:922–929.

Rocha, LA, DR Robertson, J Roman & BW Bowen. 2005. Ecological speciation in tropical reef fishes. *Proceedings of the Royal Society of London. Series B* 272. DOI: 10.1098/2004.3005.

Rundle, HD & P Nosil. 2005. Ecological speciation. *Ecology Letters* 8:336–352.

Salzburger, W, T Mack, E Verheyen & A Meyer. 2005. Out of Tanganyika: Genesis, explosive speciation, key-innovations and phylogeography of the haplochromine cichlid fishes. *BMC Evolutionary Biology* 5:17. DOI: 10.1186/1471-2148-5-17.

Schwarz, D, BM Matta, ML Shakir-Botteri & BA McPheron. 2005. Host shift to an invasive plant triggers rapid animal hybrid speciation. *Nature* 436:546–549.

Scriber, JM & GJ Ording. 2005. Ecological speciation without host plant specialization; possible origins of a recently described cryptic *Papilio* species. *Entomologia Experimentalis et Applicata* 115:247–263.

Sikorski, J & E Nevo. 2005. Adaptation and incipient sympatric speciation of *Bacillus simplex* under microclimatic contrast at "Evolution Canyons" I and II, Israel. *Proceedings of the National Academy of Sciences USA* 102:15924–15929.

Skarstein, F, I Folstad, HP Rønning. 2005. Spawning colouration, parasites and habitat selection in *Salvelinus alpinus*: Initiating speciation by sexual selection? *Journal of Fish Biology* 67:969–980.

Soulier-Perkins, A. 2005. Phylogenetic evidence for multiple invasions and speciation in caves: The Australian planthopper genus *Solonaima* (Hemiptera: Fulgoromorpha: Cixiidae). *Systematic Entomology* 30:281–288.

Steele, CA, BC Carstens, A Storfer & J Sullivan. 2005. Testing hypotheses of speciation timing in *Dicamptodon copei* and *Dicamptodon aterrimus* (Caudata: Dicamptodontidae). *Molecular Phylogenetics & Evolution* 36:90–100.

Stump, AD, MC Fitzpatrick, MF Lobo, S Traoré, N Sagnon, C Costantini, FH Collins & NJ Besansky. 2005. Centromere-proximal differentiation and speciation in *Anopheles gambiae*. *Proceedings of the National Academy of Sciences USA* 102:15930–15935.

Switzer, WM, M Salemi, V Shanmugam, F Gao, M Cong, C Kuiken, V Bhullar, BE Beer, D Vallet, A Gautier-Hion, Z Tooze, F Villinger, EC Holmes & W Heneine. 2005. Ancient co-speciation of simian foamy viruses and primates. *Nature* 434:376–380.

Taylor, MS & ME Hellberg. 2005. Marine radiations at small geographic scales: Speciation in neotropical reef gobies (*Elacatinus*). *Evolution* 59:374–385.

Telschow, A, P Hammerstein & JH Werren. 2005. The effect of *Wolbachia* versus genetic incompatibilities on reinforcement and speciation. *Evolution* 59:1607–1619.

Tonnis, B, PR Grant, R Grant & K Petren. 2005. Habitat selection and ecological speciation in Galápagos warbler finches (*Certhidea olivacea* and *Certhidea fusca*). *Proceedings of the Royal Society of London. Series B* 272. DOI: 10.1098/rspb.2004.3030.

Triantafillos, L & M Adams. 2005. Genetic evidence that the northern calamary, *Sepioteuthis lessoniana*, is a species complex in Australian waters. *ICES Journal of Marine Science* 62:1665–1670.

Vallejo, B, Jr. 2005. Inferring the mode of speciation in Indo-West Pacific *Conus* (Gastropoda: Conidae). *Journal of Biogeography* 32:1429–1439.

Verzijden, MN, RF Lachlan, MR Servedio & K Shaw. 2005. Female mate-choice behavior and sympatric speciation. *Evolution* 59:2097–2108.

Walsh, HE, IL Jones & VL Friesen. 2005. A test of founder effect speciation using multiple loci in the auklets (*Aethia* spp.). *Genetics* 171:1885–1894.

2010

Abbott, RJ, MJ Hegarty, SJ Hiscock & AC Brennan. 2010. Homoploid hybrid speciation in action. *Taxon* 59:1375–1386.

Arslan, E, E Gülbahçe, H Arıkoğlu, A Arslan, EV Bužan & B Kryštufek. 2010. Mitochondrial divergence between three cytotypes of the Anatolian Mole Rat, *Nannospalax xanthodon* (Nordmann, 1840) (Mammalia: Rodentia). *Zoology in the Middle East* 50:27–34.

Ayasse, M. 2010. Chemical ecology in deceptive orchids. *Chemoecology* 20:171–178.

Bakloushinskaya, IY, SA Romanenko, AS Graphodatsky, SN Matveevsky, EA Lyapunova & OL Kolomiets. 2010. The role of chromosome rearrangements in the evolution of mole voles of the genus *Ellobius* (Rodentia, Mammalia). *Russian Journal of Genetics* 46:1143–1145.

Bao, X, N Liu, J Qu, X Wang, B An, L Wen & S Song. 2010. The phylogenetic position and speciation dynamics of the genus *Perdix* (Phasianidae, Galliformes). *Molecular Phylogenetics and Evolution* 56:840–847.

Barton, NH. 2010. What role does natural selection play in speciation? *Philosophical Transactions of the Royal Society. Series B* 365:1825–1840.

Bocak, L & T Yagi. 2010. Evolution of mimicry patterns in *Metriorrhynchus* (Coleoptera: Lycidae): The history of dispersal and speciation in Southeast Asia. *Evolution* 64:39–52.

Brown, JD & RJ O'Neill. 2010. Chromosomes, conflict, and epigenetics: Chromosomal speciation revisited. *Annual Review of Genomics and Human Genetics* 11:291–316.

Burbrink, FT & RA Pyron. 2010. How does ecological opportunity influence rates of speciation, extinction, and morphological diversification in New World ratsnakes (tribe Lampropeltini)? *Evolution* 64:934–943.

Butlin, RK. 2010. Population genomics and speciation. *Genetica* 138:409–418.

Camargo, A, B Sinervo & JW Sites. 2010. Lizards as model organisms for linking phylogeographic and speciation studies. *Molecular Ecology* 19:3250–3270.

Carneiro, M, JA Blanco-Aguiar & R Villafuerte. 2010. Speciation in the European rabbit (*Oryctolagus cuniculus*): Islands of differentiation on the X chromosome and autosomes. *Evolution* 64:3443–3460.

Chase, MW, O Paun & MF Fay. 2010. Hybridization and speciation in angiosperms: A role for pollinator shifts? *BMC Biology* 8:45. DOI: 10.1186/1741-7007-8-45.

Ciplak, B, KG Heller & F Willemse. 2010. Phylogeny and biogeography *Eupholidoptera* Marăn (Orthoptera, Tettigoniidae): Morphological speciation in correlation with the geographical evolution of the eastern Mediterranean. *Systematic Entomology* 35:722–738.

Clarke, A & JA Crame. 2010. Evolutionary dynamics at high latitudes: Speciation and extinction in polar marine faunas. *Philosophical Transactions of the Royal Society. Series B* 365. DOI: 10.1098/rstb.2010.0270.f

Cook, JM & ST Segar. 2010. Speciation in fig wasps. *Ecological Entomology* 35(s1):54–66.

Cordero-Rivera, A & MO Lorenzo-Carballa. 2010. Three sisters in the same dress: Cryptic speciation in African odonates. *Molecular Ecology* 19:3840–3841.

Corl, A, AR Davis, SR Kuchta & B Sinervo. 2010. Selective loss of polymorphic mating types is associated with rapid phenotypic evolution during morphic speciation. *Proceedings of the National Academy of Sciences USA* 107:4254–4259.

Crawford, DJ. 2010. Progenitor-derivative species pairs and plant speciation. *Taxon* 59:1413–1423.

Crow, KD, H Munehara & G Bernardi. 2010. Sympatric speciation in a genus of marine reef fishes. *Molecular Ecology* 19:2089–2105.

Daniels, SR & H Ruhberg. 2010. Molecular and morphological variation in a South African velvet worm *Peripatopsis moseleyi* (Onychophora, Peripatopsidae): Evidence for cryptic speciation. *Journal of Zoology* 282:171–179.

Dapporto, L. 2010. Speciation in Mediterranean refugia and post-glacial expansion of *Zerynthia polyxena* (Lepidoptera, Papilionidae). *Journal of Zoological Systematics & Evolutionary Research* 48:229–237.

Dasmahapatra, KK, G Lamas & F Simpson. 2010. The anatomy of a 'suture zone' in Amazonian butterflies: A coalescent-based test for vicariant geographic divergence and speciation. *Molecular Ecology* 19:4283–4301.

Davis, ALV & CH Scholtz. 2010. Speciation and evolution of eco-climatic ranges in the intertropical African dung beetle genus, *Diastellopalpus* van Lansberge. *Biological Journal of the Linnean Society* 99:407–423.

den Tex, RJ, R Thorington & JE Maldonado. 2010. Speciation dynamics in the SE Asian tropics: Putting a time perspective on the phylogeny and biogeography of Sundaland tree squirrels, *Sundasciurus*. *Molecular Phylogenetics & Evolution* 55:711–720.

Devier, B, G Aguileta, ME Hood & T Giraud. 2010. Using phylogenies of pheromone receptor genes in the *Microbotryum violaceum* species complex to investigate possible speciation by hybridization. *Mycologia* 102:689–696.

Drosopoulos, S, A Maryańska-Nadochowska & VG Kuznetsova. 2010. The Mediterranean: Area of origin of polymorphism and speciation in the spittlebug *Philaenus* (Hemiptera, Aphrophoridae). *Zoosystematics & Evolution* 86:125–128.

Drummond, CS, HJ Xue, JB Yoder & O Pellmyr. 2010. Host-associated divergence and incipient speciation in the yucca moth *Prodoxus coloradensis* (Lepidoptera: Prodoxidae) on three species of host plants. *Heredity* 105:183–196.

Druzhinina, IS, CP Kubicek, M Komón-Zelazowska, TB Mulaw & J Bissett. 2010. The *Trichoderma harzianum* demon: Complex speciation history resulting in coexistence of hypothetical biological species, recent agamospecies and numerous relict lineages. *BMC Evolutionary Biology* 10. DOI: 10.1186/1471-2148-10-94.

Elmer, KR & A Meyer. 2010. Sympatric speciation without borders? *Molecular Ecology* 19:1991–1993.

Etges, WJ, CC De Oliveira, MAF Noor & MG Ritchie. 2010. Genetics of incipient speciation in *Drosophila mojavensis*. III. Life-history divergence in allopatry and reproductive isolation. *Evolution* 64:3549–3569.

Faria, R & A Navarro. 2010. Chromosomal speciation revisited: Rearranging theory with pieces of evidence. *Trends in Ecology and Evolution* 25:660–669.

Feder, JL & AA Forbes. 2010. Sequential speciation and the diversity of parasitic insects. *Ecological Entomology* 35:67–76.

Feder, JL & P Nosil. 2010. The efficacy of divergence hitchhiking in generating genomic islands during ecological speciation. *Evolution* 64:1729–1747.

Feldberg, K, J Váňa, DG Long & AJ Shaw. 2010. A phylogeny of Adelanthaceae (Jungermanniales, Marchantiophyta) based on nuclear and chloroplast DNA markers, with comments on classification, cryptic speciation and biogeography. *Molecular Phylogenetics and Evolution* 55:293–304.

Ferreira, RS, C Poteaux, JHC Delabie, D Fresneau & F Rybak. 2010. Stridulations reveal cryptic speciation in neotropical sympatric ants. *PLoS One*. DOI: 10.1371/journal.pone.0015363.

Finney, JC, DT Pettay, EM Sampayo, ME Warner, HA Oxenford & TC LaJeunesse. 2010. The relative significance of host–habitat, depth, and geography on the ecology, endemism, and speciation of coral endosymbionts in the genus *Symbiodinium*. *Microbial Ecology* 60:250–263.

Flagel, LE & JF Wendel. 2010. Evolutionary rate variation, genomic dominance and duplicate gene expression evolution during allotetraploid cotton speciation. *New Phytologist* 186:184–193.

Frías-Lasserre, D. 2010. A new species and karyotype variation in the bordering distribution of *Mepraia spinolai* (Porter) and *Mepraia gajardoi* Frías et al. (Hemiptera: Reduviidae: Triatominae) in Chile and its parapatric model of speciation. *Neotropical Entomology* 39. http://dx.doi.org/10.1590/S1519-566X2010000400017.

Funk, DJ. 2010. Does strong selection promote host specialisation and ecological speciation in insect herbivores? Evidence from *Neochlamisus* leaf beetles. *Ecological Entomology* 35:41–53.

Gentekaki, E & DH Lynn. 2010. Evidence for cryptic speciation in *Carchesium polypinum* Linnaeus, 1758 (Ciliophora: Peritrichia) inferred from mitochondrial, nuclear, and morphological markers. *Journal of Eukaryotic Microbiology* 57:508–519.

Gilles, A, C Costedoat, B Barascud & A Voisin. 2010. Speciation pattern of *Telestes souffia* complex (Teleostei, Cyprinidae) in Europe using morphological and molecular markers. *Zoologica Scripta* 39:225–242.

Giraud, T, P Gladieux & S Gavrilets. 2010. Linking the emergence of fungal plant diseases with ecological speciation. *Trends in Ecology and Evolution* 25:387–395.

Givnish TJ. 2010. Ecology of plant speciation. *Taxon* 59:1326–1366.

Goldberg, EE, JR Kohn, R Lande & KA Robertson. 2010. Species selection maintains self-incompatibility. *Science* 330:493–495.

Hardy, NB & LG Cook. 2010. Gall-induction in insects: Evolutionary dead-end or speciation driver? *BMC Evolutionary Biology* 10:257. DOI: 10.1186/1471-2148-10-257.

He, K, YJ Li, MC Brandley, LK Lin, YX Wang, Y Zhang & X Jiang. 2010. A multi-locus phylogeny of Nectogalini shrews and influences of the paleoclimate on speciation and evolution. *Molecular Phylogenetics & Evolution* 56:734–746.

Hohenlohe, PA & SJ Arnold. 2010. Dimensionality of mate choice, sexual isolation, and speciation. *Proceedings of the National Academy of Sciences USA* 107:16583–16588.

Holt, BG, IM Côté & BC Emerson. 2010. Signatures of speciation? Distribution and diversity of *Hypoplectrus* (Teleostei: Serranidae) colour morphotypes. *Global Ecology and Biogeography* 19:432–441.

Hoskin, CJ & M Higgie. 2010. Speciation via species interactions: The divergence of mating traits within species. *Ecology Letters* 13:409–420.

Hoso, M, Y Kameda, SP Wu, T Asami, M Kato & M Hori. 2010. A speciation gene for left-right reversal in snails results in anti-predator adaptation. *Nature Communications* 1:133. DOI: 10.1038/ncomms1133.

Hua, X & JJ Wiens. 2010. Latitudinal variation in speciation mechanisms in frogs. *Evolution* 64:429–443.

Hughes, AL & F Verra. 2010. Malaria parasite sequences from chimpanzee support the co-speciation hypothesis for the origin of virulent human malaria (*Plasmodium falciparum*). *Molecular Phylogenetics and Evolution* 57:135–143.

Jean, P & S Jean-Christophe. 2010. The pea aphid complex as a model of ecological speciation. *Ecological Entomology* 35:119–130.

Johannesson, K, M Panova, P Kemppainen, C André, E Rólan-Alvarez & RK Butlin. 2010. Repeated evolution of reproductive isolation in a marine snail: Unveiling mechanisms of speciation. *Philosophical Transactions of Royal Society B* 365:1735–1747.

Kao, KC, K Schwartz & G Sherlock. 2010. A genome-wide analysis reveals no nuclear Dobzhansky-Muller pairs of determinants of speciation between *S. cerevisiae* and *S. paradoxus*, but suggests more complex incompatibilities. *PLoS Genetics*. DOI: 10.1371/journal.pgen.1001038.

Kelly, LJ, AR Leitch, JJ Clarkson, RB Hunter, S Knapp & MW Chase. 2010. Intragenic recombination events and evidence for hybrid speciation in *Nicotiana* (Solanaceae). *Molecular Biology & Evolution* 27:781–799.

Kim, C, H Shin, YT Chang, et al. 2010. Speciation pathway of *Isoëtes* (Isoëtaceae) in East Asia inferred from molecular phylogenetic relationships. *American Journal of Botany* 97:958–969.

Kisel, Y & TG Barraclough. 2010. Speciation has a spatial scale that depends on levels of gene flow. *American Naturalist* 175:316–334.

Klemetsen, A. 2010. The charr problem revisited: Exceptional phenotypic plasticity promotes ecological speciation in postglacial lakes. *Freshwater Reviews* 2010 (3):49–74.

Knudsen, R, R Primicerio, PA Amundsen, et al. 2010. Temporal stability of individual feeding specialization may promote speciation. *Journal of Animal Ecology* 79:161–168.

Kölsch, G & BV Pedersen. 2010. Can the tight co-speciation between reed beetles (Col., Chrysomelidae, Donaciinae) and their bacterial endosymbionts, which provide cocoon material, clarify the deeper phylogeny of the hosts. *Molecular Phylogenetics and Evolution* 54:810–821.

Kopp, M. 2010. Speciation and the neutral theory of biodiversity. *BioEssays* 32:546–570.

Kraaijeveld, K. 2010. Genome size and species diversification. *Evolutionary Biology* 37:227–233.

Kreier, HP, K Feldberg, F Mahr & A Bombosch. 2010. Phylogeny of the leafy liverwort *Ptilidium*: Cryptic speciation and shared haplotypes between the Northern and Southern Hemispheres. *Molecular Phylogenetics and Evolution* 57:1260–1267.

Lago-Lestón, A, C Mota, L Kautsky & GA Pearson. 2010. Functional divergence in heat shock response following rapid speciation of *Fucus* spp. in the Baltic Sea. *Marine Biology*. DOI: 10.1007/s00227-009-1348-1.

Lancaster, LT. 2010. Molecular evolutionary rates predict both extinction and speciation in temperate angiosperm lineages. *BMC Evolutionary Biology* 10:162. DOI: 10.1186/1471-2148-10-162.

Larmuseau, MHD, T Huyse, K Vancampenhout, JKJ Van Houdt & FAM Vokkaert. 2010. High molecular diversity in the rhodopsin gene in closely related goby fishes: A role for visual pigments in adaptive speciation? *Molecular Phylogenetics and Evolution* 55:687–698.

Lawniczak, MKN, SJ Emrich, AK Holloway, AP Regier, M Olson, B White, S Redmond, L Fulton, E Applebaum, J Godfrey, C Farmer, A Chinwalla, S-P Yang, P Minx, J Nelson, K Kyung, BP Walenz, E Garcia-Hernandez, M Aguiar, LD Viswanatham, Y-H Rogers, RL Strausberg, CA Saski, D Lawson, FH Collins, FC Katatos, GK Christophides, SW Clifton, EF Kirkness & NJ Besansky. 2010. Widespread divergence between incipient *Anopheles gambiae* species revealed by whole genome sequences. *Science* 330:512–514.

Lawrence, JG & AC Retchless. 2010. The myth of bacterial species and speciation. *Biology & Philosophy* 25:569–588.

Leaché, AD & MK Fujita. 2010. Bayesian species delimitation in West African forest geckos (*Hemidactylus fasciatus*). *Proceedings of Royal Society B* 282. DOI: 10.1098/rspb.2010.0662.

Leal, M & JB Losos. 2010. Evolutionary biology: Communication and speciation. *Nature* 467:159–160.

Leray, M, R Beldade, SJ Holbrook, RJ Schmitt, S Planes & G Bernardi. 2010. Allopatric divergence and speciation in coral reef fish: The three-spot Dascyllus, *Dascyllus trimaculatus*, species complex. *Evolution* 64:1218–1230.

Li, J-W, CKL Yeung, P Tsai & RC Lin. 2010. Rejecting strictly allopatric speciation on a continental island: Prolonged postdivergence gene flow between Taiwan (*Leucodioptron taewanus*, Passeriformes Timaliidae) and Chinese (*L. canorum canorum*) hwameis. *Molecular Ecology* 19:494–507.

Linder, HP. 2010. Gradual speciation in a global hotspot of plant diversity. *Molecular Ecology* 19:4583–4585.

Lindström, K, M Murwira, A Willems & N Altier. 2010. The biodiversity of beneficial microbe-host mutualism: The case of rhizobia. *Research in Microbiology* 161:453–463.

Lo, EYY, S Stefanović & TA Dickinson. 2010. Reconstructing reticulation history in a phylogenetic framework and the potential of allopatric speciation driven by polyploidy in an agamic complex in *Crataegus* (Rosaceae). *Evolution* 64:3593–3608.

Ma, JX, YN Li, C Vogl & F Ehrendorfer. 2010. Allopolyploid speciation and ongoing backcrossing between diploid progenitor and tetraploid progeny lineages in the *Achillea millefolium* species complex: Analyses of single-copy nuclear genes and genomic AFLP. *BMC Evolutionary Biology* 10 100. DOI: 10.1186/1471-2148-10-100.

Malay, MCMD & G Paulay. 2010. Peripatric speciation drives diversification and distributional pattern of reef hermit crabs (Decapoda: Diogenidae). *Evolution* 64:634–662.

Maley, JM & K Winker. 2010. Diversification at high latitudes: Speciation of buntings in the genus *Plectrophenax* inferred from mitochondrial and nuclear markers. *Molecular Ecology* 19:785–797.

Martínez, JJ, RE González-Ittig, GR Theiler, R Ojeda, C Lanzone, A Ojeda & CN Gardenal. 2010. Patterns of speciation in two sibling species of *Graomys* (Rodentia, Cricetidae) based on mtDNA sequences. *Journal of Zoological Systematics and Evolutionary Research* 48:159–166.

Matsubayashi, KW, I Ohshima & P Nosil. 2010. Ecological speciation in phytophagous insects. *Entomologia Experimentalis et Applicata* 134:1–27.

McCormack, JE, AJ Zellmer & LL Knowles. 2010. Does niche divergence accompany allopatric divergence in *Aphelocoma* jays as predicted under ecological speciation?: Insights from tests with niche models. *Evolution* 64:1231–1244.

McDaniel, SF, M Von Stackelberg, S Richardt, RS Quatrano, R Reski & SA Rensing. 2010. The speciation history of the *Physcomitrium-Physcomitrella* species complex. *Evolution* 64: 217–231.

Michel, AP, S Sim, THQ Powell, MS Taylor, P Nosil & JL Feder. 2010. Widespread genomic divergence during sympatric speciation. *Proceedings of the National Academy of Sciences USA* 107:9724–9729.

Miller, WJ, L Ehrman & D Schneider. 2010. Infectious speciation revisited: Impact of symbiont-depletion on female fitness and mating behavior of *Drosophila paulistorum*. *PLoS Pathogens* 6(12): e1001214. DOI: 10.1371/journal.ppat.1001214.

Miyamoto, H, RJ Machida & S Nishida. 2010. Genetic diversity and cryptic speciation of the deep sea chaetognath *Caecosagitta macrocephala* (Fowler, 1904). *Deep Sea Research II* 57:2211–2219.

Mladineo, I, NJ Bott, BF Nowak & BA Block. 2010. Multilocus phylogenetic analyses reveal that habitat selection drives the speciation of Didymozoidae (Digenea) parasitizing Pacific and Atlantic bluefin tunas. *Parasitology* 137:1013–1025.

Morgan, K, Y Linton, P Somboon & P Saikia. 2010. Inter-specific gene flow dynamics during the Pleistocene-dated speciation of forest-dependent mosquitoes in Southeast Asia. *Molecular Ecology* 19:2269–2285.

Muratovicć, E, O Robin, F Bogunić, D Šoljan & S Siljak-Yakovlev. 2010. Karyotype evolution and speciation of European lilies from *Lilium* sect. *Liriotypus*. *Taxon* 59:165–175.

Nicholls, JA, S Preuss, A Hayward & G Melika. 2010. Concordant phylogeography and cryptic speciation in two Western Palaearctic oak gall parasitoid species complexes. *Molecular Ecology* 19:592–609.

Nolte, AW & D. Tautz. 2010. Understanding the onset of hybrid speciation. *Trends in Genetics* 26:54–58.

Novo, M, A Almodóvar, R Fernández & D Trigo. 2010. Cryptic speciation of hormogastrid earthworms revealed by mitochondrial and nuclear data. *Molecular Phylogenetics and Evolution* 56:507–512.

Nyman, T, V Vikberg & DR Smith. 2010. How common is ecological speciation in plant-feeding insects? A 'higher' Nematinae perspective. *BMC Evolutionary Biology* 10: DOI: 10.1186/1471-2148-10-266.

Ording, GJ, RJ Mercader, ML Aardema & JM Scriber. 2010. Allochronic isolation and incipient hybrid speciation in tiger swallowtail butterflies. *Oecologia* 162:523–531.

Pagán, I & EC Holmes. 2010. Long-term evolution of the Luteoviridae: Time scale and mode of virus speciation. *Journal of Virology* 84:6177–6187.

Payseur, BA. 2010. Using differential introgression in hybrid zones to identify genomic regions involved in speciation. *Molecular Ecology Resources* 10:806–820.

Peakall, R, D Ebert, J Poldy, RA Barrow, W. Francke, CC Bower & FP Schiestl. 2010. Pollinator specificity, floral odour chemistry and the phylogeny of Australian sexually deceptive *Chiloglottis* orchids: Implications for pollinator-driven speciation. *New Phytologist* 188:437–458.

Perrie, LR, LD Shepherd & PJ De Lange. 2010. Parallel polyploid speciation: Distinct sympatric gene-pools of recurrently derived allo-octoploid *Asplenium* ferns. *Molecular Ecology* 19:2916–2932.

Pfenning, DW, MA Wund, EC Snell-Rood, T Cruickshank, CD Schlichting & AP Moczek. 2010. Phenotypic plasticity's impacts on diversification and speciation. *Trends in Ecology and Evolution* 25:457–467.

Pfenninger, M, E Véla, R Jesse & MA Elejalde. 2010. Temporal speciation pattern in the western Mediterranean genus *Tudorella* P. Fischer, 1885 (Gastropoda, Pomatiidae) supports the Tyrrhenian vicariance hypothesis. *Molecular Phylogenetics and Evolution* 54:427–436.

Pigot, AL, AB Phillimore, IPF Owens & CDL Orme. 2010. The shape and temporal dynamics of phylogenetic trees arising from geographic speciation. *Systematic Biology* 59:660–673.

Ploch, S, YJ Choi, C Rost, HD Shin, E Schilling & M Thines. 2010. Evolution of diversity in *Albugo* is driven by high host specificity and multiple speciation events on closely related Brassicaceae. *Molecular Phylogenetics and Evolution* 57:812–820.

Poisot, T & Y Desdevises. 2010. Putative speciation events in *Lamellodiscus* (Monogenea: Diplectanidae) assessed by a morphometric approach. *Biological Journal of the Linnean Society* 99:559–569.

Presgraves, DC. 2010. Speciation genetics: Search for the missing snowball. *Current Biology* 20:R1073–R1074.

Puillandre, N, AV Sysoev & BM Olivera. 2010. Loss of planktotrophy and speciation: Geographical fragmentation in the deep-water gastropod genus *Bathytoma* (Gastropoda, Conoidea) in the western Pacific. *Systematics and Biodiversity* 8:371–394.

Pyron, RA & FT Burbrink. 2010. Hard and soft allopatry: Physically and ecologically medi-ated modes of geographic speciation. *Journal of Biogeography* 37:2005–2015.

Qvarnström, A, AM Rice & H Ellegren. 2010. Speciation in *Ficedula* flycatchers. *Proceedings of the Royal Society of London Series: B Genomics of Speciation*. DOI: 10:1098/rstb.2009.0306.

Rabosky, DL & RE Glor. 2010. Equilibrium speciation dynamics in a model adaptive radiation of island lizards. *Proceedings of the National Academy of Sciences USA* 107:22178–22183.

Ren, J, X Liu, F Jiang, X Guo & B Liu. 2010. Unusual conservation of mitochondrial gene order in *Crassostrea* oysters: Evidence for recent speciation in Asia. *BMC Evolutionary Biology* 10:394. DOI: 10.1186/1471-2148-10-394.

Retchless, AC & JG Lawrence. 2010. Phylogenetic incongruence arising from fragmented speciation in enteric bacteria. *Proceedings of the National Academy of Sciences USA* 107:11453–11458.

Rice, AM & DW Pfennig. 2010. Does character displacement initiate speciation? Evidence of reduced gene flow between populations experiencing divergent selection. *Journal of Evolutionary Biology* 23:854–856.

Richards, PM & A Davison. 2010. Adaptive radiations: Competition rules for Galápagos gas-tropods. *Current Biology* 20:R28–R30.

Rieseberg, LH & BK Blackman. 2010. Speciation genes in plants. *Annals of Botany* DOI: 10.1093/aob/mcq126.

Robertson, A, TCG Rich, AM Allen, L Houston, C Roberts, JR Bridle, SA Harris & SJ Hiscock. 2010. Hybridization and polyploidy as drivers of continuing evolution and speciation in *Sorbus*. *Molecular Ecology* 19:1675–1690.

Rosindell, J, SJ Cornell, SP Hubbell & RS Etienne. 2010. Protracted speciation revitalizes the neutral theory of biodiversity. *Ecology Letters* 13:716–727.

Rymer, PD, JC Manning & P Goldblatt. 2010. Pollinator behaviour and plant speciation: Can assortative mating and disruptive selection maintain distinct floral morphs in sympatry? *New Phytologist* 188:426–436.

Rymer, PD, SD Johnson & V Savolainen. 2010. Evidence of recent and continuous speciation in a biodiversity hotspot: A population genetic approach in southern African gladioli (*Gladiolus*; Iridaceae). *Molecular Ecology* 19:4765–4782.

Sætre, GP & SA Saether. 2010. Ecology and genetics of speciation in *Ficedula* flycatchers. *Molecular Ecology* 19:1091–1106.

Salazar C, SW Baxter, C Pardo-Diaz, G Wu, A Surridge, M Linares, E Bermingham & CD Jiggins. 2010. Genetic evidence for hybrid trait speciation in *Heliconius* butterflies. *PLoS Genetics* 6(4): e1000930. DOI: 10.1371/journal.pgen.1000930.

Sanders, KL & MSY Lee. 2010. Uncoupling ecological innovation and speciation in sea snakes (Elapidae, Hydrophiinae, Hydrophiini). *Journal of Evolutionary Biology* 23:2685–2693.

Seidel, RA, BK Lang & DJ Berg. 2010. Salinity tolerance as a potential driver of ecological speciation in amphipods (*Gammarus* spp.) from the northern Chihuahuan Desert. *Journal of the North American Benthological Society* 29:1161–1169.

Seifert, B. 2010. Intranidal mating, gyne polymorphism, polygyny, and supercoloniality as factors for sympatric and parapatric speciation in ants. *Ecological Entomology* 35:33–40 Issue Supplement.

Sobel, JW, GF Chen, LR Watt & DW Schemske. 2010. The biology of speciation. *Evolution* 64:295–315.

Sorhannus, U, JD Ortiz, M Wolf & MG Fox. 2010. Microevolution and speciation in *Thalassiosira weissflogii* (Bacillariophyta). *Protist* 161:237–249.

Storchová, R, J Reif & MW Nachman. 2010. Female heterogamety and speciation: Reduced introgression of the Z chromosome between two species of nightingales. *Evolution* 64:456–471.

Stukenbrock, EH, FG Jørgensen, M Zala & TT Hansen. 2010. Whole-genome and chromosome evolution associated with host adaptation and speciation of the wheat pathogen *Mycosphaerella graminicola*. *PLoS Genetics* DOI: 10.1371/journal.pgen.1001189.

Techow, N, C O'Ryan, RA Phillips, R Gales, M Marin, D Patterson-Fraser, F Quintana, MS Ritz, DR Thompson, RM Wanless, H Weimerskirch & PG Ryan. 2010. Speciation and phylogeography of giant petrels *Macronectes*. *Molecular Phylogenetics and Evolution* 54:472–487.

Teterina, VI, LV Sukhanova & SV Kirilchik. 2010. Molecular divergence and speciation of Baikal oilfish (Comephoridae): Facts and hypotheses. *Molecular Phylogenetics and Evolution* 56:336–342.

Thorpe, RS, Y Surget-Groba & H Johansson. 2010. Genetic tests for ecological and allopatric speciation in anoles on an island archipelago. *PLoS Genetics* DOI: 10.1371/journal.pgen.1000929.

Turner, TL & MW Hahn. 2010. Genomic islands of speciation or genomic islands and speciation? *Molecular Ecology* 19:848–850.

Ujvárosi, L, M Bálint, T Schmitt & N Mészáros. 2010. Divergence and speciation in the Carpathians area: Patterns of morphological and genetic diversity of the crane fly *Pedicia occulta* (Diptera: Pediciidae). *Journal of the North American Benthological Society* 29:1075–1088.

Van Bocxlaer, I, SP Loader, K Roelants, SD Biju, M Menegon & F Bossuyt. 2010. Gradual adaptation toward a range-expansion phenotype initiated the global radiation of toads. *Science* 327:679–682.

VanderWerf, LC Young & NW Yeung. 2010. Stepping stone speciation in Hawaii's flycatchers: Molecular divergence supports new island endemics within the elepaio. *Conservation Genetics* 11:1283–1298.

Venail, J & A Dell'Olivo. 2010. Speciation genes in the genus *Petunia*. *Philosophical Transactions of the Royal Society. Series B* 365: DOI: 10.1098/rstb.2009.0242.

Venditti, C & M Pagel. 2010. Speciation as an active force in promoting genetic evolution. *Trends in Ecology & Evolution* 25:14–20.

Venditti, C, A Meade & M Pagel. 2010. Phylogenies reveal new interpretation of speciation and the Red Queen. *Nature* 463:349–352.

Vereecken, NJ, A Dafni & S Cozzolino. 2010. Pollination syndromes in Mediterranean orchids—implications for speciation, taxonomy and conservation. *Botanical Review* 76:220–240.

Voelker, G, RK Outlaw & RCK Bowie 2010. Pliocene forest dynamics as a primary driver of African bird speciation. *Global Ecology and Biogeography* 19:111–121.

Wiens, JJ, CA Kuczynski & P Stevens. 2010. Discordant mitochondrial and nuclear gene phylogenies in emydid turtles: Implications for speciation and conservation. *Biological Journal of the Linnean Society* 99:445–461.

Winkler, K. 2010. On the origin of species through heteropatric differentiation: A review and a model of speciation in migratory animals. *Ornithological Monographs* 69:1–30.

Wolf, JBW, J Lindell & N Backström. 2010. Speciation genetics: Current status and evolving approaches. *Philosophical Transactions of the Royal Society. Series B* 365. DOI: 10.1098/rstb.2010.0023.

Xavier, JR, PG Rachello-Dolmen, F Parra-Velandia, CHL Schonberg, JAJ Breeuwer & RWM van Soest. 2010. Molecular evidence of cryptic speciation in the "cosmopolitan" excavating sponge *Cliona celata* (Porifera, Clionaidae). *Molecular Phylogenetics and Evolution* 56:13–20.

Yuan, J, Z He, X Yuan, X Jiang, X Sun & S Zou. 2010. Speciation of polyploid Cyprinidae fish of common carp, crucian carp, and silver crucian carp derived from duplicated Hox genes. *Journal of Experimental Zoology Part B: Molecular and Developmental Evolution* 314B:445–456.

Zimkus, BM, MO Rödel & A Hillers. 2010. Complex patterns of continental speciation: Molecular phylogenetics and biogeography of sub-Saharan puddle frogs (*Phrynobatrachus*). *Molecular Phylogenetics & Evolution* 55:883–900.

Zulliger, DE & HA Lessios. 2010. Phylogenetic relationships in the genus *Astropecten* Gray (Paxillosida: Asteropectinidae) on a global scale: Molecular evidence for morphological convergence, species-complexes and possible cryptic speciation. *Zootaxa* 2504:1–19.

PAPERS WITH COMPETITION & SPECIATION IN TITLE

Almaça, C. 1988. Competition, available space, and speciation in Mediterranean crabs. *Italian Journal of Zoology* 22:477–497.

Atamas, N, MS Atamas, F Atamas & SP Atamas. 2012. Non-local competition drives both rapid divergence and prolonged stasis in a model of speciation in populations with degenerate resource consumption. *Theoretical Biology and Medical Modeling* 9:56.

Brigatti, E, JS Sá Martins & I Roditi. 2006. Evolution of polymorphism and sympatric speciation through competition in a unimodal distribution of resources. arXiv:q-bio/0505017 [q-bio.PE].

Brigatti, E, JS Sá Martins & I Roditi. 2015. Evolution of biodiversity and sympatric speciation through competition in a unimodal distribution of resources. *Physica A: Statistical Mechanics and Its Applications* 376:378–386.

Dijkstra, PD, O Seehausen, BLA Gricar, ME Maan & TGG Groothuis. 2006. Can male-male competition stabilize speciation? A test in Lake Victoria haplochromine cichlid fish. *Behavior, Ecology, and Sociobiology* 59:704–713.

Dijkstra, PD, O Seehausen, MER Pierotti & TGG Groothuis. 2006. Male–male competition and speciation: Aggression bias towards differently coloured rivals varies between stages of speciation in a Lake Victoria cichlid species. *Journal of Evolutionary Biology* 20. DOI: 10.1111/j.1420-9101.2006.01266.x.

Doebeli, M. 1996. A quantitative genetic competition model for sympatric speciation. *Journal of Evolutionary Biology* 9:893–909.

Hood, GR, SP Egan & JL Feder. 2012. Interspecific competition and speciation in endoparasitoids. *Evolutionary Biology* 39:219–230.

Maruvka, YE, T Kalisky & NM Shnerb. 2008. Nonlocal competition and the speciation transition on random networks. *Physical Review E* 78, 031920. *Behavior, Ecology, and Sociobiology* 59:704–713.

Pellissier, L. 2015. Stability and the competition-dispersal trade-off as drivers of speciation and biodiversity gradients. *Frontiers in Ecology & Evolution* 3:52. DOI: 10.3389/fevo.2015.00052.

Pellmyr, O. 1986. Three pollination morphs in *Cimicifuga simplex*; incipient speciation due to inferiority in competition. *Oecologia* 68:304–307.

Polechová, J & NH Barton. 2005. Speciation through competition: A critical review. *Evolution* 59:1194–1210.

Qvarnström, A, N Vallin & A Rudh. 2012. The role of male contest competition over mates in speciation. *Current Zoology* 58:493–509.

Rabosky, DL. 2013. Diversity-dependence, ecological speciation, and the role of competition in macroevolution. *Annual Review Ecology & Systematics* 44:481–502.

Walter, GH, PE Hulley & AJFK Craig. 1984. Speciation, adaptation and interspecific competition. *Oikos* 43:246–248.

West-Eberhard, MJ. 1983. Sexual selection, social competition and speciation. *The Quarterly Review of Biology* 58:155–183.

Winkelmann, K, MJ Genner, T Takahashi & L Rüber. 2014. Competition-driven speciation in cichlid fish. *Nature Communications* 5. DOI: 10.1038/ncomms4412.

EVOLUTION IN ACTION PAPERS

Ayasse, M, J Gögler & J Stökl. 2010. Pollinator-driven speciation in sexually deceptive orchids of the genus *Ophrys*. In: *Evolution in Action*, M. Glaubrecht (Ed.), Springer-Verlag, Berlin, pp 101–118.

Blattner, FR, T Pleines & SS Jakob. 2010. Rapid radiation in the barley genus *Hordeum* (Poaceae) during the Pleistocene in the Americas. In: *Evolution in Action*, M. Glaubrecht (Ed.), Springer-Verlag, Berlin, pp 17–33.

Feldhaar, H, J Gadau & B Fiala. 2010. Speciation in obligately plant-associated *Cremastogaster* ants: Host distribution rather than adaptation towards specific hosts drives the process. In: *Evolution in Action*, M. Glaubrecht (Ed.), Springer-Verlag, Berlin, pp 193–213.

Herder, F & UK Schliewen. 2010. Beyond sympatric speciation: Radiation of sailfin silverside fishes in the Malili Lakes (Sulawesi). In: *Evolution in Action*, M. Glaubrecht (Ed.), Springer-Verlag, Berlin, pp 465–483.

Johannesen, J, T Diegisser & A Seitz. 2010. Speciation via differential host-plant use in the tephritid fly *Tephritis conura*. In: *Evolution in Action*, M. Glaubrecht (Ed.), Springer-Verlag, Berlin, pp 239–260.

Kilian, B, W Martin & F Salamini. 2010. Genetic diversity, evolution and domestication of wheat and barley in the Fertile Crescent. In: *Evolution in Action*, M. Glaubrecht (Ed.), Springer-Verlag, Berlin, pp 137–166.

Köhler, F, S Panha & M Glaubrecht. 2010. Speciation and radiation in a river: Assessing the morphological and genetic differentiation in a species flock of viviparous gastropods (Cerithioidea: Pachychilidae). In: *Evolution in Action*, M. Glaubrecht (Ed.), Springer-Verlag, Berlin, pp 513–550.

Kohnen, A, R Brandl, R Fricke, F Gallenmüller, K Klinge, I Köhnen, W Maier, F Oberwinkler, C Ritz, T Speck, G Theissen, T Tscharntke, A Vaupel & V Wissemann. 2010. Radiation, biological diversity and host-parasite interactions in wild roses, rust fungi and insects. In: *Evolution in Action*, M. Glaubrecht (Ed.), Springer-Verlag, Berlin, pp 215–238.

Liebers-Helbig, D, V Sternkopf, AJ Helbig & P de Knijff. 2010. The Herring Gull complex (*Larus argentatus - fuscus - cachinnans*). In: *Evolution in Action*, M. Glaubrecht (Ed.), Springer-Verlag, Berlin, pp 351–371.

Marten, A, M Kaib & R Brandl. 2010. Are cuticular hydrocarbons involved in speciation of fungus-growing termites (Isoptera: Macrotermitinae)? In: *Evolution in Action*, M. Glaubrecht (Ed.), Springer-Verlag, Berlin, pp 283–306.

Mayer, F, D Berger, B Gottsberger & W Schulze. 2010. Non-ecological radiations in acoustically communicating grasshoppers? In: *Evolution in Action*, M. Glaubrecht (Ed.), Springer-Verlag, Berlin, pp 451–464.

Nitz, B, G Falkner & G Haszprunar. 2010. Inferring multiple Corsican *Limax* (Pulmonata: Limacidae) radiations: A combined approach using morphology and molecules. In: *Evolution in Action*, M. Glaubrecht (Ed.), Springer-Verlag, Berlin, pp 405–435.

Paetsch, M, S Mayland-Quellhorst, H Hurka & B Neuffer. 2010. Evolution of the mating system in the genus *Capsella* (Brassicaceae). In: *Evolution in Action*, M. Glaubrecht (Ed.), Springer-Verlag, Berlin, pp 77–100.

Plötner, J, T Uzzell, P Beerli, Ç Akin, CC Bilgin, C Haefeli, T Ohst, F Köhler, R Schreiber, G-D Guex, SN Litvinchuk, R Westaway, H-U Reyer, N Pruvost & H Hotz. 2010. Genetic divergence and evolution of reproductive isolation in eastern Mediterranean water frogs. In: *Evolution in Action*, M. Glaubrecht (Ed.), Springer-Verlag, Berlin, pp 373–403.

Sauer, J & B Hausdorf. 2010. Palaeogeography or sexual selection: Which factors promoted Cretan land snail radiations? In: *Evolution in Action*, M. Glaubrecht (Ed.), Springer-Verlag, Berlin, pp 437–450.

Schneider, H, H-P Krier, T Janssen, E Otto, H Muth & J Henrichs. 2010. Key innovations versus key opportunities: Identifying causes of rapid radiations in derived ferns. In: *Evolution in Action*, M. Glaubrecht (Ed.), Springer-Verlag, Berlin, pp 61–75.

Schneider, H, T Janssen, N Byrstiakova, J Heinrichs, S Hennequin & F Rakotondrainibe. 2010. Rapid radiations and neoendemism in the Madagascan Biodiversity Hotspot. In: *Evolution in Action*, M. Glaubrecht (Ed.), Springer-Verlag, Berlin, pp 3–15.

Schubart, CD, T Weil, JT Stenderup, KA Crandall & T Santl. 2010. Ongoing phenotypic and genotypic diversification in adaptively radiated freshwater crabs from Jamaica. In: *Evolution in Action*, M. Glaubrecht (Ed.), Springer-Verlag, Berlin, pp 323–349.

Stephan, W & T Städler. 2010. Population genetics of speciation and demographic inference under populations subdivision: Insights from studies on wild tomatoes (*Solanum* sect. *Lycopersicon*). In: *Evolution in Action*, M. Glaubrecht (Ed.), Springer-Verlag, Berlin, pp 119–135.

Thiv, M, K Esfeld & M Koch. 2010. Studying adaptive radiation at the molecular level: A case study in the Macaronesian Crassulaceae-Sempervivoideae. In: *Evolution in Action*, M. Glaubrecht (Ed.), Springer-Verlag, Berlin, pp 35–59.

Tiedemann, R, PGD Feulner & F Kirschbaum. 2010. Electric organ discharge divergence promotes ecological speciation in sympatrically occurring African weakly electric fish (*Campylomormyrus*). In: *Evolution in Action*, M. Glaubrecht (Ed.), Springer-Verlag, Berlin, pp 307–321.

von Rintelen, T, K von Rintelen & M Glaubrecht. 2010. The species flocks of the viviparous freshwater gastropod *Tylomelania* (Mollusca: Cerithioidea: Pachychilidae) in the ancient lakes of Sulawesi, Indonesia: The role of geography, trophic morphology and color as driving forces in adaptive radiation. In: *Evolution in Action*, M. Glaubrecht (Ed.), Springer-Verlag, Berlin, pp 485–512.

Wägele, H, MJ Raupach, I Burghardt, Y Grzymbowski & K Händeler. 2010. Solar powered seaslugs (Ophistobranchia, Gastropoda, Mollusca): Incorporation of photosynthetic units: A key character enhancing radiation? In: *Evolution in Action*, M. Glaubrecht (Ed.), Springer-Verlag, Berlin, pp 263–282.

Weising, K, D Guicking, C Fey-Wagner, T Kröger-Killan, T Wöhrmann, W Dorstewitz, G Bänfer, U Moog, M Vogel, C Baier, FR Blattner & B Fiala. 2010. Mechanisms of speciation in Southeast Asian ant-plants of the genus *Macaranga* (Euphorbiaceae). In: *Evolution in Action*, M. Glaubrecht (Ed.), Springer-Verlag, Berlin, pp 169–191.

Wilke, T, M Benke, M Brändle, C Albrecht & J-M Bichain. 2010. The neglected side of the coin: Non-adaptive radiations in the spring snails (*Bythinella* spp.). In: *Evolution in Action*, M. Glaubrecht (Ed.), Springer-Verlag, Berlin, pp 551–578.

APPENDIX 8.2 PATTERNS OF ACCUMULATION OF STUDY SYSTEMS FOR THE 504 PAPERS ON SPECIATION IN APPENDIX 8.1

1965
Theory/Models
Bacteria
Protists
Plants
Insects
Arachnids
Crustaceans
Fishes
Amphibians
Mammals

1970
Fungi
Crustaceans
Reptiles[a]
Birds
Animals
Coevolution
Ecosystem
Life/Organisms

1975
Gastropods
Tunicates

1980
Nematodes
Parasites

1985
Algae
Flatworms
Conservation Biology

2000
Echinoderms
Hawaiian Biota

2005
Viruses
Jellyfishes
Rotifers

2010
Sponges
Arrow Worms
Annelids
Onychophora
Vertebrates
Marine Fauna

---

[a] Includes squamates + turtles for this sampling effort.

APPENDIX 8.3 PATTERNS OF ACCUMULATION OF MODES OF SPECIATION IDENTIFIED FOR THE 504 PAPERS ON SPECIATION IN APPENDIX 8.1

1965
No Mechanism
Models
Genetics
Allopatric Speciation
Climate/Geology
Coevolution
Incipient Speciation
Natural Selection
Premating Isolation
Transformation

1970
Multiple Mechanisms
Character Displacement
Dispersal
Ecological Adjustment
Phyletic Change
Sympatric Speciation

1975
Adaptive Differentiation
Sexual Selection
Species Evolution
Vicariance

1980
Invasion
Isolation
Parapatric Speciation
Parasitism

1985
Competition
Ecological Speciation
Fragmentation
Geographic Speciation
Habitat Specialization
Insular Speciation

Organization Theory
Punctuated Equilibria
Random Morphological Change
Replacement Pattern Speciation
Stasipatic Speciation

1995
Rapid Speciation

2000
Allochronic Speciation
Alloxenic Speciation
Autochthonous Speciation
Centrifugal Speciation
Determinism
Environmental Selection
Life Cycle Change
Reproductive Isolation
Speciation in situ

2005
Adaptive Radiation
Elevational Speciation
Force of Distance
Frequency Dependent Selection
Historic Factors
Humans as Selective Agents
Reinforcement
Symbiogenesis

2010
Heteropatric Speciation
Infectious Speciation
Morphic Speciation
Phenotypic Plasticity
Phylogenetic Differentiation
Predation
Sensory Drive
Sequential Speciation

APPENDIX 8.4   JOURNALS THAT PUBLISHED THE
504 PAPERS IN APPENDIX 8.1

1965
American Journal of Medicine
Annual Review of Entomology
Annual Review of Phytopathology
Bulletin of the World Health Organization
Canadian Journal of Genetics & Cytology
Chromosoma
Cytologia
Ecological Monographs
Euphytica
Evolution
Genetics
Hydrobiologia
Isis
Japanese Journal of Genetics
Journal of Bacteriology
Journal of General Microbiology
Journal of the New York Entomological Society
Journal of Parasitology
Journal of Zoology
Oikos
Proceedings of the Royal Society London Ser. B
Silvae Genetica
Systematic Zoology/Systematic Biology

1970
Adansonia
American Naturalist
American Zoologist
Ann Rev of Ecology & Systematics
Annals de Genetique
Antonie van Leewenhoek
Arquivos de Zoologia
Australian Journal of Entomology
Bacteriological Review
Biotropica
Brittonia
Bulletin of the Torrey Botanical Club
Canadian Entomologist
Genetica
Genetika
Great Basin Naturalist
Journal of the Fishery Res Brd Canada
Journal of Theoretical Biology
Mammal Review
Natural History Bulletin of the Siam Soc.

Taxon
Trans. & Proc. Botanical Soc. Edinburgh
Trans. American Fisheries Society
Trans. of the New York Academy of Sci
Wilson Bulletin

1975
Anal Inst Bot AJ Cav.
Ann de Science Forest
Ann Missouri Botanical Garden
Beaufortia
Botanical Review
Bryologist
Herpetologica
New Phytologist
Oecologia
Proc. National Acad. Sciences
Rev. Chil de Entomol.
Zool. Journal of Linnean Soc.

1980
American Birds
American Journal of Botany
Bioscience
Bolletino di Zoologia
Bonn Zool Beitrage
Botanical Journal Linnean Society
Copeia
Deutsche Entom Zeits
Environmental Biology of Fishes
Heredity
Ibis
Italian Journal of Zoology
Journal of Biogeography
Journal of Mammalogy
Journal of the Roy Soc of New Zealand
Perspectives in Biology & Medicine
Systematic Entomology

1985
Auk
Australian Journal of Botany
Biochemical Systematics and Evolution
Biometrical Journal
Entomological Gener.
Japanese Journal of Entomology

Journal of the Kansas Entomologial Soc.
Journal. Zool. Systematics & Evolution
Ornithological Monographs
Plant Systematics and Evolution
Proc. Roy Soc. Edinburgh Sect. B
Systematic Botany
Transvaal Museum Monographs
Zoological Science

1990
Bulletin of Marine Science
Canadian Journal of Zoology
Journal of the Arnold Arboretum

1995
Canadian Journal of Botany
Canadian J Fish & Aquatic Sci.
Current Biology
Experientia
Nature

2000
Adv. In Ecological Research
BioEssays
Cladistics
Ekológia
Evolutionary Biology
Evolutionary Eco Res.
Genes & Genetic Systems
Intern. J. Parasitology
Intern. J. Primatology
Journ. Evolutionary Biology
Journal of Fish Biology
Lethaia
Molecular Ecology
Mol. Phylogenetics & Evolution
Plant Species Biology
Science
Trends in Microbiology
Turkish Journal of Zoology

2005
Behavior Genetics
Biochem. Systemat & Evolution
BMC Evolutionary Biology
Ecology Letters
Entomol Expet App
Evolutionary Ecology
Fish & Fisheries
Gene

Genetic Resources & Crop Evolution
Infection & Immunity
Insectes Sociaux
Journal of Arachnology
Journal of Heredity
Journal of Mammalian Evolution
Journal of Marine Science
Marine Biology
Marine Ecology Progress Series
Proceed. Royal Irish Academy
Scientia Marina
Theory in Biosciences
Theory in Parasitology

2010
Ann Rev. Genomics & Hum Genet.
Annals of Botany
Biolog. Journ. Linnean Society
Biology & Philosophy
Chemoecology
Conservation Genetics
Deep Sea Research
Ecological Entomology
Freshwater Reviews
Global Ecology & Biogeography
Journal of Animal Ecology
Journal of Eukaryotic Microbiology
Journal of Experimental Zoology
Jour. North Amer. Benthologic. Soc.
Journal of Virology
Molecular Biology & Evolution
Molecular Ecology Resources
Mycologia
Nature Communications
Neotropical Entomology
Parasitology
Philos. Trans. Roy Soc. Ser. B
Phylogenetics & Evolution
PLoS Genetics
PLoS One
PLoS Pathogens
Protist
Research in Microbiology
Russian Journal of Genetics
Systematics & Biodiversity
Trends in Ecology & Evolution
Trends in Genetics
Zoologia Scripta
Zoology in the Middle East
Zoosystematics & Evolution
Zootaxa

REFERENCES

Bronstein, J. L. 2009. The evolution of facilitation and mutualism. *Journal of Ecology* 97:1160–1170.

Brown, J. H. 1981. Two decades of homage to Santa Rosalia: Towards a general theory of diversity. *American Zoologist* 21:877–888.

Bruno, J. F., J. J. Stachowicz and M. D. Bertness. 2003. Inclusion of facilitation into ecological theory. *Trends in Ecology and Evolution* 18:119–125.

Callaway, R. M. 1995. Positive interactions among plants. *The Botanical Review* 61:306–349.

Callaway, R. M. 2007. *Positive interactions and interdependence in plant communities.* Dordrecht: Springer.

Connell, J. H. 1980. Diversity and the coevolution of competitors, or the ghost of competition past. *Oikos* 131–138.

Connell, J. H. 1983. On the prevalence and relative importance of interspecific competition: Evidence from field experiments. *American Naturalist* 122:661–696.

Connell, J. H., T. P. Hughes, C. C. Wallace, J. E. Tanner, K. E. Harms and A. M. Kerr. 2004. A long-term study of competition and diversity of corals. *Ecological Monographs* 74:179–210.

Den Boer, P. J. 1986. The present status of the competitive exclusion principle. *Trends in Ecology and Evolution* 1:25–28.

Glaubrecht, M. (ed.). 2010. *Evolution in action.* Heidelberg: Springer.

Hacker, S. D. and S. D. Gaines. 1997. Some implications of direct positive interactions for community species richness. *Ecology* 78:1990–2003.

Hood, G. R., S. P. Egan and J. L. Feder. 2012. Interspecific competition and speciation in endoparasitoids. *Evolutionary Biology* 39:219–230.

Hutchinson, G. E. 1959. Homage to Santa Rosalia or why are there so many kinds of animals? *American Naturalist* 93:145–159.

Kaplan, I. and R. F. Denno. 2007. Interspecific interactions in phytophagous insects revisited: A quantitative assessment of competition theory. *Ecology Letters* 10:977–994.

Lotka, A. J. 1920. Analytical note on certain rhythmic relations in organic systems. *Proceedings of the National Academy of Sciences USA* 6:410–415.

Lotka, A. J. 1925. *Elements of physical biology.* Baltimore: Williams & Wilkins Co.

McIntire, E. J. B. and A. Fajardo. 2014. Facilitation as a ubiquitous driver of biodiversity. *New Phytologist* 201:403–416.

Menge, B. A. and J. P. Sutherland. 1976. Species diversity gradients: Synthesis of the roles of predation, competition, and temporal heterogeneity. *American Naturalist* 110:351–369.

Peters, R. H. 1976. Tautology in evolution and ecology. *American Naturalist* 110:1–12.

Peters, R. H. 1991. *A critique for ecology.* New York: Cambridge University Press.

Ricklefs, R. and R. Relyea. 2014. *Ecology: The economy of nature.* New York: W.H. Freeman and Co.

Schoener, T. W. 1982. The controversy over interspecific competition. *American Scientist* 70:586–595.

Schoener, T. W. 1983. Field experiments on interspecific competition. *The American Naturalist* 122:240–285.

Simberloff, D. 1982. The status of competition theory in ecology. *Annales Zoologici Fennici* 19:241–253.

Simberloff, D. 1983. Competition theory, hypothesis testing and other community ecological buzzwords. *American Naturalist* 122:626–635.

Stachowicz, J. J. 2001. Mutualism, facilitation, and the structure of ecological communities. *BioScience* 51:235–246.

Stiling, P. 2012. *Ecology: Global insights and investigations.* New York: McGraw Hill.

Strong, D. R., Jr. 1980. Null hypotheses in ecology. *Synthese* 43:271–285.

Strong, D. R., Jr., L. A. Szyska and D. S. Simberloff. 1979. Tests of community-wide character displacement against null hypotheses. *Evolution* 33:897–913.

Valiente-Banuet, A. and M. Verdú. 2007. Facilitation can increase the phylogenetic diversity of plant communities. *Ecology Letters* 10:1029–1036.

Volterra, V. 1926a. Fluctuations in the abundance of a species considered mathematically. *Nature* 118:558–560.

Volterra, V. 1926b. Variazioni e fluttuazioni del number d'individuiin specie animali conviventi. *Memoire della R. Accademia Nazionale dei Lincei*, series 6, 2:31–112.

Volterra, V. 1928. Variations and fluctuation of the number of individuals in animal species living together. *Journal du Conseil Permanent International pour l'Exploration de la Mer* 3:3–51.

Wake, D. B. and V. T. Vredenburg. 2008. Are we in the midst of the sixth mass extinction? A view from the world of amphibians. *Proceedings of the National Academy of Sciences USA* 105:11466–11473.

Wiens, J. A. 1977. On competition and variable environments. *American Scientist* 65:590–597.

9 Epistemological Concern for Estimating Extinction
Introducing a New Model for Comparing Phylogenies

Prosanta Chakrabarty and Subir Shakya
Louisiana State University

CONTENTS

> Instead of choosing the best among a set of models, no matter how insufficient they all
> are, one could identify situations where there is no existing adequate model.
>
> **Rabosky and Goldberg (2015)**

9.1   INTRODUCTION

Extinctions, being generally unobservable in the fossil record and in the lifetime
of a scientist, are a mystery. Most species that have gone extinct left no trace of
their existence. For most species that have lived on Earth, all that remains are their
descendants, who themselves have evolved into other forms and perhaps have also
gone extinct. Unfortunately, reliably measuring extinction by relying on the relationships among extant taxa and a scant fossil record is problematic, if not impossible (Rabosky 2010). The most widely used birth–death models estimate extinction
rates that are close to zero (Nee 2006; Höhna et al. 2011) even though extinction

rates were more likely to be relatively high (Rosenblum et al. 2012). A lack of a robust measure of extinction is unfortunate because knowledge of the true amount of extinction that has taken place in a clade could be extremely informative (Simpson 1944). Extinction rates could help us better understand speciation and ecological turnover in a clade. Knowing the total number of extinct species in a group can tell us about the total diversity once contained within that clade (He and Hubbell 2011). Past extinction events can even provide relevant information on current ones (Régnier et al. 2015): for example, did past climate change result in a lineage more or less resilient to current levels of change? Are extinction events correlated across groups in space and time?

Here, we propose a new model for estimating extinction events using a simple formula that can be applied across any phylogeny. Like other models of extinction, this new model cannot be verified or falsified (because we cannot know the true amount of extinction) outside of a simulation. Our goal is for this new model to be a useful exercise to compare phylogenies and to help us rethink our assumptions about measuring extinction. We attempt with this method to harken back to the earliest days of studying biological diversification by relying on tree symmetry as a proxy for understanding diversification and extinction (see discussion in Pennell and Harmon 2013) and in using a sister clade comparison to investigate diversification within the tree (as previously proposed, perhaps first by Slowinski and Guyer 1993). The major assumption of this model is that a constant rate of origination without extinction will lead to a symmetrical tree. The pragmatic justification for this assumption is based on some empirical evidence that time-constant diversification tends to produce more balanced trees (Nee et al. 1994b; Chan and Moore 2005; Ricklefs 2007; Morlon 2014), although never a symmetrical one. This assumption allows us to easily make comparisons between different kinds of asymmetrical trees versus the baseline of symmetry. The symmetrical "expected" tree will have two sides (sister clades), when viewed in two dimensions, with the same total number of species. As shown in Figure 9.1, one can then calculate the number of extinct taxa by subtracting the number of species in the "expected" tree from the number of taxa in the original "observed" tree.

The symmetrical, expected tree could be treated as the null hypothesis for any given phylogeny. This null hypothesis allows groups of different sizes and phylogenetic histories to be easily compared against their expected tree, which allows us to measure extinction as a fraction of the total number of expected species.

In this model, extinction is the only cause of asymmetry in a phylogeny (see Pearson 1998 for similar thoughts, also Heard and Mooers 2002 for a discussion). We note that other sources of noise, including uneven sampling, rate variation across lineages, and phylogenetic error (Huelsenbeck and Kirkpatrick 1996) may also contribute to tree asymmetry; however, these other sources of noise are beyond the scope of this manuscript, which focuses on the birth–death process assuming no other sources of error. However, we do propose a way that incomplete sampling, including from the fossil record, can be incorporated into our model of extinction.

One advantage of this new method is that measures of extinction in trees with complete taxon sampling are less error prone than those with incomplete sampling

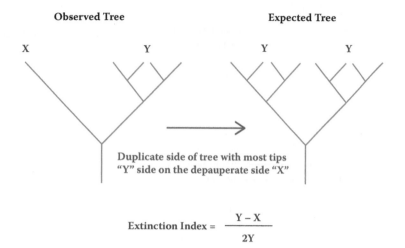

$$\text{Extinction Index} = \frac{Y - X}{2Y}$$

FIGURE 9.1 *Using the Complete Tree Null Model:* The phylogeny obtained by the user, that is, the "observed tree," should be pruned to the ingroup or section where extinction is to be measured. The phylogeny (which must not have a basal polytomy) is then made symmetrical by duplicating the species-rich side of the phylogeny ("Y" side) on to the species-poor side "X" side to obtain the "expected" (zero extinction) tree. The percentage of extinction is then measured by the equation $(Y - X)/2Y$. Where X is the number of species on the species-poor side of the tree, and Y is the number of species on the species-rich side of the tree.

(Rabosky 2010). Under this method, which we call the "Complete Tree Null Model" (CTNM), one need only know if missing (unsampled) taxa belong to the species-rich ("Y" clade or side) or species-poor clade ("X" clade or side) of a phylogeny to achieve a measure of complete taxon sampling as shown in Figure 9.1. In addition, Figure 9.2 shows an instance where a phylogeny of a family of organisms is represented by only one species of each genus in that family; by plugging in the numbers of species for each genus, one can then know the number of taxa expected on each sister clade of the tree. These "missing taxa" may also include fossils that can be added to the extinction total as shown in Figure 9.3. Even in the absence of major clades, if one knows the relative position of missing taxa within the tree (either right side or left), then estimates can be made about the total number of taxa in the ingroup. Maximum likelihood models for adding taxa to incomplete trees are also available for particularly difficult to place clades or extinct lineages (Revell et al. 2015) and these can easily be applied to the decision-making process for placing missing taxa.

This minimalist, or parsimonious approach (in terms of minimizing ad hoc assumptions) to measure extinction is an alternative to the many perhaps over-parameterized estimations of diversification. To our knowledge, no model currently exists that focuses solely on extinction. Instead, most models focus on speciation and extinction simultaneously to better understand diversification. In most of these cases, extinction is modeled strictly for the sake of understanding rates of speciation and overall diversification. Because speciation is as difficult to measure as extinction,

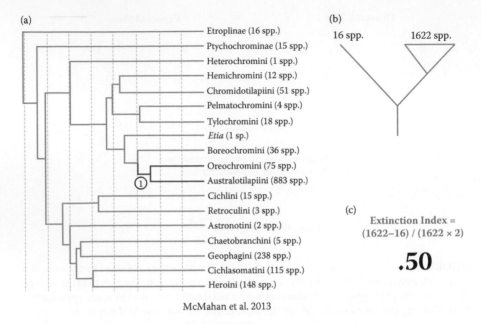

McMahan et al. 2013

FIGURE 9.2 *Adding missing (unsampled) taxa to the CTNM*: Cichlid phylogeny from McMahan et al. (2013) (a). Number of total species is included in the X and Y sides of the tree (b), even though they were not sampled in the original phylogeny, missing taxa can be added simply by counting them in the totals for each side of the tree. The totals with the missing taxa can then be used to measure the percentage of extinction following the CTNM (c).

measuring both simultaneously could be thought of as impractical as simultaneously measuring the velocity and position of a Higgs Boson.

Extinction, perhaps because of the difficulty in finding evidence for it, typically plays second fiddle to speciation (although obviously extinction plays a major role in diversification). Extinction rates are often held constant (e.g., Heath et al. 2015) or if estimated, are typically underestimated, that is, close to zero (Nee 2006; Morlon et al. 2011; Höhna et al. 2011).

Many diversification models have so many parameters at play that it is difficult to measure any one parameter independently (Rabosky and Goldberg 2015) and certainly extinction cannot be accurately measured when extinction is assigned an *a priori* arbitrary constant rate. The point of many diversification models is not to measure extinction at all, perhaps because of the difficulty of measuring extinction with an incomplete fossil record and extrapolating the death of species solely on the pattern of diversification of extant species. Rabosky (2010) goes as far as to state that "extinction should not be estimated on molecular phylogenies" and that "extinction rates should not be estimated in the absence of fossil data"—a stand at odds with other researchers (see Nee et al. 1994a; Paradis 2003). Morlon et al. (2011) who have noted that many researchers purported to find an extinction rate of zero based on likelihood-based molecular phylogenies despite the frequency of known extinction events. Höhna et al. (2011:2586) show that the commonly used birth–death models

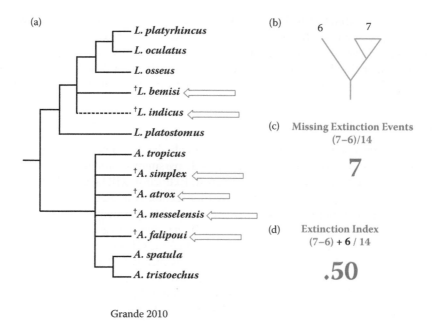

Grande 2010

FIGURE 9.3 *Adding fossil information to the CTNM*: Phylogeny of Lepisosteidae from Grande 2010 that includes placement of fossil gars (green arrows) (a). The fossil species are included in the "Expected" tree just as with extant taxa (b). The percentage of missing extinct taxa is then calculated in the same way as previously discussed (c) with the exception that the fossil taxa are then added to the numerator again because they too are extinct species (d). Note that groups with a large fossil record can have extinction over 0.50 or even over 1 depending on the number of fossil species known. (From Wright, J.J., et al. *Mol. Phylogeneti. Evol.*, 63, 848–856, 2012.)

often underestimate extinction and that they "…often show an estimated death rate close to zero even though it is well known that species go extinct during evolution."

The greatest source of empirical evidence of extinction and speciation is the fossil record (Rosenblum et al. 2012). Yet, fossils also only provide an incomplete record of extinction. We continue to struggle with incorporating evidence from the fossil record into our understanding of the phylogenetic history of a group (Rabosky 2010; Morlon et al. 2011; Pennell and Harmon 2013). Unfortunately, many groups lack a known fossil record altogether, but that does not mean their extinction rate should be considered zero as many birth–death models estimate. In the CTNM method, fossils can easily be incorporated into our measure of extinction.

In the CTNM fossil taxa are at first treated like extant taxa; that is, they are placed as tips in the phylogeny similar to how missing (unsampled) taxa were added as described above and as shown in Figure 9.3. The percentage of missing taxa is then calculated as before with the exception that these fossil taxa are then re-added to the numerator of the extinction equation. Therefore, the CTNM provides an easy and intuitive way to add evidence of extinct species from the fossil record to a phylogeny and incorporates that evidence to estimate the amount of extinction in a group.

How overall extinction is determined in many studies is not always clear. Many models of speciation and divergence assume a constant rate of extinction (see above) or one that changes based on a fixed set of branching points based on a likelihood function and/or fossil evidence (Morlon et al. 2011). In the CTNM, we assume a constant rate for speciation in order to measure the variable of interest: extinction.

9.2   METHODS

To carry out the complete tree null model approach, the user obtains an empirically derived phylogeny (from maximum likelihood, parsimony, simulations, etc.) that should be pruned to the ingroup taxa of interest. Extinction will be calculated on the remaining (nonpruned) section of the tree; this section is treated as the "observed tree." The two sides of the observed tree are then made symmetrical by duplicating the species-rich side of the phylogeny ("Y-side") on to the opposite species-poor side ("X-side") to create the "expected tree," which under the CTNM is assumed to be the result of zero extinction. The index of extinction is then calculated by the fraction: $(Y - X)/2Y$; where Y equals the number of species on the species-rich half of the tree and X equals the number of species on the species-poor side of the tree. A schematic representation of this method is shown in Figure 9.1. The result from the calculation, although presented as a proportion, should not be considered a literal proportion of extinct taxa (again this is unknowable). Rather we ask the user to interpret this value as an "index of extinction" without units that can be used to compare across groups.

We examined several independent phylogenies and used our CTNM to compare the index of extinction across groups. We sampled phylogenies with either robust taxon sampling (Near et al. 2011; Collins et al. 2015) or with a tree structure that permits easy addition of unsampled taxa. For instance, Thompson et al. (2014) sampled less than 40% of the known species of piranhas (Serrasalmidae) but included examples of all genera. Assuming the unsampled species remain with their congeners, it was relatively easy to add the missing taxa to these trees by simply adding the number of unsampled taxa to those existing branches. After adding those unsampled taxa the side of the phylogeny (either right or left) with the most taxa was again designated "Y" and the side with the least species was designated "X." $Y - X$ is the number of species "missing" from the X side of the tree (given the assumption of constant origination); 2Y is the total number of expected species under the CTNM. A schematic representation of this method is shown in Figure 9.2. Similarly, one can add information about known fossils in the same way unsampled taxa were added. The fossil taxa are added to the expected tree just as extant taxa are and the extinction index is calculated in the same manner as discussed above with the exception that the number of fossil taxa are added to the numerator $(Y - X)$ again (because they too are extinct species). A schematic representation of this method is shown in Figure 9.3.

We also created a custom Python script, in Appendix 9.1, to run the CTNM and calculate an extinction index automatically. This Python script will accept a newick file for any phylogenetic tree and provide an extinction index as calculated by the CTNM. This script also calculates a p-value for a chi-squared test assuming the degrees of freedom equals 1 (because there are two variables, the X and Y side of the tree). We remind the user to input the newick file with the ingroup taxa of interest only. An anonymous reviewer also generously provided a draft R script (Appendix 9.2) to examine the distribution of the CTNM on simulated trees. Each analysis ran a different extinction rate to generate 1000 simulated trees to calculate the extinction index on each tree.

9.3   RESULTS AND DISCUSSION

Our comparison of multiple groups using the CTNM is shown in Table 9.1. We find a somewhat narrow range of extinction indices across groups. The highest value achievable is 0.50 (unless additional fossils can be added as shown in Figure 9.3), the lowest 0. This narrow range was also achieved under simulations from a custom R script as shown in Figure 9.4, simulating trees with extinction rates from 0 to 0.8 and showing a heavy skewing towards an extinction index of 0.50, the higher the extinction rate. The results from the simulation also differ from the conclusions of Slowinski and Guyer (1989) in showing that tree probabilities do change with random extinction. Notably, some variation was found with our empirical data particularly when the contribution from the fossil record was included.

The percentage of total extinction for a group as measured by the CTNM is essentially a measure of tree imbalance or asymmetry. The more asymmetrical (many species on one side versus the other) the greater the amount of extinction that will be assessed. The most symmetrical lineage we discovered belonged to the Crocodylia (Oaks 2011), a group that was found to have an extinction index of 0.26.

We recovered several groups with very high asymmetry and thus a very high extinction index (0.49). These groups include: Vireonidae (Vireos), Hirundinidae

---

TABLE 9.1
List of Groups Compared Using the CTNM

Taxon	Reference	Number of Living Species	Extinction Index
Crocodylia	Oaks (2011)	23	0.26
Etheostomatinae	Near et al. (2011)	247	0.36
Monarchidae	Andersen et al. (2014)	99	0.37
Polycentridae	Collins et al. (2015)	5	0.38
Cotingidae	Berv and Prum (2014)	66	0.41
Serrasalmidae	Thompson et al. (2014)	99	0.42
Falconidae	Fuchs et al. (2015)	64	0.43
Leiognathidae	Chakrabarty et al. (2011)	43	0.45
Vireonidae	Slager et al. (2014)	52	0.49
Hirundinidae	Sheldon et al. (2005)	83	0.49
Cichlidae	McMahan et al. (2013)	1638	0.50
Lepisosteiformes	Grande (2010)	6	0.50[a]
Notothenioidei	Near et al. (2015)	114	0.50
Ostariophysi	Fink and Fink (1981)	10,237	0.50

[a] Use of fossils in measuring the percentage of total extinct in a group.

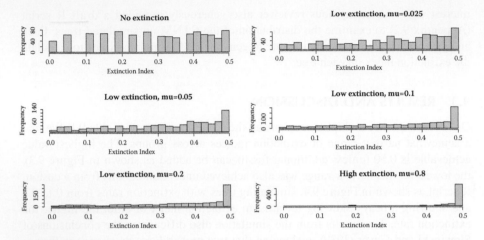

FIGURE 9.4 Results from R script from Appendix 9.2 showing the frequency distribution of the extinction index (as calculated by CTNM) with 1000 simulated trees with varying extinction rates.

(swallows), Cichlidae (cichlids), Lepisosteiformes (gars), Notothenioidei (icefishes), and Ostariophysi (the speciose superorder that includes catfishes, minnows, tetras, electric knifefishes, among other freshwater fishes).

This simple method to estimate extinction may be too simple for the palate of many modern researchers. We know, for instance, that extinction and speciation rates vary among clades (Jetz et al. 2011; Rabosky et al. 2013) and also that a constant speciation rate will not necessarily lead to a symmetric tree (Losos and Adler 1994). Our analysis does not address any rates of extinction but shows only an overall fraction of extinction (which we refer to as an index), a more conservative approach. This CTNM may even be of use to those that prefer more complex models; for instance, the conservative estimate of extinction derived from the CTNM can be used in choosing a simulated rate of extinction being applied in another method.

The CTNM examines just one set of sister-clade contrasts, the one between the left and right side of the tree within the ingroup. More complex models could be developed in which each clade in a tree is examined from the base to the tips to get a "rate" based on the CTNM approach. Slowinski and Guyer (1993) proposed a similar method using multiple sister-clade contrasts, except these contrasts were modeled with random speciation and extinction. More recent constant rate models, such as the equal rate Markov (ERM), allow stochasticity and will result in some random tree imbalance (Mooers and Heard 1997). These stochastic models make adding missing (unsampled) taxa more complex than in the CTNM.

Knowledge of the diversification rate of a clade is highly sought after, potentially informing us about adaptive radiations, ecological shifts, and key adaptations, among other evolutionary phenomena. Unfortunately, knowledge about diversification shifts may be misleading due to assumptions about extinction. Differences in extinction rate can lead to two groups with the same diversification rate having phylogenetic trees of different shapes (Nee et al. 1994b; Rabosky 2006b).

A popular model by Magallón and Sanderson (2001) that searches for evidence of exceptional diversification rates uses estimates of extinction between 0% and 99%: if the speciation rate falls beyond the error bars of an assumed 0%–99% extinction, the diversification is assumed to have a significant shift in rate. This broad estimate of extinction (0%–99%) likely keeps us from capturing many of the other possible significant shifts. The focus of the CTNM on extinction specifically is not novel, but is intended to shift the discussion back to simpler criteria or assumption sets of measuring extinction.

The need for a simpler model is clear. In some commonly used programs, even arbitrary characters (like the length of a taxon name) can be found to have a (obviously spurious) correlation with a given rate of speciation under a constant birth–death assumption (see Rabosky and Goldberg 2015; also see discussion of birth–death models in Pennell and Harmon 2013). Models to estimate tree balance exist for both maximum likelihood (Chan and Moore 2005) and Bayesian approaches (Moore and Donoghue 2009) that can be compared with results from our parsimony approach. Despite some concerns, birth–death models remain a useful and popular tool in estimating diversification (Nee 2006; Höhna et al. 2011; Frost et al. 2015; Heath et al. 2015). Here, we propose an alternative that may also be of value and that is philosophically rooted in Hennig's (1966) model of speciation.

In Hennig's (1966) model, every speciation event results in the origin of two sister taxa with the simultaneous extinction of the ancestor. In this model, every node in a phylogenetic tree represents an ancestor and also an extinction event. The number of nodes in a resolved tree is always equal to $n-1$, with the number of total taxa in the phylogeny $= n$. Therefore, the index of extinction events is always $(n-1)/[n + (n-1)]$; or, in other words, the index of extinction events is equal to the total number of nodes in a tree $(n-1)$ divided by the total number of extinct and extant taxa $[(n-1) + n]$. In a phylogenetic tree with 25 taxa represented as tips, there will be 24 internal nodes; the background extinction in this case is then $24/(25 + 24)$ or 0.49. As shown in Figure 9.4, for most trees the number will approach 0.50. This baseline index of 0.50 extinction, paired with a constant rate of speciation, is effectively a high turnover birth–death model (see more on that in Höhna et al. 2011) that some argue is improbable (Rabosky and Lovette 2008). In the CTNM, the extinction index will never go above 0.50 in the absence of additional evidence from the fossil record, because extinction is effectively only measured on one side of the tree (the depauperate side). As shown in the case of Figure 9.3, the addition of fossils can increase this percentage infinitely.

In addition to Hennig, others have proposed that ancestral species are more likely to become extinct than their descendants (the so-called "Simpsonian step-series"; Simpson 1953; Pearson 1998). Rosenblum et al. (2012) argue that measuring the persistence of species is more important than speciation itself, since speciation rates appear to be higher than the number of taxa that persist. These "failed speciations" of incipient species discussed in Rosenblum et al. (2012) might explain the high background extinction rates of the Hennigian model (but see their caveats about high turnover in birth–death models) (Figure 9.5).

MEDUSA (Alfaro et al. 2009) and BAMM (Rabosky 2014) are two of the most widely used programs to examine shifts in diversification rate using more dynamic

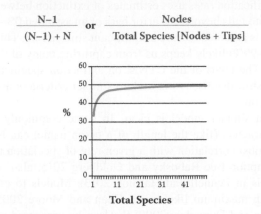

# of nodes is (N−1); where n is total number of species

$$\frac{N-1}{(N-1)+N} \quad \text{or} \quad \frac{\text{Nodes}}{\text{Total Species [Nodes + Tips]}}$$

FIGURE 9.5 *The Hennig background extinction rate graphed*: The Hennig (1966) speciation model predicts that every extinct ancestor is a node on a phylogeny. The number of nodes is equal to "*n*−1," where "*n*" is the number of tips or species that are sampled in a phylogeny. The percentage of nodes (or extinctions) approaches but never reaches 50% as the number of species sampled increases. This 50% baseline from the Hennig model is a philosophical basis for the extinction bar set in the CTNM with the exception that evidence from the fossil record will allow the percentage to increase above this 50% bar.

birth–death models. MEDUSA uses a likelihood-based step-wise Akaike information criterion to find the best-fit rate shifts in birth and death models but assumes a constant rate of speciation and extinction through clades. BAMM 2.0, a Bayesian approach, allows time-constant and time-varying diversification modes and varies extinction rates across clades. The CTNM proposed here can be considered the parsimony alternative to these Bayesian and likelihood approaches, and one that focuses solely on extinction (and not diversification as a whole) by keeping speciation constant.

The assumption that a constant rate of speciation will result in species arising equally across all branches is certainly wrong (empirically proven so by Losos and Adler 1994). And as a rough estimate of extinction, the CTNM is certainly underestimating extinction on the speciose side of the tree ("Y" side—where it is zero) and overestimating it on the depauperate side of the tree ("X" side—where all the extinction is counted). Despite that, averaged over the entire tree, the number of extinct taxa calculated might be closer to reality than in the constant rate models that find extinction to be near zero or that are based on arbitrary or assumption-laden parameters. The CTNM model has the advantage of easily incorporating missing taxa (the largest source of error for measuring extinction) and for incorporating the known fossil record for a group (the only empirical evidence of extinction that we have). Certainly better models to explain imbalance in empirical trees are still needed. This new approach is far from a perfect measure of extinction, but it can be a first step in rethinking how we measure the most enigmatic of biological processes: extinction.

ACKNOWLEDGMENTS

This work was inspired by Donn Rosen, whose clear thinking approach often found the simplest ways to explain complex patterns in natural history. Although we are obviously incapable of similar clear thinking, we hope Dr. Rosen would have considered our approach interesting if not useful. We would like to thank the 2015 Systematics Discussion Group at Louisiana State University that helped hammer out the details of this idea through tough questions, parallel dialogue, perplexed looks, and a sprinkling of mockery and ridicule—often over beers. Also to those that commented on both our poster presentation in the 2015 Evolution meetings in Brazil and the oral presentation at the 2015 Joint Meetings of Ichthyologists and Herpetologists in Reno and two anonymous reviewers of this chapter. PC also thanks the Albert lab at the University of Louisiana, Lafayette, for discussing this topic between trawls on the Amazon. Lastly, we thank the organizers of the JMIH Rosen ("Donn Rosen and the Assumptions that Inhibit Scientific Progress in Comparative Biology") Symposium, Lynne Parenti and Brian Crother, for the invitation to present and publish our work. This work was supported by NSF grant DEB 1354149 to PC.

APPENDIX 9.1   PYTHON SCRIPT FOR AUTOMATING THE PROCESS OF MEASURING THE EXTINCTION INDEX CALCULATED BY THE CTNM

```
GITHUB link: https://github.com/subirshakya/Epi_Con_Est_
Extinction/blob/master/Ext_index_estimate.py
from __future__ import (division, print_function)
importnumpy
fromscipy.stats import chisquare as chi
```

class Node:

```
    """This holds the tree data and creates node class"""
    def __init__(self,name="",parent=None,children=None,
branchlength = 0):
        self.name = name #Name of node
        self.parent = None #Name of parent (initially set to
None, but can be changed)
        if children is None:
          self.children = [] #List to hold children nodes
        else:
          self.children = children #List to append children nodes
        self.brl = branchlength #Branch lengths
```

class Tree:

```
    """
    Defines a class of phylogenetic tree, consisting of linked
Node objects.
    """

    def __init__(self, data):
        self.root = Node("root") #Define root
```

```python
        self.newicksplicer(data, self.root) #Splice newick data
    defnewicksplicer(self, data, base):
        """
        Splices newick data to create a node based tree. Takes a
base argument which is the root node.
        Make sure the file has a single line of newick code. Any
miscellaneous characters will throw the program off. Also make
sure it does not end in ;"
        Newick should start with ( and end in ). Can modify
script to do more.
        Note: The script handles every node as if there is a
bifurcation. Any polytomy will not be correctly programmed.
        """

        data = data.replace(" ", "")[1: len(data)]    #Get rid of
all spaces and removes first and last parenthesis
        n = 0
        ifdata.count(",") !=0: #While there is no more comma
separated taxa
            for key in range(len(data)): #Find the corresponding
comma for a given parenthesis (n will be 0 for the correct
comma)
                if data[key] == "(":
                    n += 1 #Increase index of n by 1 for 1 step into
new node
                elif data[key] == ")":
                    n -= 1 #Decrease index of n by 1 for 1 step out node
                elif data[key] == ",":
                    if n == 0: #To check for correct comma
                        vals = (data[0:key], data[key+1:len(data)-1])
#Break newick into left and right datasets
                        for unit in vals: #For each entry of dataset
                            if ":" in unit: #For cases with branch lengths
                                data = unit[0:unit.rfind(":")] #get rid of
trailing branchlength if provided
                                node_creater = Node(data, parent = base)
#Create node entry
                                node_creater.brl = float(unit[unit.
rfind(":")+1:]) #Append branch length of that branch
                                base.children.append(node_creater) #Create
children
                                self.newicksplicer(data, node_creater)
#Recursive function
                            else: #For case with no branch lengths
                                data = unit
                                node_creater = Node(data, parent = base)
                                base.children.append(node_creater)
                                self.newicksplicer(data, node_creater)
                        break #Terminate loop, we don't need to look any
further
```

```
    defcountsym(self, node):
        """
        Breaks tree into two halves at the root node and
passes a command to each node.
        """
        list = [] #List to hold output values
        for child in node.children:
                val = self.countsym2(child)
                nodes = self.nodecount (child)
                list.append((val, nodes))
        return list

    defnodecount(self, node):
        """
        Count number of nodes on tree
        """
        start = 0
        ifnode.children == []:
                return 0 #Terminal node returns 0 as there
are no nodes aboveit
        else:
                start += 1 #Any non-terminal node will add
1 to the total
                for child in node.children:
                        start += self.nodecount(child)
                return start

    def length(self, node):
        """
        Count length to each bifurcation
        """
        list = []
        ifnode.children == []: #Terminal branch returns
branch length
                returnnode.brl #Not sure about this part,
but without it the numbers are skewed
        else:
                try:
                        list.extend(node.brl)
                except:
                        list.append(node.brl)
                for child in node.children:
                        try:
                                list.extend(self.
length(child))
                        except:
                                list.append(self.
length(child))
                list = filter(lambda x:x != 0, list)
#Filter to remove zeros from the list
                return list
```

```
        defdivcount (self):
            """
            Tallies the lengths to give total diversification
rate
            """
            list = self.length(self.root)
            val = numpy.mean(list)
            returnval

    def summarize(self):
            """
            Summarize data
            """
            list = self.countsym(self.root)
            obs = list[0][0]+list[1][0]
            exp = max(list[0][0],list[1][0])*2
            print ("Number of observed species: ", obs)
            print ("Number of extinct species: ", exp-obs)
            print ("% extinct: ", (exp-obs)/(exp)*100)
            print ("chi: ", chi(f_obs = obs, f_exp = exp,
ddof = -(obs-1)))

"""
data =          #Provide newick here
sim = Tree(data) #Loads tree into program
sim.summarize() #Prints summary stats for the data
"""
```

APPENDIX 9.2   DRAFT (NOT ANNOTATED) R SCRIPT PROVIDED BY AN ANONYMOUS REVIEWER TO CALCULATE THE EXTINCTION INDEX OF THE CTNM ON SIMULATED TREES WITH VARYING EXTINCTION RATES (0, 0.025, 0.05, 0.1, 0.2, AND 0.8)

```
GITHUB link: https://github.com/subirshakya/Epi_Con_Est_
Extinction/blob/master/Graph_test.R
## function to compute asymmetry of root node
library(ips)
library(ape)
library(TreeSim)

# root asymmetry (ra) function
# 'tr' is a object of class 'phylo' from the ape library
ra<- function(tr)
{
  N<- Ntip(tr)
  rn<- N+1
  ee<- tr$edge
```

```
   ees<- ee[ee[,1]==rn, ]  # extract edges that stem from root
node
   desc.node<- ees[,2]
   if(any(desc.node< N)) { RS<- 1; LS<- N-1  # if either
descendant of root is a terminal tip
   } else {
     RN<- descendants(tr, node=desc.node[1], type="t")  # get
descdendants
     LN<- descendants(tr, node=desc.node[2], type="t")
     RS<- length(RN)
     LS<- length(LN)
   }

   ext<- abs(RS - LS) / (2* max(c(RS, LS))) # compute asymmetry
metric
   res<- c(ext, max(c(RS, LS)), min(c(RS, LS)))
   names(res)<- c("asym", "right", "left")

   # side with more tips is considered the "right" side of tree
   return(res)

}

nrep<- 1000
layout(1:3)

# no extinction
tr<- sim.bd.taxa(n=32, numbsim=nrep, lambda=1, mu=0)
rav<- t(sapply(tr, ra))
hist(rav[,1], 25, col="pink", main="No extinction")

# low extinction
tr<- sim.bd.taxa(n=32, numbsim=nrep, lambda=1, mu=0.2)
rav<- t(sapply(tr, ra))
hist(rav[,1], 25, col="pink", main="Low extinction, mu=0.2")

# high extinction
tr<- sim.bd.taxa(n=32, numbsim=nrep, lambda=1, mu=0.8)
rav<- t(sapply(tr, ra))
hist(rav[,1], 25, col="pink", main="High extinction, mu=0.8")
```

REFERENCES

Alfaro, M. E., F. Santini, C. Brock, H. Alamillo, A. Dornburg, D. L. Rabosky, G. Carnevale and L. J. Harmon. 2009. Nine exceptional radiations plus high turnover explain species diversity in jawed vertebrates. *Proceedings of the National Academy of Sciences USA* 106: 13410–13414.

Andersen, M. J., P. A. Hosner, C. E. Filardi and R. G. Moyle. 2015. Phylogeny of the monarch flycatchers reveals extensive paraphyly and novel relationships within a major Australo-Pacific radiation. *Molecular Phylogenetics and Evolution* 83: 118–136.

Berv, J. S. and R. O. Prum. 2014. A comprehensive multilocus phylogeny of the Neotropical cotingas (Cotingidae, Aves) with a comparative evolutionary analysis of breeding system and plumage dimorphism and a revised phylogenetic classification. *Molecular Phylogenetics and Evolution* 81: 120–136.

Chakrabarty, P., M. P. Davis, W. L. Smith, Z. H. Baldwin and J. S. Sparks. 2011. Is sexual selection driving diversification of the bioluminescent ponyfishes (Teleostei: Leiognathidae)? *Molecular Ecology* 20: 2818–2834.

Chan, K. M. A. and B. R. Moore. 2005. SYMMETREE: Whole-tree analysis of differential diversification rates. *Bioinformatics* 21: 1709–1710.

Collins, R. A., R. Britz and L. Rüber. 2015. Phylogenetic systematics of leaffishes (Teleostei: Polycentridae, Nandidae). *Journal of Zoological Systematics and Evolutionary Research* 53: 259–272.

Fink, S. V. and W. L. Fink. 1981. Interrelationships of the ostariophysan fishes (Teleostei). *Zoological Journal of the Linnean Society* 72: 297–353.

Frost, S. D. W., O. G. Pybus, J. R. Gog, C. Viboud, S. Bonhoeffer and T. Bedford. 2015. Eight challenges in phylodynamic inference. *Epidemics* 10: 88–92.

Fuchs, J., J. A. Johnson and D. P. Mindell. 2015. Rapid diversification of falcons (Aves: Falconidae) due to expansion of open habitats in the Late Miocene. *Molecular Phylogenetics and Evolution* 82: 166–182.

Grande, L. 2010. An empirical synthetic pattern study of gars (Lepisosteiformes) and closely related species, based mostly on skeletal anatomy. The resurrection of Holostei. *Copeia* 10(2A): 1.

He, F. and S. P. Hubbell. 2011. Species–area relationships always overestimate extinction rates from habitat loss. *Nature* 473: 368–371.

Heard, S. B. and A. O. Mooers. 2002. Signatures of random and selective mass extinctions in phylogenetic tree balance. *Systematic Biology* 51: 889–897.

Heath, T. A., J. P. Huelsenbeck and T. Stadler. 2015. The fossilized birth–death process: A coherent model of fossil calibration for divergence time estimation. *Proceedings of the National Academy of Sciences USA* 111: 2957–2966.

Hennig, W. 1966. *Phylogenetic Systematics*. Urbana, Illinois: University of Illinois Press.

Höhna, S., T. Stadler, F. Ronquist and T. Britton. 2011. Inferring speciation and extinction rates under different sampling schemes. *Molecular Biology and Evolution* 28: 2577–2589.

Huelsenbeck, J. P. and M. Kirkpatrick. 1996. Do phylogenetic methods produce trees with biased shapes? *Evolution* 50: 1418–1424.

Jetz, W., G. H. Thomas, J. B. Joy, K. Hartmann and A. Mooers. 2012. The global diversity of birds in space and time. *Nature* 491: 444–448.

Losos, J. B. and F. R. Adler. 1994. Stumped by trees? A generalized null model for patterns of organismal diversity. *American Naturalist* 145: 329–342.

Magallón, S. and M. J. Sanderson. 2001. Absolute diversification rates in angiosperm clades. *Evolution* 55: 1762–1780.

McMahan, C., P. Chakrabarty, W. L. M. Smith, J. S. Sparks and M. P. Davis. 2013. Temporal patterns of diversification across global cichlid biodiversity (Acanthomorpha: Cichlidae). *PLoS One* 8(e71162): 1–9.

Mooers, A. O. and S. B. Heard. 1997. Inferring evolutionary process from phylogenetic tree shape. *Quarterly Review of Biology* 72: 31–54.

Moore, B. R. and M. J. Donoghue. 2009. A Bayesian approach for evaluating the impact of historical events on rates of diversification. *Proceedings of the National Academy of Sciences USA* 106: 4307–4312.

Morlon, H. 2014. Phylogenetic approaches for studying diversification. *Ecology Letters* 17: 508–525.

Morlon, H., T. L. Parsons and J. B. Plotkin. 2011. Reconciling molecular phylogenies with the fossil record. *Proceedings of the National Academy of Sciences USA* 108: 16327–16332.

Near, T. J., A. Dornburg, R. C. Harrington, C. Oliveira, T. W. Pietsch, C. E. Thacker, T. P. Satoh, E. Katayama, P. C. Wainwright, J. T. Eastman and J. M. Beaulieu. 2015. Identification of the notothenioid sister lineage illuminates the biogeographic history of an Antarctic adaptive radiation. *BMC Evolutionary Biology* 15: 109.

Near, T. J., C. M. Bossu, G. S. Bradburd, R. L. Carlson, R. C. Harrrington, P. R. Hollingsworth Jr., B. P. Keckand and D. A. Etnier. 2011. Phylogeny and temporal diversification of darters (Percidae: Etheostomatinae) *Systematic Biology* 60: 565–595.

Nee, S. 2006. Birth–death models in macroevolution. *Annual Review of Ecology Evolution and Systematics* 37: 1–17.

Nee, S., E. C. Holmes, R. M. May and P. H. Harvey. 1994a. Extinction rates can be estimated from molecular phylogenies. *Philosophical Transactions of the Royal Society of London B* 344: 77–82.

Nee, S., R. M. May and P. H. Harvey. 1994b. The reconstructed evolutionary process. *Philosophical Transactions of the Royal Society of London B* 344: 305–311.

Oaks, J. R. 2011. A time-calibrated species tree of Crocodylia reveals a recent radiation of the true crocodiles. *Evolution* 65: 3285–3297.

Paradis, E. 2003. Analysis of diversification: combining phylogenetic and taxonomic data. *Proceedings of the Royal Society of London B* 270: 2499–2505.

Pearson, P. N. 1998. Speciation and extinction asymmetries in paleontological phylogenies: Evidence for evolutionary progress. *Paleobiology* 24: 305–335.

Pennell, M. W. and L. J. Harmon. 2013. An integrative view of phylogenetic comparative methods: Connections to population genetics, community ecology, and paleobiology. *Annals of the New York Academy of Sciences* 1289: 90–105.

Rabosky, D. L. 2006. Likelihood methods for detecting temporal shifts in diversification rates. *Evolution* 60: 1152–1164.

Rabosky, D. L. 2010. Extinction rate should not be estimated from molecular phylogenies. *Evolution* 64: 1816–1824.

Rabosky, D. L. 2014. Automatic detection of key innovations, rate shifts, and diversity dependence on phylogenetic trees. *PLoS One* 9: e89543.

Rabosky, D. L. and E. E. Goldberg. 2015. Model inadequacy and mistaken inferences of trait-dependent speciation. *Systematic Biology* 64: 340–355.

Rabosky, D. L. and I. J. Lovette. 2008. Density dependent diversification in North American wood-warblers. *Proceedings of the Royal Society of London B* 276: 995–997.

Rabosky, D. L., F. Santini, J. T. Eastman, S. A. Smith, B. Sidlauskas, J. Chang and M. E. Alfaro. 2013. Rates of speciation and morphological evolution are correlated across the largest vertebrate radiation. *Nature Communications* 4: 1–8.

Régnier, C., G. Achaz, A. Lambert, R. H. Cowie, P. Bouchet and B. Fontaine. 2015. Mass extinction in poorly known taxa. *Proceedings of the National Academy of Sciences USA* 112: 7761–7766.

Revell, L. J., D. L. Mahler, R. G. Reynolds and G. J. Slater. 2015. Placing cryptic, recently extinct, or hypothesized taxa into an ultrametric phylogeny using continuous character data: A case study with the lizard *Anolisroosevelti*. *Evolution* 69: 1027–1035.

Ricklefs, R. E. 2007. Estimating diversification rates from phylogenetic information. *Trends in Ecology and Evolution* 22(11): 601–610.

Rosenblum, E. B., B. A. J. Sarver, J. W. Brown, S. D. Roches, K. M. Hardwick, T. D. Hether, J. M. Eastman, M. W. Pennell and L. J. Harmon. 2012. Goldilocks meets Santa Rosalia: An ephemeral speciation model explains patterns of diversification across time scales. *Evolutionary Biology* 39: 255–261.

Sheldon, F. H., L. A. Whittingham, R. G. Moyle, B. Slikas, and D. W. Winkler. 2005. Phylogeny of swallows (Aves: Hirundinidae) estimated from nuclear and mitochondrial DNA sequences. *Molecular Phylogenetics and Evolution* 35: 254–270.

Simpson, G. G. 1944. *Tempo and Mode of Evolution.* New York, NY: Columbia University Press.

Simpson, G. G. 1953. *The Major Features of Evolution.* New York: Columbia University Press.

Slager, D. L., C. J. Battey, R. W. Bryson, G. Voelker and J. Klicka. 2014. A multilocus phylogeny of a major New World avian radiation: The Vireonidae. *Molecular Phylogenetics and Evolution* 80: 95–104.

Slowinski, J. B. and C. Guyer. 1989. Testing the stochasticity of patterns of organismal diversity: An improved null model. *American Naturalist* 134: 907–921.

Slowinski, J. B. and C. Guyer. 1993. Testing whether certain traits have caused amplified diversification—An improved method based on a model of random speciation and extinction. *American Naturalist* 142: 1019–1024.

Thompson, A. W., R. R. Betancur, H. Lopez-Fernandez and G. Orti. 2014. A time calibrated, multi-locus phylogeny of piranhas and pacus (Characiformes: Serrasalmidae) and a comparison of species tree methods. *Molecular Phylogenetics and Evolution* 81: 242–257.

Wright, J. J., S. R. David and T. J. Near. 2012. Gene trees, species trees, and morphology converge on a similar phylogeny of living gars (Actinopterygii: Holostei: Lepisosteidae), an ancient clade of ray-finned fishes. *Molecular Phylogenetics and Evolution* 63: 848–856.

10 Donn Rosen and the Perils of Paleontology

Lance Grande
The Field Museum

CONTENTS

10.1 INTRODUCTION

My career was hugely influenced by Donn Rosen and his close colleagues, Colin Patterson and Gary Nelson. Like Lynne Parenti, I was in the PhD program at the American Museum of Natural History (AMNH) during the height of the cladistics revolution in systematic biology. Rosen and Nelson were my graduate co-advisors. It was an exciting time to be in systematic biology, and New York was where it was all happening in the late 1970s and early 1980s. Donn was the paternal figure in the Ichthyology Department of the AMNH to both students and staff. He died far too young, but he left an impressive legacy of students who took curatorial positions in major natural history museums across the world.

In the spring of 2015, Lynne sent me a copy of Donn's list of 32 "Assumptions that inhibit scientific progress in comparative biology" that serves as a core to this volume and is reproduced in Chapter 1. After reading the list I was struck by how clearly I could remember Donn making these points in his lectures and seminars, particularly as they pertained to systematics and phylogenetic research. As a student, I came to appreciate his subversive challenge of dogmatic principles, particularly those regarding paleontology. As a professional paleontologist, I continued to appreciate these ideas even more. Here I discuss five of the points Donn thought to inhibit progress in comparative biology, particularly with regard to evolutionary theory. These are: *fossils specify the age of their including taxon* (Rosen inhibiting assumption 7), *the stratigraphic sequence never lies* (Rosen inhibiting assumption 6), *punctuated equilibrium rescues the imperfections of the fossil record from criticism and is testable* (Rosen inhibiting assumption 4), *it is important to search for ancestors* (Rosen inhibiting assumption 5), and *pure paleontology provides essential information about the history of life* (Rosen inhibiting assumption 3).

These inhibiting assumptions fall into two basic categories: those that overestimate the quality and completeness of the fossil record (4, 6, and 7), and those that over embellish the power of paleontology for studies of biohistory (3 and 5).

10.2   THE EARLY DEVELOPMENT OF PALEONTOLOGY AS A SCIENCE

Over the last several hundred years, paleontology emerged as an important field of science. As with all scientific specialties, its fundamental tenets evolved over time. The breakthrough concept of one period contained problematic elements for the next. All scientific progress must be evaluated within its own time in history to be appreciated. Prior to the seventeenth century, the nature of fossils was unclear. Fossils had been known for thousands of years, but they had been interpreted mainly as the remains of extant species, or even geologic features of inorganic origin. During the seventeenth and eighteenth centuries, the nature of fossils to past life developed gradually. Georges Cuvier (1769–1832) made a huge advance by establishing the reality of extinction. Fossils were finally recognized as species that were no longer living on Earth (e.g., Cuvier 1813). Although Cuvier's work led to the emergence of Paleontology as a science, he was a Creationist. He believed that extinct species in the fossil record were the result of periodic catastrophic extinction events followed by successive creation events. He did not accept the possibility that species could evolve or be modified in form over time. To him, fossil species demonstrated only that there were alternating cycles of mass extinction and creation. In fact, he saw his concept of extinction as in opposition to any concept of biological evolution (Rudwick 1972). This was far from the later views of Charles Darwin, who would integrate extinction into a theory of evolution. The history of paleontology between Cuvier and Darwin is complex, involving many steps and influential paleontologists. This history is covered in detail by Rudwick (1976) so I will jump ahead to post-Darwinian time.

Once Darwin brought evolutionary theory into broad acceptance within the scientific community, fossils were seen as part of the evolutionary tree of life. This was another important advance in paleontology. Through comparative anatomy, fossils were now seen as forming a natural pattern with living species that could be explained by evolution. It did not matter whether species were extinct or living; they could all be tied together, as in Darwin's first known sketch of a phylogenetic tree, reproduced here in Figure 10.1. But the expectations of fossils and an extrapolation of their significance to evolutionary studies grew over time. Fossils were eventually seen by many paleontologists of the mid to late twentieth century as necessary for the effective study of evolutionary relationships and superior to extant taxa for such purposes. This was particularly true for paleomammalogists. Simpson (1961:83) stated that "fossils provide the soundest basis for evolutionary classification," and that mammal classifications expressing evolutionary relationships "have come to depend more on fossils than on recent animals." Fossils were thought to offer the way to identify specific ancestors. Gingrich (1979:58) stated "Paleontology is uniquely situated for study of speciation in natural populations in a natural environment because of the time dimension contributed by the fossil record." He proposed

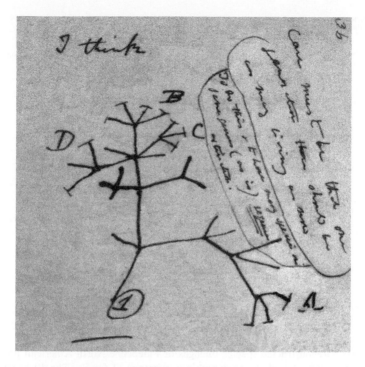

FIGURE 10.1 A sketch from Darwin's 1837 "Notebook B" where he makes his first branching diagram of evolutionary relationship. The species are clustered into groups of genera, and above the sketch he wrote the words "I think".

the *stratophenetic approach* to identify ancestors using fossil species from a single locality that are morphologically similar (at least with regard to those fragments that were preserved) and interpret the older species as the ancestor of the younger species (Gingrich 1979:56). Some of the fossil species in his example using fossil primates as in Figure 10.2 were represented only by teeth and jaw fragments. Other paleontologists also began to look for ancestor–descendant lineages in the fossil record. Some saw fossils as graphic representations of evolution from which evolutionary lineages could be read, like some sort of stop motion movie recorded in the rocks as in Figure 10.3.

10.3   COMING TO GRIPS WITH THE LIMITATIONS OF THE FOSSIL RECORD

While fossils were being promoted as essential to studies of phylogeny and organismal classification by some, they were seen as far less critical by others. Løvtrup (1977:21) expressed the opinion that "the discovery of a new fossil has no impact on classification" and "if, with the wealth of information … from living animals we cannot arrange [vertebrate animals] in a correct phylogenetic system, then no amount of fossil data will ever help us approach this goal." Colin Patterson, recognized as

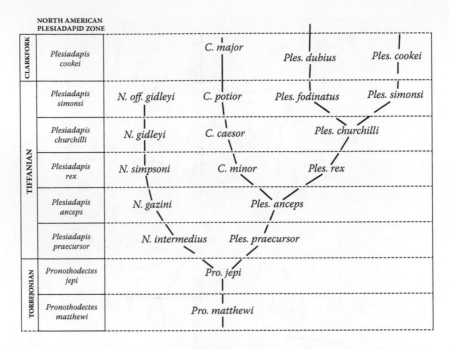

FIGURE 10.2 Reconstructed species lineages of Paleocene North American primates of the family Plesiapidae based on stratigraphic occurrence of fossils. The 19 fossil species are in the genera *Pronothodectes*, *Plesiadapis*, *Nannodectes*, and *Chironyoides*, and most are represented only by isolated teeth and jaw elements. Gingrich proposed this method of lineage reconstruction, which is called stratophenetics. (From Gingrich, P., *Univ. Michigan Pap. Paleontol.* 15, 1976, 1–140.)

the leading fish paleontologist of the twentieth century (Bonde 2000), concluded that "instances of fossils overturning theories of relationships based on Recent organisms are very rare" (Patterson 1981; Grande 2000). In his book *Evolution* (1978:133), Patterson stated "Fossils can tell us many things, but one thing they can never tell us is whether they were ancestors of anything else," a sentiment echoed by Donn Rosen in many presentations and publications (Rosen et al. 1981). As someone who has been a professional paleontologist for more than three decades, I too am convinced of this.

One problem with trying to reconstruct detailed lineages from the fossil record is that the record is far too incomplete. Logistically, this will always be true. It has been estimated that well over five billion species have become extinct on this planet (Raup 1991; Kunin and Gaston 1997). There are only about 250,000 extinct species described as fossils (e.g. Prothero 1999), which translates to less than 0.05% of the Earth's extinct species (i.e., less than 1 out of every 2,000 extinct species) known from the fossil record. The problem of missing taxa is compounded by missing data on many of the fossils that do exist. Many fossil species are very poorly preserved leaving most of their anatomy unknown.

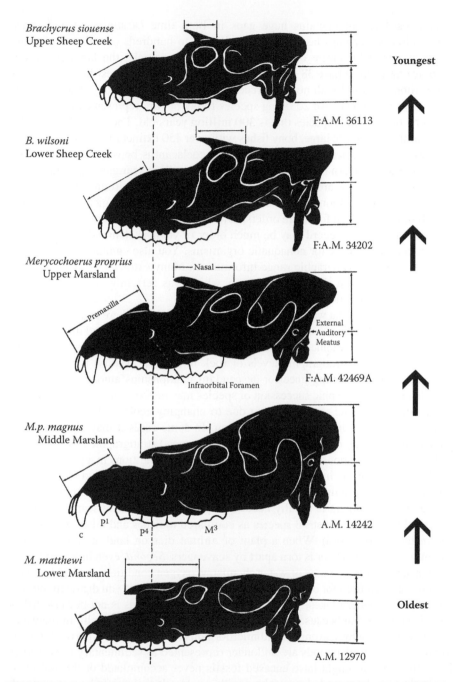

FIGURE 10.3 "Phylogenetic development" of oreodont mammals through the Miocene Marsland and Sheep Creek Formations of Nebraska, based primarily on stratigraphy. Schultz and Falkenbach called this their "stratigraphic approach" to systematics. (From Schultz, C. and Falkenbach, C., *Bull. Am. Mus. Nat. Hist.* 139, 427, 1968, with my arrows added for clarity.)

The fossil record contains huge gaps in both time (stratigraphy) and space (geography). Major gaps lasting tens of millions to hundreds of millions of years are documented for entire orders of animals. For example, the lampreys (order Petromyzontiformes) have 38 living species today but the youngest fossil species is 120 million years old with no fossil record between then and today. The hagfishes (order Myxiniformes) have 20 living species today but are represented in the fossil record by only a single species that is 300 million years old. The coelacanths (order Coelacanthiformes) are large, bony fishes with over 150 extinct fossil species known plus two living species. There are no known coelacanths between the end of the Cretaceous and modern times, leaving a classic 65 million year gap. Hagfishes, lampreys, and coelacanths are all aquatic animals; they live in environments even more conducive to fossilization than those of animals on land. The logical consequence of such time gaps, even for entire orders, is that many taxa (even species) of animals or plants in the fossil record may be much older or younger than indicated by their fossil record, especially for nonaquatic organisms. You also cannot trust the fossil record for detailed geographic range information. Many extinct species are known by fossils from only a single rock outcrop, which was certainly only a fraction of their former geographic ranges.

The fossil record that we know is only a minute sample of where and when extinct plant and animal species existed through time, or what species existed through time. Consequently, stratigraphic succession of species (i.e., a sequence of species moving up through the stratigraphic levels of rock in a given locality) is of relatively little use in reconstructing ancestor-descendant relationships among species. In a given locality, stratigraphic succession of species may be the result of closely related species moving in and out of the area due to changing ecological conditions over time rather than evolutionary transformation. In some cases it may also represent inadequate sampling of variation within some (or most) stratigraphic levels, or even cyclic changes in response to changing environmental conditions in a given area.

The reasons for the incompleteness of the fossil record are many, and the conditions for a dead organism to become fossilized are both rare and complex. *First*, an aquatic environment is usually needed (explaining the relative scarcity of well-preserved fossilized terrestrial species as compared to fishes and other aquatic species in the fossil record). When a plant or animal dies on land, its body usually decomposes completely or is torn apart by scavengers. *Second*, even in aquatic environments, rare combinations of water chemistry, oxygen conditions, and sedimentation rates are required for dead organisms to become fossils. If you dig deep into the mud of many modern lakes, you will find no fish bones. This is not because fishes never lived there, but because the skeletons either dissolved chemically in the water or were consumed by scavengers and microorganisms. *Third*, at any particular location today, most time periods are no longer represented as sedimentary rock. Either the sediments that might have encased fossils never accumulated or the sedimentary rocks were later eroded away along with any fossils they might have contained. *Lastly*, many extinct species that became fossils may never be discovered due to their inaccessibility.

Clearly, fossils do not specify the absolute age range for the taxon they represent (pertaining to Rosen's inhibiting assumption 7), because fossils do not indicate anything close to the complete record of their range or existence through time. Gaps of 100 million years or more are possible for taxa. Also, because fossil localities represent only a small fraction of the complete geographic range of the species they represent, their absence should not be used to reject biological theories of area relationship. Another inescapable conclusion we come to, based on the incompleteness of the fossil record, is that the stratigraphic sequence can lie (pertaining to Rosen's inhibiting assumption 6). I mean this in the sense that the stratigraphic sequence can be over-interpreted. Just because the fossil record of one lineage is older than that of another lineage, it does not dictate a phylogenetic branching sequence, lineage, or even the extinction of a taxon. For a given fossil locality, the stratigraphic sequence may be showing a succession of species migrating in and out of an area in response to changing ecological conditions or artifacts of inadequate sampling.

Finally, punctuated equilibrium does not rescue the imperfections of the fossil record from criticism (pertaining to Rosen's inhibiting assumption 4). Punctuated equilibrium (Eldredge and Gould 1972, after a similar theory by Tremaux 1865) is a theory in evolutionary biology—which proposes that evolution occurs in relatively sudden jumps rather than gradually, thus explaining the lack of gradual transition forms. Once species appear in the fossil record they show little net evolutionary change for most of their geological history. When significant evolutionary change occurs, the theory proposes that it is generally restricted to rare and geologically rapid events of branching speciation called cladogenesis. Cladogenesis is the process by which a species splits into two distinct species, rather than one species gradually transforming into another. Punctuated equilibrium may very well have been the reason for some of the gaps, but how could you distinguish it from the many other possible explanations for a given gap?

Even in an imaginary world where there was a complete fossil record of all species leading to our present day biodiversity, I cannot imagine how it would be possible to compile and interpret such a massive volume of data. That would mean billions of morphotypes mapped over millions of different geographic ranges and a project of infinite complexity. Today we have not yet even managed to compile data for all *living* species.

10.4   THE POWER OF FOSSILS TO IDENTIFY ANCESTORS

As explained above, fossils do not show us specific ancestors (pertaining to Rosen's inhibiting assumption 5). There are even practical limitations on the comparability of fossil and living taxa at the "species" level. The biological species concept is the most commonly used to demark a species among living taxa. We do not consider certain types of morphological variation to indicate species differences (e.g., the vastly different morphotypes of the domestic dog, changing morphotypes due to cyclic ecological pressures on local populations, changing morphotypes due to ontogenetic development, and differences due to sexual dimorphism). These types of non-taxonomic morphological differences are difficult to impossible to distinguish in the fossil record, particularly for species based on incomplete fragments. There is an empirical gap that must be considered when comparing the validity of extant species to fossil species. So how then do we make an equivalent unit of comparison between fossil and living taxa for phylogenetic analysis? And considering the uncertainty of species validity for fossil taxa together with the incompleteness of the fossil record, how can identification of ancestors based on stratigraphy and taxa represented by highly fragmentary material be more than speculation? We can still integrate fossil and living organisms at a more general level to look at general interrelationships, but slight changes in morphology over time from a few fossil localities does not prove anagenesis. They *may* be the result of anagenesis, but there is no convincing way to demonstrate that scientifically.

Thanks to the early influence of Brundin (1966) and Hennig (e.g., 1966), and later systematists who promoted and expanded these ideas (e.g., Nelson and Platnick 1981) the focus of biohistory and paleontology is no longer a search for the ancestor, it is the search for relative relationships (the sister group), whether fossil or living. Not being able to identify specific ancestor–descendent lineages of species is basically irrelevant to modern evolutionary research, although a few still debate that today. Many ancestral species no doubt exist among the fossils that we know of today, but we simply have no satisfactory way to identify which ones they are. I know for a fact that my daughter's direct ancestor is her mother, because I saw her birth. But if you never observed that process, and 50 million years from now you had only the adult fossil of my daughter and the adult fossil of her mother, you would have no way of knowing who was whose ancestor (e.g., which one was the mother and which one was the daughter). Nevertheless, even without knowledge of the exact ancestor–descendant connection you could still hypothesize a pattern of relationships among my daughter, her mother, and a fossil chimp. My daughter and her mother share unique characteristics not shared by the chimp indicating that they are more closely related to each other than either is to the chimp. The problem is similar for identification of an ancestor–descendent species pair in the fossil record. Although we can use scientific methods to hypothesize that a species is more closely related to one species than to another, identification of specific ancestral species with their descendent species is a speculative guess at best.

Although pure paleontology provides valuable information about the history of life, it is not "essential" to evolutionary classification. Phylogenetic research is perfectly able to proceed without it (pertaining to Rosen's inhibiting assumption 3). Although not essential in that regard, paleontology is nevertheless valuable in providing missing pieces in the history of life on this planet and enriching our understanding of past biodiversity.

10.5   THE CONTRIBUTIONS OF PALEONTOLOGY TO THE STUDY OF BIOHISTORY

I do not want to leave the impression that I believe paleontology is unimportant in the study of biohistory. We have come a long way in the last 30 years in evaluating the importance of paleontology in the study of phylogeny. Although the power of paleontology has been over embellished at times with regard to its power to identify ancestors, it has much to offer. The fossil record still constitutes evidence supporting the theory of evolution at a general level. Paleontology tells us that life originated billions of years ago on this planet, diversified over time, and that millions of species did not survive to the present day. Also, there is basic order to broad levels of evolutionary complexity. The oldest fossils are only single-celled organisms. Simple multiple-celled organisms are found in rocks much older than any rocks containing multiple-celled organisms with heads. Organisms with heads are found in rocks much older than any rocks containing animals with heads and limbs. Animals with heads and limbs are found in rocks much older than any rocks containing animals with heads, limbs, and hair. Animals with heads, limbs, and hair are found in rocks older than any rocks containing animals with heads, limbs, and hair that walk on two legs. Many such examples of increasing general complexity exist in the fossil record. As Neil Shubin (2008) puts it, "If, digging in 600-million-year-old rocks, we found the earliest jellyfish lying next to the skeleton of a woodchuck, then we would have to rewrite our textbooks." We will never find a 600-million-year-old woodchuck, and I stake my reputation on that prediction.

Fossil species also provide many new character combinations that add richness and depth to our understanding about the history of life. Occasionally, a fossil taxon may even overturn a previous phylogenetic pattern based on extant taxa because of a unique character combination that changes a former most-parsimonious conclusion (e.g., Grande 2010:825–827). Fossils frequently extend the geographic range of extant taxa that have become regionally extinct (e.g., Grande 1985; Grande and Bemis 1998:636–643; Grande 2010:812–815). And sometimes fossils produce character combinations that further validate phylogenetic predictions (e.g., "missing links" such as *Archaeopteryx* and feathered dinosaurs).

I will always appreciate Donn's lessons of self-criticism in science, as well as those of Patterson and Nelson. It is necessary to reassess our instances of overreaching on occasion. Recognizing the limitations of our work in science is part of progress and part of what gives our work strength in the end.

REFERENCES

Bonde, N. 2000. Colin Patterson: The greatest fish paleobiologist of the 20th century. *The Linnaean Society*, Special Issue 2:33–38.

Brundin, L. 1966. Transantarctic relationships and their significance, as evidenced by chironomid midges with a monograph of the subfamilies Podonominae and Aphroteniinae and the austral Heptagyiae. *Kungliga Svenska Vetenskapakademiens Handlingar*, Fjarde Serien 11:1–472.

Cuvier, G. 1813. *Essay on Theory of the Earth*. London: Natural History Museum.

Darwin, C. 1859. *On the Origin of Species by Means of Natural Selection, or the Preservation of Favored Races in the Struggle for Life*. London: John Murray Publisher. 502pp.

Eldredge, N. 2005. *Darwin: Discovering the Tree of Life*. New York: W.W. Norton and Company. 256pp.

Eldredge, N. and S. Gould. 1972. Punctuated equilibrium: An alternative to phyletic gradualism. In: *Models in Paleobiology*. Schopf, T. J. M. (ed.). San Francisco: Freeman, Cooper and Co. pp. 82–115.

Gingrich, P. 1976. Cranial anatomy and evolution of early tertiary plesiadapidae. *University of Michigan Papers on Paleontology* 15:1–140.

Gingrich, P. 1979. The stratophenetic approach to phylogenetic reconstruction in vertebrate paleontology. In: *Phylogenetic Analysis and Paleontology*. Cracraft, J. and N. Eldredge (eds.). New York: Columbia University Press. pp. 113–163.

Grande, L. 1985. The use of paleontology in systematics and biogeography, and a time control refinement for historical biogeography. *Paleobiology* 11:234–243.

Grande, L. 2000. Fossils, phylogeny, and Patterson's rule. In: *Colin Patterson (1933–1998): A Celebration of His Life*. Forey, P., B. Gardiner and C. Humphries (eds.) The Linnaean, Special Issue No. 2. The Linnean Society of London. London: Academic Press. pp. 24–32.

Grande, L. 2010. *An Empirical Synthetic Pattern Study of Gars and Closely Related Species (Lepisosteiformes) Based Mostly on Skeletal Anatomy: The Resurrection of Holostei*. American Society of Ichthyologists and Herpetologists. Special Publication 7, Lawrence, Kansas: Allen Press, 874pp.

Grande, L. and W. E. Bemis. 1998. A comprehensive phylogenetic study of amiid fishes (Amiidae) based on comparative skeletal anatomy. An empirical search for interconnected patterns of natural history. Society of Vertebrate Paleontology Memoir 4:i–x, 1–690; supplement to *Journal of Vertebrate Paleontology* 18(1).

Hennig, W. 1966. *Phylogenetic Systematics*. Urbana: University of Illinois Press. 263pp. [Translated book summarizing methods published in earlier papers by Hennig that influenced Brundin.]

Kunin, W. and K. Gaston (eds.) 1997. *The Biology of Rarity: Causes or Consequences of Rare-Common Differences*. London: Chapman and Hall. 279pp.

Løvtrup, S. 1977. *The Phylogeny of Vertebrata*. New York: Wiley. 330pp.

Nelson, G. and N. Platnick. 1981. *Systematics and Biogeography: Cladistics and Vicariance*. New York: Columbia University Press. 567pp.

Patterson, C. 1978. *Evolution*. London: British Museum (Natural History) and Routledge & Kegan Paul. 197pp. [Second edition published in 1999 by the Natural History Museum, London, 166pp.]

Patterson, C. 1981. The significance of fossils in determining evolutionary relationships. *Annual Review of Ecology and Systematics* 12:195–223.

Prothero, D. 1999. *Fossil Record. Encyclopedia of Paleontology*. Chicago: Fitzroy Dearborn Publishers. pp. 490–492.

Raup, D. 1991. *Extinction: Bad Genes or Bad Luck?* New York: W.W. Norton.

Rosen, D., P. Forey, B. Gardiner and C. Patterson. 1981. Lungfishes, tetrapods, paleontology, and plesiomorphy. *Bulletin American Museum of Natural History* 167:159–276.

Rudwick, M. 1976. *The Meaning of Fossils. Episodes in the History of Paleontology*. Chicago: The University of Chicago Press. 287pp.

Schultz, C. and C. Falkenbach. 1968. The phylogeny of oreodonts. *Bulletin of the American Museum of Natural History* 139:1–498.

Shubin, N. 2008. *Your Inner Fish. A Journey into the 3.5-Billion-Year-Old History of the Human Body*. New York: Pantheon Books. 230pp.

Simpson, G. G. 1961. *Principles of Animal Taxonomy*. New York: Columbia University Press.

Tremaux, P. 1865. *Origine et Transformations de l'Homme et des Autres Etres*. Paris: Hachette.

11 Biogeographic Origin of Mainland *Norops* (Squamata: Dactyloidae)

Kirsten E. Nicholson
Central Michigan University

Craig Guyer
Auburn University

John G. Phillips
University of Tulsa

CONTENTS

11.1 INTRODUCTION

Over a career that challenged many concepts held as conventional wisdom to other biologists, Donn Rosen formulated a number of insightful ideas, the impact of which have inspired and laid the groundwork for many contemporary scientists. His ideas are still relevant and controversial, as evidenced by the content of this volume. One topic of particular interest to Rosen was the complex biogeographic history of Central America and the Caribbean, an interest that was influenced by the writings of Leon Croizat (e.g., Croizat 1962), and that eventually led to publication of Rosen's seminal paper in biogeographic theory and practice (Rosen 1976). Rosen's (1985) visit to the University of Miami, where he presented the list of axioms guiding contributions to this volume, and an earlier visit to the University of Southern California were crucial to the development of ideas for a vicariance explanation for the distribution of anoles (Dactyloidae) presented in Guyer and Savage (1986). Here, we return to the same group of Neotropical squamates and the explanation of vicariance versus dispersal for distribution patterns of the genus *Norops* across Central and South America. Our contribution evaluates Rosen's notions that the axioms "all distributions result from vicariance" and "all distributions result from dispersal" inhibit scientific progress.

Anoles are a species-rich family of squamates found throughout the Neotropics and have held a principal place in ecological and evolutionary studies (see Losos 2009 and sources within). Many biologists have demonstrated an affinity toward this group, providing an impressive breadth of literature on anole biology, tackling major concepts such as ecomorphology and adaptive radiations (Losos 1994, 2009 and references therein; Losos et al. 1998; Williams 1972, 1983), biogeographic hypotheses (Nicholson et al. 2005), and genomic advances (Alföldi et al. 2011). However, the majority of this work has focused on Caribbean species, leaving the origin and distribution of mainland anoles poorly understood, and their biogeographic history confused. Anole biologists have provided discussion on the origin of mainland *Norops* (Guyer and Savage 1986; Nicholson et al. 2012), but no studies have explicitly tested the biogeographic origin of the group. Nicholson et al. (2012) reconstructed a phylogeny across anoles that included many *Norops* species, but performed no additional tests to determine any information regarding biogeographic histories within this clade or in regards to anoles as a whole. However, Nicholson et al. did make inferences about the origin of *Norops* using their reconstructed phylogeny.

Mainland *Norops* refers to a monophyletic assemblage of species primarily found within continental South and Central America and belonging to the *N. auratus* species group (defined in Nicholson et al. 2012), excluding the clades of *Norops* found on Cuba (*N. sagrei* series) and Jamaica (*N. valencienni* series; Figure 11.1). The *N. auratus* group comprises more than 150 species distributed throughout the mainland and is resolved as a monophyletic lineage in all recent studies (Jackman et al. 1999; Poe 2004; Nicholson et al. 2005, 2012; Alföldi et al. 2011; Pyron et al. 2013). To date, only three studies have investigated mainland biogeographic relationships of *Norops* (Vanzolini and Williams 1970; Glor et al. 2001; Phillips et al. 2015), but all three have had taxon sampling restricted to smaller subclades within the group, the observed patterns of which may not be relevant to the broader group. Phillips et al. (2015) investigated phylogeographic patterns within the *N. humilis* species complex (widespread across Mesoamerica), while Vanzolini and Williams (1970) and Glor et al. (2001) investigated patterns within South American groups. Additional studies have conducted phylogeographic reconstructions of other clades of mainland anoles (*Norops* and *Dactyloa*; e.g., Castañeda and de Queiroz 2013; Prates et al. 2015; Guarnizo et al. 2015), but did not conduct heuristic analyses in order to investigate biogeographic patterns within mainland anoles. The combined results of these studies could be used as a potential model for testing biogeographic patterns in other species complexes.

Older biogeographic studies concerning all anoles (Guyer and Savage 1986, 1992; Williams 1989) proposed that mainland *Norops* originated in the northern part of its current range and colonized southwards, ultimately invading South America after the closure of the Panamanian Portal (Figure 11.2), a biogeographic event that was believed, at that time, to have occurred ~3.5 mya. However, subsequent studies presented results inconsistent with a north-to-south colonization pattern. Guyer and

FIGURE 11.1   The distribution of *Norops* anoles. Colors indicate three recognized subclades: red = *N. sagrei* clade on Cuba, green = *N. valencienni* clade on Jamaica, and blue = *N. auratus* mainland clade.

Savage (1986, 1992) first noted that mainland *Norops* forms a monophyletic group, and inferred that this clade resulted from sequential vicariance of the *N. sagrei* (eastern Cuba separates from nuclear Central America) and *N. valencienni* (Jamaica separates from nuclear Central America) groups, leaving the *N. auratus* group in nuclear Central America. Later, Glor et al. (2001) proposed an alternative hypothesis to a Central American origin of the *N. auratus* group finding instead a basal split within the *N. auratus* species group, yielding one lineage of South American origin and another of Central American origin, a pattern confirmed by Nicholson et al. (2005, 2012) with more extensive taxon sampling. More importantly, Glor et al. (2001) estimated the date of origin for *N. nitens*, a species complex of endemic

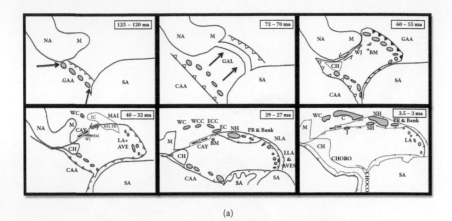

(a)

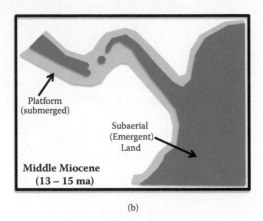

(b)

FIGURE 11.2 Characterizations of the geologic history of Central America. (a) Older reconstructions used by previous anole biogeographers (Guyer and Savage 1986; Nicholson et al. 2012). (b) Recent reconstructions by Montes et al. (2015) reproduced here.

South American mainland anoles, at ~15 mya, too old to be consistent with the traditional hypothesis that South American *Norops* must have originated after the closure of the Panamanian Portal ~3.5 mya (but see below). In short, the observed patterns in accumulating recent information (i.e., placement of endemic South American lineages at the base of the tree; date of origin for them as ~15 mya) are unexpected if (a) the general pattern is expected to be a north-to-south distribution and (b) if South American endemic species could not have originated prior to a relatively recent closure of the Panamanian Portal.

To explain better these inconsistent biogeographic interpretations, Nicholson et al. (2012) proposed an alternative hypothesis: that *Norops* originated earlier than previously hypothesized, and established a widespread distribution prior to the separation of North and South America by the eastward movement of the Greater Antillean Land bridge (hereafter GAL). When the GAL moved eastward and separated the

Americas (beginning ~55 mya, see Figure 11.2), some members of the ancestral *Norops* lineage remained in North and in South America. They also proposed that northern *Norops* members moved southwards as blocks moved into place reconnecting North and South America (Figure 11.2a). Recent geologic work provides evidence for much earlier reconnections between North and South America, pushing back initial modern connections to 23 mya with a final closure of the Panamanian Portal being 15 mya (Montes et al. 2012a,b, 2015). These older dates help to explain some of the inconsistent biogeographic conclusions of previous studies, but focused studies specifically testing these ideas are still lacking.

Given the discordance between the traditional paradigm (north-to-south movement, invasion of South America after closure of the Panamanian Portal) and recent evidence regarding patterns of dispersal in mainland *Norops*, we sought to test further the timing and patterns hypothesized by previous biogeographic hypotheses for mainland anoles. Our specific objectives were to estimate the dates of major clades within the mainland anole tree to determine if: (a) an endemic South American lineage predated the closure of the Panamanian Portal (any of the proposed dates from the literature) and (b) mainland *Norops* were present in North and South America prior to the proposed date of separation of North and South America by the eastward plate movement and separation from the GAL. In addition, we investigated the north-to-south colonization pattern purported to have occurred as the blocks moved into place reconnecting North and South America.

11.2   DO ENDEMIC SOUTH AMERICAN *NOROPS* PREDATE THE CLOSURE OF THE PORTAL?

To examine this question we generated a time-calibrated phylogenetic tree based on accumulated information from seven mtDNA genes (see Appendix 11.1 for methods) that allowed for the broadest coverage of species with no missing data. Our taxon sampling included a total of 117 lineages, 80 ingroup, and 37 outgroup (10 non-dactyloid, 8 representing other genera of anoles, and 19 representing species of Caribbean *Norops*). Sequences for 87 species were available from GENBANK and we present new data for 30 taxa (see Appendix 11.2 for a list of species and sources of data). The phylogeny estimated from these data was then used to reconstruct ancestral areas (LaGrange analysis). We used eight geographic areas, three in Central America, four in South America, and one connecting Central and South America (Figure 11.3). As in previous studies, our phylogenetic tree recovered three main lineages of mainland *Norops* (Figure 11.4). Unfortunately, recent phylogenetic reconstructions (e.g., Nicholson et al. 2005, 2012) have found little concordance between traditional recommendations of beta taxonomy within *Norops* (Etheridge 1959; Williams 1976) and lineages recovered by recent phylogenetic estimations such that taxonomic names for major lineages are lacking. To rectify this problem, we follow Nicholson et al. (2012) in recognizing all mainland *Norops* as belonging to a monophyletic *N. auratus* species group and erect three formal series within this species group. The *N. chrysolepis* series represents a monophyletic lineage of *Norops* largely restricted to South America (encompassing the *N. nitens* group described above). The remaining two series are sister taxa (Figures 11.4 and 11.5) that are distributed

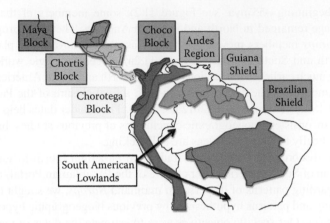

FIGURE 11.3 The geologic blocks used in ancestral area estimations to test the biogeographic history of mainland *Norops* (see Appendix for details).

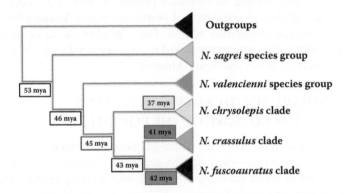

FIGURE 11.4 Simplified Bayesian phylogenetic reconstruction with estimated dates of origin derived from BEAST. Major groups are noted (defined in text).

from Central America to South America and comprise the *N. crassulus* and *N. fuscoauratus* series. In designating these three series, we only include species known from molecular work or inferred from recent taxonomy to belong to each lineage.

The split between Jamaican (*Norops valencienni*) and *N. auratus* series anoles occurred ca. 46 mya (95% HPD: 42.6–50.3 mya), with divergence between *N. chrysolepis* and the ancestor of *N. fuscoauratus* and *N. crassulus* series anoles occurring ca. 45 mya (95% HPD: 41.7–49.1 mya). Divergence between *N. fuscoauratus* and *N. crassulus* series anoles is estimated to have occurred 43 mya (95% HPD: 40.4–47.6 mya). Divergence within the *N. crassulus* and *N. fuscoauratus* series occurred rapidly, starting ca. 41–42 mya, with divergence within the *N. chrysolepis* series occurring slightly later (37.5 mya) (95% HPD listed in figure caption of Figure 11.4). Accounting for uncertainty, our dates indicate an earlier occupation of South America by *N. chrysolepis* series anoles than estimated by Glor et al. (2001). In fact, the dates estimated by Glor et al. (2001) are consistent with recent estimates

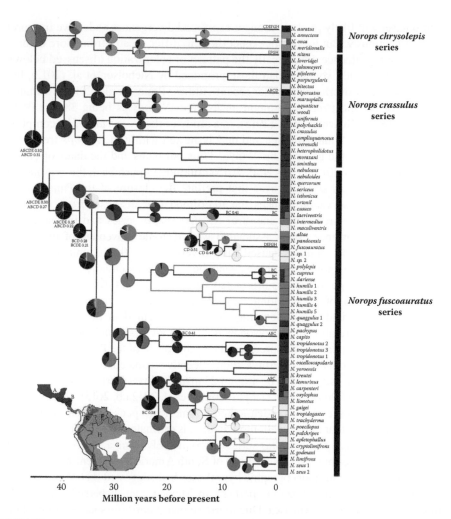

FIGURE 11.5 Ancestral area estimation for mainland *Norops* species. Regions used are as follows: Maya Block/Mexico (A), Chortis Block (B), Chorotega Block (C), Choco Block (D), Andes Mountains and surrounding region (E), Guyana Shield (F), Brazilian Shield (G), and South American Lowlands (H). Any ancestral ranges that have a probability <1 (multiple possible ranges) are placed on the internal branch using pie charts to express percentages. See map inset for color key. Branches and ancestral ranges that occupy multiple areas ("widespread") simultaneously are in black, and significant widespread ranges are listed with corresponding probabilities.

of closure of the Panamanian Portal at 15 mya, a finding that might have rescued Vanzolini and Williams's (1970) hypothesis that the *N. chrysolepis* series invaded South America after the closure of the Panamanian Portal. We recognize the limitation of using only mitochondrial genes to estimate divergence times and the risk of overestimating these dates (summarized in Prates et al. 2015 and references therein, but also see Mulcahy et al. 2012). We anticipate the addition of several nuclear genes to our dataset will likely produce younger estimated divergences and so we interpret our dates cautiously. But we do not expect them to differ greatly from more diverse datasets, nor to influence dramatically the pattern of colonization (discussed below). Our dates are slightly younger than those proposed by Nicholson et al. (2012), but are similar to and overlap with other recent studies. Prates et al. (2015) estimated the origin of the most recent common ancestor to *Norops* as being 38 mya (95% HPD 28–50 mya). Guarnizo et al. (2015), in a study of the *N. meridionalis* complex, a group endemic to South America and deeply nested within the *Norops* tree, estimated the origin of the group to be 21 mya. Thus our dates and the dates of others (e.g., Guarnizo et al. 2015; Prates et al. 2015) continue to push back the origination for the *N. chrysolepis* series to ages that predate the closure of the Panamanian Portal. Such an observation reduces the likelihood that dispersal from northern ancestors was the source of the ancestral *N. chrysolepis* series and increases the likelihood that vicariance events associated with tectonic movement of the South American and Caribbean plates caused early divergences within mainland *Norops*. Our dates support the notion that these events occurred at essentially the same time that Caribbean lineages of *Norops* originated via dissection of GAL.

The combination of node dating and ancestral area reconstruction analyses (discussed below) yields results contrary to ideas presented in the literature, but supportive of interpretation in Nicholson et al. (2012). We first sought to determine if endemic South American species predated the final closure of the Panamanian Portal (in early literature thought to be 3.5 mya, more recently estimated to be around 15 mya with first connection estimated to be around 23 mya [3.5 mya: Iturralde-Vinent 2006, Pindell and Kennan 2009; 15 and 23 mya: Montes et al. 2012a,b, 2015]. The *N. chrysolepis* clade, composed of 9–13 endemic South American species and one additional species (*N. auratus*) that ranges well into Panama, is estimated to have diverged from the *N. crassulus* and *N. fuscoauratus* series approximately 45 mya, well before any of the estimated closures or initial connections of South America with Central American tectonic blocks. Previous biogeographic hypotheses had suggested that *Norops* originated in northern Central America, perhaps in Mexico, and dispersed southwards ultimately invading South America. Due to the stepwise north-to-south reconnection resulting from the movement of tectonic blocks into place forming current Central America, the assumption was that endemic South American *Norops* species could not have originated until fairly recently, but that is clearly not the case. Our dates suggest that the clade originated much earlier than originally hypothesized.

11.3   HOW LONG HAS *NOROPS* OCCUPIED THE MAINLAND?

Our second objective was to investigate whether mainland members of *Norops* were present in North or South America prior to the separation of the continents by the eastward movement of the Greater Antillean Landblock (GAL, estimated to have completely separated from mainland areas by 55 mya; summarized in Nicholson et al. 2012). Nicholson et al. (2012) hypothesized that *Norops* originated much earlier than previously thought and became widespread via dispersal across the GAL. The origin of *Norops* in our analysis is estimated to have occurred ca. 58 mya (95% HPD: 52.8–64.7 mya), while mainland *Norops* is estimated to have originated ca. 45 mya

(95% HPD: 42.6–50.3 mya), a date close to, but 5–12 million years after the estimated separation of the GAL from the mainland (~55 mya). We conclude that these dates support the hypothesis that ancestral *Norops* was present before the GAL completely separated from the mainland and that the separation of the GAL from the mainland may have led to the origin of the mainland *Norops* clade. However, what is not yet clear from these analyses is exactly where *Norops* originated. Examination of the pattern of ancestral areas estimated for members of the *N. crassulus* and *N. fuscoauratus* clades reveals a general north-to-south pattern over a very short time frame and originating at effectively the same time as the *N. chrysolepis* clade. These results suggest that a rapid radiation of mainland anoles may have occurred subsequent to their origin and separation from their Caribbean ancestors. Given an observed north-to-south pattern (discussed further below), the implication is that the origin of *Norops* was somewhere in currently recognized Mexico. But our results are inconclusive on this point, as the ancestor to the *N. crassulus* + *N. fuscoauratus* clades is estimated to be widespread, and at a time when the GAL was separated from the continents. Additional data and analysis are required to derive conclusive results on this point.

11.4   WHERE DID MAINLAND *NOROPS* ORIGINATE AND WHAT IS THE PATTERN OF DISPERSAL ACROSS THE PANAMANIAN PORTAL?

Our Lagrange ancestral area reconstruction estimates two major centers of origin for mainland anoles (Figure 11.5). The *N. chrysolepis* series is estimated to have originated in the South American lowlands, and the ancestor of the *N. fuscoauratus* + *N. crassulus* series is estimated to have occupied a widespread Central and South American distribution (Figure 11.5, Regions A–E). However, the reconstruction of a widespread ancestor spanning the gap between North and South America may be an artifact of the program. The timing for the presence of the *N. fuscoauratus* + *N. crassulus* series ancestor (discussed above) may be problematic, because the final connection between present-day northern Central America with South America was not made until ~23–15 mya (reviewed in Townsend 2014). However, there was a series of islands spanning the two land masses after the departure of the GAL (Figure 11.2), so it is conceivable that mainland *Norops* occupied the estimated range through island-hopping, as has been hypothesized for Caribbean anoles (Glor et al. 2005) and more recently for several mainland *Norops* not included in this analysis (e.g., *N. townsendi* on Cocos Island and *N. medemi* and *N. biporcatus* on Gorgona Island, both in the Pacific Ocean). With a high number of species in present-day northern Central America, the *N. crassulus* series was estimated to have originated on the Chortis Block. From that ancestral area, our results indicate movement both north and south by clade members, some towards the Mayan block, others towards the Chorotegan block. Also, a single taxon within this clade (*N. bitectus*), endemic to the Choco block, indicates at least one southward colonization of South America and subsequent speciation in that region.

A general pattern of southward movement within the *N. fuscoauratus* series is indicated by estimated ancestral areas, with Chocoan taxa generally originating 15 mya or earlier, Chorotegan taxa appearing 30 mya or earlier, and Mayan/Chortis taxa appearing 45 mya or earlier. Within this series of anoles at least two north-to-south crossings of the Panamanian Portal are indicated. Thus, both *N. crassulus* and *N. fuscoauratus* series anoles diversified in a way that appears to have allowed invasion of South America from ancestors of Central American origin. These ancestors appear to have crossed from Central America into South America independently, but these colonization events did not definitively occur prior to 15 mya, supporting the earlier closure of the Portal proposed by Montes et al. (2012a).

Taken together, our data indicate that the Panamanian Portal has played a lengthy role in establishing biogeographic patterns of mainland *Norops*. Despite this lengthy time frame, relatively few extant taxa are found on both sides of this former barrier and relatively few ancestors crossed it and later speciated. Of these cases, the Portal has been crossed more frequently by *Norops* dispersing in the north-to-south direction (seven instances) than in the south-to-north direction (two instances). This directional disparity is expressed also in the distances that dispersers have penetrated South America (at least as far south as Bolivia in the case of *Norops ortoni* dispersing from a Central American origin; no farther north than southeastern Costa Rica in the case of *N. auratus* dispersing from South America). Some of this disparity is likely related to the physical area available to dispersers. Those lineages moving south across the Portal reach an enormous continent with complex physiography created by the Northern Andes. *N. fuscoauratus* appears to represent the only case of a recent arrival to South America dispersing widely across the expanse of new space. Species moving north across the Portal must reach a narrow connection with complex physiography created by the Talamancan Mountains, but that gradually widens as dispersers move farther north. We find no case representing a counter example to that of *N. fuscoauratus*. Thus, most of the diversity of mainland *Norops* stems from an ancient *N. chrysolepis* series that diversified across South America, an ancient ancestor to the *N. crassulus* + *N. fuscoauratus* series that diversified across Central America, and relatively few taxa crossing the Panamanian Portal from one continent to the other. The fact that few species managed to cross the Portal suggests that niche space had already been filled by each of the ancestral lineages. This view of Neotropical biogeography differs strikingly from the faunal exchanges described by older versions of mainland *Norops* history (Williams 1976) or other taxa (Marshall et al. 1982).

ACKNOWLEDGMENTS

We graciously thank Drs. Brian Crother and Lynne Parenti for inviting us to participate in the Rosen Symposium (JMIH 2015) where these ideas could be shared and discussed. We are very grateful for the assistance of several field and lab collaborators and those that shared tissues: Sarah Burton, Kevin de Queiroz, Jenny Gubler, Sara Huetteman, Gunther Köhler, David Laurencio, Holly Maddox, Lenin Obando, Javier Sunyer, and Jay Walton. This project was funded by a National Science Foundation grant to KEN (DEB 0949359), and internal grant funds from Central Michigan University. All animals were treated humanely and under the appropriate collection, importation, and IACUC permits.

APPENDIX 11.1   PHYLOGENETIC SAMPLING METHODS

We combined previously published data for seven mtDNA genes (a continuous section spanning NADH-ubiquinone oxidoreductase chain 2 (ND2), tRNATrp, tRNAAla, tRNAAsn, tRNACys, tRNATyr, origin of light strand replication, and partial CO1) with the same data for 30 new taxa (see Appendix 2). These combined data represent the entire geographic range of the mainland *N. auratus* species group as well as many other lineages variously recognized as full species.

The resulting dataset totaled 1423 base pairs aligned manually in MacClade (Maddison and Maddison 2005). We included eight Dactyloid outgroups representing each of the major lineages of anoles, as well as 19 Caribbean *Norops* species (see Appendix 2). We also included 10 non-Dactyloid outgroup species for their close relationship to Dactyloidae, because they are members of clades associated with fossil calibration dates that have been successfully used by others (e.g., Guarnizo et al. 2015; Prates et al. 2015) for dating nodes using the BEAST package (Drummond et al. 2012).

Bayesian phylogenetic analyses were performed employing the GTR + I + G model selected following the AIC criterion implemented in PartitionFinder (Lanfear et al. 2012). PartitionFinder did not recommend partitioned analysis of these data. An initial analysis running four chains for one-hundred-million (1.0×10^8) generations indicated convergence of parameters by 2.5 million generations with no subsequent changes to the plateau. Therefore, three independent runs were performed for 40 million generations each and each run was examined in Tracer (Rambaut and Drummond 2007) for convergence of parameters. Because all three runs performed similarly, they were combined in LogCombiner (BEAST package, Drummond et al. 2012) to produce the resultant consensus phylogeny, and TreeAnnotator was used to remove 20% burn in. Node support was evaluated via significant posterior probabilities.

Dates for clades throughout the tree were estimated via BEAST (Drummond et al. 2012) analyses. To calibrate dates, we used both the 0.65%/mya molecular calibration rate of Macey et al. (1998) as well as three node calibration dates for the non-Dactyloid outgroups as extracted from Prates et al. (2015): with upper and lower 95% HPD; 1. For the node defining Pleurodonta: 82.256 mya (71.331–98.585); 2. For the node defining Corytophanidae: 61.231 mya (57.350–67.207); 3. For the node defining *Leiosaurus* and *Urostrophus*: 53.789 mya (50.638–58.955). We employed a Yule Speciation prior and a relaxed lognormal clock, and used default settings for parameters relative to substitution rates, nucleotide frequencies, and the Yule tree prior.

Pattern of dispersal was examined via ancestral area estimation using Lagrange (Ree and Smith 2008). Eight areas were coded for this analysis, representing the major geologic blocks of Central America and important geological features of South America relevant to the movement of mainland anoles during their evolution: Maya Block, Chortis Block, Chorotega Block, Choco Block, Andes Mountains Region, Guiana Shield, Brazilian Shield, and South American Lowlands (Figure 11.3; see Nicholson et al. 2012 for a review of these details). Dispersal was constrained to occur between adjacent areas, and island taxa were removed from this analysis because (a) ours and previous analyses (unpublished) indicate that islands have been colonized recently and directly from mainland locations and thus are not informative regarding the testing of major mainland biogeographic patterns, and (b) Lagrange allows a maximum of eight ancestral areas to be designated.

APPENDIX 11.2   TAXON SAMPLING AND LOCATION DATA

Species	Outgroup/Ingroup	Data Source
Chalarodon madagascariensis	Outgroup (non-Dactyloid)	AF528722
Corytophanes cristatus	Outgroup (non-Dactyloid)	AF528717
Enyalioides laticeps	Outgroup (non-Dactyloid)	AF528718
Leiosaurus catamarcensis	Outgroup (non-Dactyloid)	AF528731
Plica plica	Outgroup (non-Dactyloid)	AF528748
Polychrus marmoratus	Outgroup (non-Dactyloid)	AF528738
Morunasaurus annularis	Outgroup (non-Dactyloid)	AF528720
Stenocercus guentheri	Outgroup (non-Dactyloid)	JQ687071
Uranoscodon supersiliosus	Outgroup (non-Dactyloid)	AF528749
Urostrophus vautieri	Outgroup (non-Dactyloid)	AF528734
Anolis brunneus	Outgroup (Dactyloid)	KF819779
Audantia cybotes	Outgroup (Dactyloid)	AF528723
Chamaelinorops koopmani	Outgroup (Dactyloid)	KF819783
Ctenonotus cristatellus	Outgroup (Dactyloid)	AF528724
Dactyloa frenata	Outgroup (Dactyloid)	AY909752
Dactyloa transversalis	Outgroup (Dactyloid)	AF337769
Deiroptyx chlorocyanus	Outgroup (Dactyloid)	AY296163
Xiphosurus barbatus	Outgroup (Dactyloid)	AY296146
Norops ahli	Outgroup (Dactyloid)	AF055941
N. allogus	Outgroup (Dactyloid)	AY296152
N. bremeri	Outgroup (Dactyloid)	AY296157
N. confusus	Outgroup (Dactyloid)	AY909787
N. guafe	Outgroup (Dactyloid)	AY909788
N. homolechis	Outgroup (Dactyloid)	AY296179
N. imias	Outgroup (Dactyloid)	AF294314
N. jubar	Outgroup (Dactyloid)	AY296182
N. mestrei	Outgroup (Dactyloid)	AF337779
N. ophiolepis	Outgroup (Dactyloid)	AF055942
N. rubribarbus	Outgroup (Dactyloid)	AY909789
N. sagrei	Outgroup (Dactyloid)	AY655172
N. quadriocellifer	Outgroup (Dactyloid)	AY655168
N. conspersus	Outgroup (Dactyloid)	AF294304
N. garmani	Outgroup (Dactyloid)	AF294289
N. grahami	Outgroup (Dactyloid)	AF294303
N. lineatopus	Outgroup (Dactyloid)	AF055937
N. opalinus	Outgroup (Dactyloid)	AF294309
N. reconditus	Outgroup (Dactyloid)	AY296198

Species	Outgroup/Ingroup	Data Source	Location
N. altae	Ingroup	AY909735	Costa Rica
N. amplisquamosus	Ingroup	This study	Honduras
N. annectens	Ingroup	AY909736	Venezuela
N. apletophallus	Ingroup	This study	Panama
N. aquaticus	Ingroup	AY909738	Costa Rica
N. auratus	Ingroup	AY909740	Panama
N. bicaorum	Ingroup	AY909741	Honduras
N. biporcatus	Ingroup	AF294286	Nicaragua
N. bitectus	Ingroup	This study	Colombia
N. capito	Ingroup	AY909744	Costa Rica
N. carpenteri	Ingroup	AY296160	Nicaragua
N. crassulus	Ingroup	AY909748	Mexico
N. cryptolimifrons	Ingroup	This study	Panama
N. cupreus	Ingroup	AY909750	Costa Rica
N. cusuco	Ingroup	This study	Honduras
N. dariense	Ingroup	This study	Honduras
N. fuscoauratus	Ingroup	AF337792	Brazil
N. gagei	Ingroup	This study	Panama
N. godmani	Ingroup	This study	Costa Rica
N. heteropholidotus	Ingroup	This study	Honduras
N. humilis_1	Ingroup	KJ954109	Panama
N. humilis_2	Ingroup	KJ953951	Costa Rica
N. humilis_3	Ingroup	KJ953995	Costa Rica
N. humilis_4	Ingroup	KJ953940	Costa Rica
N. humilis_5	Ingroup	KJ954008	Costa Rica
N. intermedius	Ingroup	AY909755	Costa Rica
N. isthmicus	Ingroup	AY909762	Mexico
N. johnmeyeri	Ingroup	This study	Honduras
N. kemptoni	Ingroup	AY909770	Panama
N. kreutzi	Ingroup	This study	Honduras
N. laeviventris	Ingroup	AY909756	Guatemala
N. lemurinus	Ingroup	AF294283	
N. limifrons	Ingroup	AF055943	
N. lineatus	Ingroup	AF055935	Aruba
N. lionotus	Ingroup	AY909757	Panama
N. loveridgei	Ingroup	AY909759	Honduras
N. maculiventris	Ingroup	This study	Colombia
N. marsupialis	Ingroup	KJ964103	Costa Rica
N. medemi	Ingroup	KJ953921	Colombia
N. meridionalis	Ingroup	AY909760	Paraguay
N. morazani	Ingroup	This study	Honduras
N. nebuloides	Ingroup	AY909763	Mexico
N. nebulosus	Ingroup	This study	Mexico
N. nitens	Ingroup	AF337800	Brazil
N. onca	Ingroup	AY909765	Venezuela

(Continued)

Appendix 11.2 (*Continued*)

Species	Outgroup/Ingroup	Data Source	Location
N. ortonii	Ingroup	AF294288	Brazil
N. ocelloscapularis	Ingroup	AY909767	Honduras
N. oxylophus	Ingroup	AY909768	Honduras
N. pachypus	Ingroup	AY909769	Costa Rica
N. pijolense	Ingroup	This study	Honduras
N. poecilopus	Ingroup	AY909771	Panama
N. polylepis	Ingroup	AY909772	Costa Rica
N. polyrhachis	Ingroup	AY909773	Mexico
N. pulchripes	Ingroup	This study	Costa Rica
N. purpurgularis	Ingroup	AY909774	Honduras
N. quaggulus 1	Ingroup	KJ954026	Costa Rica
N. quaggulus 2	Ingroup	KJ954052	Nicaragua
N. quercorum	Ingroup	AY909775	Mexico
N. sericeus	Ingroup	AY909778	Costa Rica
N. sminthus	Ingroup	AY909779	Honduras
N. townsendi	Ingroup	KJ953922	Costa Rica
N. trachyderma	Ingroup	AF294285	
N. transversalis	Ingroup	This study	Brazil
N. tropidogaster	Ingroup	AY909782	Panama
N. tropidonotus 1	Ingroup	This study	Mexico
N. tropidonotus 2	Ingroup	This study	Honduras
N. tropidonotus 3	Ingroup	This study	Nicaragua
N. uniformis	Ingroup	AY909784	Belize
N. utilensis	Ingroup	AY909785	Honduras
N. wermuthi	Ingroup	This study	Nicaragua
N. woodi	Ingroup	AF337780	Costa Rica
N. woodi	Ingroup	This study	Costa Rica
N. yoroensis	Ingroup	This study	Honduras
N. zeus 1	Ingroup	This study	
N. zeus 2	Ingroup	AY909786	Honduras

REFERENCES

Alföldi, J., F. D. Palma, M. Grabherr, M. Williams, L. Kong, E. Mauceli, P. Russell, C. B. Lowe, R. E. Glor, J. D. Jaffe, D. A. Ray, S. Boissinot, A. M. Shedlock, C. Botka, T. A. Castoe, J. K. Colbourne, M. K. Fujita, R. G. Moreno, B. F. ten Hallers, D. Haussler, A. Heger, D. Heiman, D. E. Janes, J. Johnson, P. J. de Jong, M. Y. Koriabine, M. Lara, P. A. Novick, C. L. Organ, S. E. Peach, S. Poe, D. D. Pollock, K. de Queiroz, T. Sanger, S. Searle, J. D. Smith, Z. Smith, R. Swofford, J. Turner-Maier, J. Wade, S. Young, A. Zadissa, S. V. Edwards, T. C. Glenn, C. J. Schenider, J. B. Losos, E. S. Lander, M. Breen, C. P. Ponting and K. Lindblad-Toh. 2011. The genome of the green anole lizard and a comparative analysis with birds and mammals. *Nature* 477:587–591.

Castañeda, M. D. R. and K. de Queiroz. 2013. Phylogeny of the Dactyloa clade of Anolis lizards: New insights from combining morphological and molecular data. *Bulletin of the Museum of Comparative Zoology* 160(7):345–398.

Croizat, L. 1962. *Space, Time, Form: The Biological Synthesis*. Published by the author, Caracas.

Drummond, A. J., M. A. Suchard, D. Xie and A. Rambaut. 2012. Bayesian phylogenetically with BEAUti and the BEAST 1.7. *Molecular Biology and Evolution* 29(8):1969–1973.

Etheridge, R. 1959. *The Relationships of the Anoles (Reptilia: Sauria: Iguanidae): An Interpretation Based on Skeletal Morphology*. PhD Dissertation, University of Michigan, Ann Arbor, 236pp.

Glor, R. E., L. J. Vitt and A. Larson. 2001. A molecular phylogenetic analysis of diversification in Amazonian *Anolis* lizards. *Molecular Ecology* 10:2661–2668.

Glor, R. E., J. B. Losos and A. Larson. 2005. Out of Cuba: Overwater dispersal and speciation among lizards in the *Anolis carolinensis* subgroup. *Molecular Ecology* 14:2419–2432.

Guarnizo, C. E., F. P. Werneck, L. G. Giugliano, M. G. Santos, J. Fenker, L. Souse, A. B. D'Angiolella, A. R. dos Santos, C. Strüssmann, M. R. Rodrigues, R. F. Dorado-Rodrigues, T. Gamble and G. R. Colli. 2015. Cryptic lineages and diversification of an endemic anole lizard (Squamata, Dactyloidae) of the Cerrado hotspot. *Molecular Phylogenetics and Evolution* 94:279–289.

Guyer, C. and J. M. Savage. 1986. Cladistic relationships among anoles (Sauria: Iguanidae). *Systematic Zoology* 35:509–531.

Guyer, C. and J. M. Savage. 1992. Anole systematics revisited. *Systematic Zoology* 41:89–110.

Iturralde-Vinent, M. A. 2006. Meso-Cenozoic Caribbean paleogeography: Implications for the historical biogeography of the region. *International Geology Review* 48:791–827.

Jackman, T. R., A. Larson, K. de Queiroz and J. B. Losos. 1999. Phylogenetic relationships and tempo of early diversification of *Anolis* lizards. *Systematic Biology* 48:254–285.

Lanfear, R., B. Calcott, S. Y. W. Ho and S. Guindon. 2012. Partition Finder: Combined selection of partitioning schemes and substitution models of phylogenetic analyses. *Molecular Biology and Evolution* 29:1695–1701.

Losos, J. B. 1994. Integrative approaches to evolutionary ecology: *Anolis* lizards as model systems. *Annual Review of Ecology and Systematics* 25:467–493.

Losos, J. B. 2009. *Lizards in an Evolutionary Tree: Ecology and Adaptive Radiation of Anoles*. Berkeley: University of California Press, 528 pages. ISBN: 0520269845.

Losos, J. B., T. R. Jackman, A. Larson, K. de Queiroz and L. Rodríguez-Schettino. 1998. Contingency and determinism in replicated adaptive radiations of island lizards. *Science* 279:2115–2118.

Macey J. R., J. A. Schulte II, N. B. Anajeva, A. Larson, N. Rastegar-Pouyani, S. M. Shammakov and T. J. Papenfuss. 1998. Phylogenetic relationships among agamid lizards of the *Laudakia caucasia* species group: Testing hypotheses of biogeographic fragmentation and an area cladogram for the Iranian Plateau. *Molecular Phylogenetics and Evolution* 10:118–131.

Maddison, W. P. and D. R. Maddison. 2005. *Analysis of Phylogeny and Character Evolution Using the Computer Program MacClade*. Sunderland, MA: Sinauer Associates.

Marshall, L. G., S. D. Webb, J. J. Sepkoski and D. M. Raup. 1982. Mammalian evolution and the great American interchange. *Science* 215:1351–1357.

Montes, C., G. Bayona, A. Cardona, D. M. Buchs, C. A. Silva, S. Morón, N. Hoyos, D. A. Ramírez, C. A. Jaramillo and V. Valencia. 2012a. Arc-continent collision and orocline formation: Closing of the Central American seaway. *Journal of Geophysical Research* 117: B04105, doi:10.1029/2011JB008959.

Montes, C., A. Cardona, R. McFadden, S. W. Morón, C. A. Silva, S. Restrepo-Moreno, D. A. Ramírez, N. Hoyos, J. Wilson, D. Farris, G. A. Bayona, C. A. Jaramillo, V. Valencia, J. Bryan and J. A. Flores. 2012b. Evidence for middle Eocene and younger land emergence in central Panama: Implications for isthmus closure. *Geological Society of America Bulletin* 124:780–799.

Montes, C., A. Cardona, C. Jaramillo, A. Pardo, J. C. Silva, V. Valencia, C. Ayala, I. C. Pérez-Angel, I. A. Rodriguez-Parra, V. Ramirez and H. Niño. 2015. Middle Miocene closure of the Central American seaway. *Science* 348:226–229.

Mulcahy, D. G., B. P. Noonan, T. Moss, T. M. Townsend, T. W. Reeder, J. W. Sites Jr. and J. J. Wiens. 2012. Estimating divergence dates and evaluating dating methods using phylogenomic and mitochondrial data in squamate reptiles. *Molecular Phylogenetics and Evolution* 65:874–991.

Nicholson, K. E., R. E. Glor, J. J. Kolbe, A. Larson, S. Blair Hedges and J. B. Losos. 2005. Mainland colonization by island lizards. *Journal of Biogeography* 32:929–938.

Nicholson, K. E., B. I. Crother, C. Guyer and J. M. Savage. 2012. It is time for a new classification of anoles (Squamata: Dactyloidae). *Zootaxa* 3477:1–108.

Phillips, J. G., J. Deitloff, C. Guyer, S. Huetteman and K. E. Nicholson. 2015. Biogeography and evolution of a widespread Central American lizard species complex: *Norops humilis* (Squamata: Dactyloidae). *BMC Evolutionary Biology* 15:143–156.

Pindell, J. L. and L. Kennan. 2009. Tectonic evolution of the Gulf of Mexico, Caribbean and northern South America in the mantle reference frame: An update. *Geological Society, London, Special Publications* 328:1–55.

Poe, S. 2004. Phylogeny of anoles. *Herpetological Monographs* 18:37–89.

Prates, I., M. R. Rodrigues, P. R. Melo-Sampaio and A. C. Carnaval. 2015. Phylogenetic relationships of Amazonian anole lizards (*Dactyloa*): Taxonomic implications, new insights about phenotypic evolution and the timing of diversification. *Molecular Phylogenetics and Evolution* 82:258–268.

Pyron, R. A., F. T. Burbrink and J. J. Wiens. 2013. A phylogeny and revised classification of Squamata, including 4161 species of lizards and snakes. *BMC Evolutionary Biology* 13:93–145.

Rambaut, A. and A. J. Drummond. 2007. *Tracer: MCMC Trace Analysis Tool. Institute of Evolutionary Biology.* Edinburgh: University of Edinburgh.

Ree, R. H. and S. A. Smith. 2008. Maximum-likelihood inference of geographic range evolution by dispersal, local extinction, and cladogenesis. *Systematic Biology* 5:4–14.

Rosen, D. E. 1976. A vicariance model of Caribbean biogeography. *Systematic Zoology* 24:431–464.

Townsend, J. H. 2014. Characterizing the chortís block biogeographic province: Geological, physiographic, and ecological associations and herpetofaunal diversity. *Mesoamerican Herpetology* 1:203–252.

Vanzolini, P. E. and E. E. Williams. 1970. South American anoles: The geographic differentiation and evolution of the *Anolis chrysolepis* species group (Sauria, Iguanidae). *Arquivos de Zoologia* 19:1–124.

Williams, E. E. 1972. The origin of faunas: Evolution of lizard congeners in a complex island fauna—a trial analysis. *Evolutionary Biology* 6:47–89.

Williams, E. E. 1976. South American anoles: A taxonomic and evolutionary summary. 1. Introduction and species list. *Breviora, Museum of Comparative Zoology* 440:1–21.

Williams, E. E. 1983. Ecomorphs, faunas, island size, and diverse end points in island radiations of *Anolis*. In: *Lizard Ecology: Studies of a Model Organism* (eds. Huey, R. B., Pianka, E. R., Schoener, T. W.). pp. 326–370, Cambridge: Harvard University Press.

Williams, E. E. 1989. A critique of Guyer and Savage (1986): Cladistics relationships among anoles (Sauria: Iguanidae): Are the data available to reclassify the anoles? In: *Biogeography of the West Indies: Past, Present, Future* (ed. Woods, C. A.). pp. 433–477, Gainesville, FL: Sandhill Crane Press.

12 Modification of a Comparative Biogeographic Method Protocol to Differentiate Vicariance and Dispersal

Mallory E. Eckstut
Southeastern Louisiana University

Brett R. Riddle
University of Nevada, Las Vegas

Brian I. Crother
Southeastern Louisiana University

CONTENTS

12.1 INTRODUCTION

Making any of the assumptions listed by Rosen (2016) when formulating or testing scientific hypotheses can inhibit progress in comparative biology. However, developing and using methods with these assumptions can be just as detrimental, especially when reconstructing historical scenarios. Two assumptions that can equally inhibit progress in comparative biogeography in particular are:

- All distributions result from vicariance
- All distributions result from dispersal

Biogeographic studies have traditionally focused on describing distributional patterns either by using phylogenetic patterns of distributions to address large spatial and deeper temporal scales, or ecologically relevant patterns of species richness that focus on smaller spatial and shallower temporal scales. More recently, there has been a drive to create methods that integrate both approaches to gain a better understanding of the dynamics of global biodiversity. This has led to the development of a variety of techniques and approaches, phylogeography (Avise 2000) being the most widely used, with the goal of revealing biological and demographic processes that have played important roles in current biotic structure and how these patterns have been shaped by a region's abiotic history (as reviewed in Brooks and McLennan 2002; Wiens and Donoghue 2004; Morrone 2009; Folinsbee and Evans 2012; Wiens 2012).

In particular, identification of spatially and temporally congruent patterns of biotic diversification is a topic that many biogeographers have addressed, but with variable levels of success. A time-honored premise is that spatial congruence in diversification patterns is the result of vicariance, where a barrier is formed within the ranges of ancestrally widespread taxa, simultaneously splitting co-distributed taxa (Croizat et al. 1974; Nelson and Platnick 1981). However, nonvicariant mechanisms can also result in congruence among taxa in biogeographic patterns. Alternative mechanisms include, for example, biotic expansion by concordant dispersal following the removal of a barrier (biotic dispersal, Platnick and Nelson 1978; or geodispersal, Lieberman and Eldredge 1996) and linear jump dispersal of taxa associated with temporally and spatially sequential development of new areas, where the oldest lineages are found in the oldest region and the youngest lineages are found in the youngest region (Hennig's progression rule; Hennig 1966; Cowie and Holland 2008).

The vicariance hypothesis in particular has been criticized because lineages often differentially respond to local abiotic and biotic factors, and many strictly cladistic (often vicariance-based) methods do not reconstruct unique responses as prevalent mechanisms of producing regional biodiversity (e.g., de Queiroz 2005; Parenti 2007). Even in scenarios where vicariance and other congruent (or general) diversification mechanisms have been argued as the prevalent mechanism (e.g., Madagascar and the North American aridlands), there are many instances of idiosyncratic (or unique) dispersal events, and it can be difficult to discern whether general or unique events contributed most substantially to a biogeographic system (e.g., Zink et al. 2000a; Yoder and Nowak 2006; Riddle et al. 2008; Hoberg and Brooks 2010). Additionally, excluding divergence times from analyses, which is common in cladistic analyses, may overpredict the number of vicariant events because of temporally pseudo-congruent patterns, which are similar biogeographic patterns produced by historically dissimilar events (Donoghue and Moore 2003; Ree and Sanmartín 2008; Eckstut et al. 2011).

In light of the complex nature of formation of regional biotas, no single analytical approach so far has allowed the teasing apart of their diversification complexity. The development of integrative, step-wise frameworks (e.g., Riddle and Hafner 2006), as shown in Figure 12.1, has been proposed as a way to optimally

---

**Step-Wise Framework
from Riddle and Hafner (2006)**

1) Identify biota, units of analysis (e.g., species or phylogroups), and distributional areas

2) Diagnose areas of endemism using distributional data (e.g., Parsimony Analysis of Endemicity)

3) Determine general divergences using multi-clade analysis (e.g., Primary Brooks Parsimony Analysis, or BPA)
   Note: PACT replaces BPA in this step and includes unique divergences

4) Resolve departure from general divergences (reticulate area relationships) using additional multi-clade analysis (e.g., Secondary BPA)
   Note: this step does not occur in PACT analysis

5) Test hypotheses of taxon and biotic distributions

FIGURE 12.1   Step-Wise framework as designed and implemented by Riddle and Hafner (2006). Notes are included to identify where the method discussed in this study, PACT, would be implemented in this step-wise framework and where this step-wise framework would not be applicable to a PACT analysis.

reveal historical biogeographic complexity (Richards et al. 2007; Morrone 2009; Riddle and Hafner 2010). Part 3 of Riddle and Hafner's (2006) proposed step-wise framework involves identifying general events using an analysis such as Primary and Secondary BPA. Secondary BPA is the precursor for a more recent developed method called phylogenetic analysis for comparing trees (PACT; Wojcicki and Brooks 2004, 2005), which is a method that incorporates all events (both general and unique, which may represent vicariance or congruent dispersal and individual dispersal events, respectively). Accordingly, step-wise frameworks serve as an integrative approach, and portions of the framework (such as using PACT instead of Secondary BPA for part 3) may be valuable to infer the relative importance of general versus unique diversification in forming regional biodiversity and reducing the prevalence of pseudo-congruence.

12.1.1   PACT AS AN INTEGRATIVE METHOD

PACT (Wojcicki and Brooks 2004, 2005) is a multiclade analytical method that can integrate historical and ecological approaches and has been shown to reveal patterns of the taxon pulse (periodic episodes of expansion, colonization, and isolation), Hennig's progression rule, and species–area relationships (Halas et al. 2005; Hoberg and Brooks 2010; Eckstut et al. 2011). PACT integrates all spatial data from each taxon–area cladogram by tracing geographic changes on a phylogenetic tree for a single taxon, as shown in Figure 12.2, to generate general area cladograms, which summarize geographic changes traced on an area cladogram that was generated using several taxon–area cladograms, as shown in Figure 12.3, and therefore assumes neither vicariance nor dispersal. Despite the proposed benefits of PACT, this algorithm was developed using cladograms that do not include a temporal calibration and without consideration of weighting or a temporal component, and as a consequence, ancestral area reconstruction is restricted to parsimony-based frameworks (Eckstut et al. 2011).

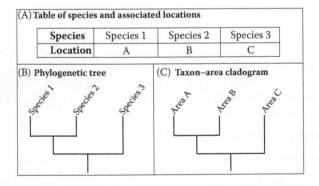

FIGURE 12.2 Generation of a taxon–area cladogram. (A) Species distributions are collected. (B) A phylogenetic tree is built. (C) The areas of occurrence for each species are overlaid onto the phylogenetic tree.

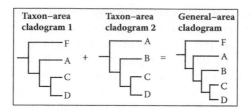

FIGURE 12.3   Generation of a general area cladogram using two input taxon–area clado-grams (1—F, A, C, D; 2—A, B, C, D). All areas are included in the output general area cladogram (F, A, B, C, D).

Excluding time can be problematic, because temporally and spatially discor-dant events can result in identical cladogram patterns (pseudo-congruence), which can then be incorrectly inferred to represent a single vicariant event (Cunningham and Collins 1994). Likewise, when dissimilar patterns emerge from similar histori-cal events, pseudo-incongruence can also occur (e.g., Donoghue and Moore 2003; Riddle and Hafner 2006). Wojcicki and Brooks (2004, 2005) suggested that the PACT algorithm inherently reduces errors in congruence, as the algorithm proceeds in a top-down fashion, and younger nodes (at the top of a tree) are presumed to be more similar to each other than older nodes (deeper in a tree), because nodes toward the top of a tree should represent younger events and deeper nodes should represent older events. Moreover, as a consequence of decreased similarity toward the base of the tree, geographic paralogy is more likely to occur at the base of a tree (Nelson and Ladiges 1996).

The top-down approach is conducted under the assumption that all phyloge-netic trees occur on the same temporal scale (i.e., the youngest node on two trees would represent recent divergences at approximately the same time, and the oldest nodes on the same two trees would represent older divergences at approximately the same time). However, the top-down approach can be problematic even if analy-ses are restricted to the same taxonomic scale (e.g., species- or genera-level infer-ences) in instances where named species, subspecies, and phylogroups are not on the same temporal scale. For example, Riddle (1995) reported vast differences between amount of divergence representing species and genera in pocket mice (*Chaetodipus* and *Perognathus*) and grasshopper mice (*Onychomys*). In pocket mice, evolution of genera dates back to at least between 9 and 10 million years (*Chaetodipus* and *Perognathus*, respectively), with divergences between species occurring from 4 mya (*C. penicillatus* and *C. intermedius*) and even approximately 8 mya for divergence within the species *P. parvus* (Riddle 1995; Riddle et al. 2014). Alternatively, the grasshopper mouse genus *Onychomys* is estimated to have evolved just over 2 mya, and divergence of the three species (*O. arenicola, O. leucogaster,* and *O. torri-dus*) occurred soon after (Riddle 1995). Thus, to correct for inconsistencies with taxonomic–temporal dynamics, molecular clocks and fossil information can each be used to reduce pseudo-congruence in general area cladograms (GACs). Lim (2008) and Folinsbee and Evans (2012) conducted molecular dating on taxon–area clado-grams that were incorporated into GACs using the PACT algorithm, but there has not been formal integration of molecular dating into the PACT framework and explicit discussion of ramifications on PACT-specific analyses.

Moreover, the current PACT algorithm is only capable of producing trees with equal branch lengths, as shown in Figure 12.4A and B. This restricts ancestral area reconstruction to parsimony-based optimization frameworks, which do not incorporate branch length into ancestral state reconstruction. Not incorporating branch lengths can produce erroneous ancestral area reconstruction results (Cunningham 1999), especially in areas where there are high rates of dispersal (Pirie et al. 2012). For example, if unique (1 taxon) and general (2+ taxa) events are not considered when optimizing the GAC, then weighting ancestral nodes can be erroneous if the optimization protocol inadvertently favors the unique event when reconstructing ancestral node state because of other patterns within the GAC. This issue can be alleviated by producing a GAC with branch lengths that can appropriately weight general and unique events for the reconstruction

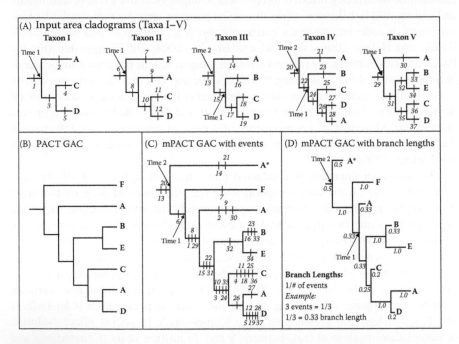

FIGURE 12.4 Deriving general area cladograms (GACs) from five input area cladograms using PACT and a modified version of PACT (mPACT). (A) The five input taxon–area clado-grams from five taxa, with areas indicated in A–F and each event numbered in italics at the hash over each branch, and timing of diversification indicated for events. (B) PACT GAC. (C) mPACT GAC with each event indicated by the hash with the associated italicized number from the input cladograms. (D) mPACT GAC with total number of events (E) converted to branch lengths by calculating $1/E$ (e.g., 3 events = $1/3$ = 0.33). These branch lengths facilitate likelihood-based optimization because ancestral nodes are more heavily weighted toward short branch lengths, and ancestral nodes are more likely similar to general (2+ congruent) events than unique (1) events. Structural differences between the mPACT and PACT GACs are indicated by A*.

of ancestral areas, wherein the ancestral area would be more heavily weighted toward general events, as shown in Figure 12.4C and D.

There are numerous cladistic biogeographic methods that are used to evaluate general area relationship patterns, including the commonly used primary and secondary Brooks parsimony analysis (BPA; Brooks 1990), component analysis (Nelson and Platnick 1981), subtree analysis (Nelson and Ladiges 1996), and tree reconciliation (Page 1994); however, none of these methods generate and incorporate branch lengths. To alleviate the issues of pseudo-congruence and absence of branch weighting, we herein develop a modification to the PACT algorithm that incorporates a temporal component to analyses by conducting molecular dating techniques using the program BEAST (Drummond and Rambaut 2007) to estimate timing of ancestral divergences (Drummond et al. 2006). Additionally, the modified PACT (mPACT) algorithm incorporates branch lengths based on frequency of events to facilitate a Maximum Likelihood-based approach for ancestral area reconstruction of the PACT GAC,[*] as shown in Figure 12.4.

12.1.2 North American Warm Deserts as a Model System

The North American warm deserts (the Sonoran, Chihuahuan, Mojave, and Peninsular deserts, as shown in Figure 12.5) are an ideal region for experimentally testing integrative historical biogeographic methods, particularly when optimizing methods that can incorporate both vicariant and dispersal events, and reduce pseudo-congruence. These deserts have been subject to an array of geologic and climatic

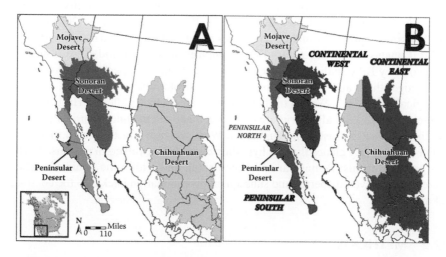

FIGURE 12.5 Map of the North American warm deserts (A), including the Continental East, Continental West, Peninsular North, and Peninsular South areas of endemism (B). (From Riddle, B.R. and Hafner, D.J., *J. Arid Environ.*, 66, 435–461, 2006.)

---

[*] We are cognizant of the problems of this approach as laid out by Siddall and Kluge (1997). We incorporate this new method as an exploratory heuristic only.

processes across various timescales (ca. 3.6 mya–present) that have produced a complex topography and diverse environmental conditions (Wells and Haragan 1983). As a consequence, unique communities of relatively distinct biotic assemblies for which a large number of phylogenetically based biogeographic datasets have been generated, including numerous vertebrates, invertebrates, and plants (e.g., Wells and Haragan 1983; Riddle 1995; Riddle et al. 2000a,b,c; Jaeger et al. 2005; Riddle and Hafner 2006; Jezkova et al. 2009; Pyron and Burbrink 2010; Graham et al. 2013). Moreover, there is potential for a disproportionate amount of pseudo-congruent diversification patterns among taxa in these regions because of the diverse timescale of geologic and climatic events that have shaped current biotic composition.

Riddle and Hafner (2006) conducted a study on the North American warm deserts where they diagnosed areas of endemism (areas producing and maintaining unique biodiversity through time) and conducted a Secondary BPA analysis (a binary-coded, matrix-based precursor to PACT for inferring general diversification events). That study provided testable hypotheses of North American warm desert biodiversity evolution and a benchmark dataset for comparing the performance of newly developed methods.

Herein, we discuss the history and development of the biogeographic technique PACT (Wojcicki and Brooks 2004, 2005), which is an integrative comparative biogeographic method that does not assume that all distributions result from either vicariance or dispersal. Moreover, we propose a modification to the PACT protocols to facilitate distinguishing hypothesized vicariant from dispersal events in a region. We test this modification on a North American warm desert biota.

To test the performance of the mPACT protocol, we compared mPACT results to the Secondary BPA analysis conducted by Riddle and Hafner (2006) and a standard PACT analysis conducted in this study. Riddle and Hafner's (2006) dataset and hypotheses provide testable hypotheses and a tool for comparison of PACT and mPACT analyses with Secondary BPA. Comparison of PACT and mPACT analyses to Secondary BPA analysis provides unique insight into the performance of PACT-based algorithms and the progression of these evolutionary methods at capturing biological complexity. We test the hypotheses set forth by Riddle and Hafner (2006), where there are four allopatric areas of endemism (Peninsular North, Peninsular South, Continental East, and Continental West, as shown in Figure 12.5) and two major evolutionary clades that are hypothesized to have evolved in response to various geologic events: Peninsular and Continental, and within the Peninsular clade, there was observed dispersal and subsequent diversification back into the continental mainland.

12.2   METHODS

12.2.1   Evaluating PACT Protocols

We used the 22 North American warm desert taxa that were previously used by Riddle and Hafner (2006), which include nine mammals, seven birds, four nonavian reptiles, one amphibian, and one plant, as shown in Figure 12.6 and Table 12.1. The use of this benchmark dataset allows us to directly compare the new methods (PACT

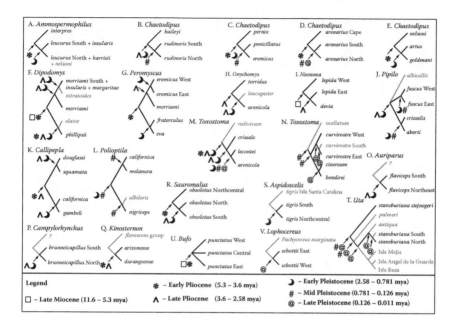

FIGURE 12.6 Cladograms and ages of divergence used in comparative analysis. Multiple symbols indicate divergence ranges that encompass numerous time periods. (From Riddle, B.R. and Hafner, D.J., *J. Arid Environ.*, 66, 435–461, 2006.)

and mPACT), as shown in Table 12.2, to Riddle and Hafner's (2006) Secondary BPA results. Moreover, including other taxa would now allow the use of areas used in Riddle and Hafner's (2006) Secondary BPA analysis, because the areas of endemism they diagnosed were based on the 22 taxa dataset. Additionally, we herein refer to Riddle and Hafner's (2006) diagnosed areas of endemism for designating regions to remain consistent with comparison of PACT and mPACT results to their Secondary BPA results, as shown in Figure 12.5.

12.2.2 Standard PACT Analysis

12.2.2.1 Step 1: Obtain Cladograms

For the standard PACT analysis, cladograms were collected from the manuscripts in which they were initially published, as shown in Table 12.1, and each was converted into a taxon–area cladogram by replacing the names of each of the terminal species with the area(s) of their respective distributions. A single lineage occurring in numerous areas was treated as a monophyletic polytomy in the taxon–area cladogram.

12.2.2.2 Step 2: GAC Construction

We combined the taxon–area cladograms by hand to form a GAC using the PACT algorithm as described by Wojcicki and Brooks (2004, 2005), because there is currently no software available to conduct PACT analysis. The PACT algorithm

TABLE 12.1

List of North American Desert Taxa as Used by Riddle and Hafner (2006) and GenBank Numbers for Sequences Used for Molecular Dating

Group	Clade	Common Name	Original Source	GenBank Accession Numbers
Plants	*Lophocereus schotti*	Whisker cactus	Nason et al. (2002)	AF328628–AF328664
Amphibians	*Bufo punctatus*	Red-spotted toad	Riddle et al. (2000), Jaeger et al. (2005)	AY010121–AY010166, DQ085629–DQ085776
Reptiles	*Uta stansburiana*	Common side-blotched lizard	Upton and Murphy (1997)	U46695–U46742
	Aspidoscelis tigris	Western whiptail	Murphy and Aguirre-Léon (2002); Radtkey et al. (1997); Reeder et al. (2002)	AF006266–AF006288
	Sauromalus species	Chuckwalla	Petren and Case (1997, 2002)	AF20223–AF20252
	Kinosternon flavescens species group	Mud turtle	Serb et al. (2001)	AF316121–AF316137
Birds	*Campylorhynchus brunneicapillus*	Cactus wren	Zink et al. (2000b)	AF291529–AF291571
	Auriparus flaviceps	Verdin	Zink et al. (2000b)	AF295205–AF295249
	Toxostoma curvirostre species group	Curve-billed thrasher	Zink and Blackwell-Rago (2000), Zink et al. (2001), Zink et al. (1999)	AF287160–AF287228, AF287496–AF2878560, AF287564–AF287629
	Toxostoma lecontei species group	LeConte's thrasher	Zink et al. (1997), Zink et al. (2001)	U75569–U75573, U75575–U75579
	Polioptila melanura species group	Black-tailed Gnatcatcher	Zink et al. (2000a), Zink and Blackwell (1998a), Zink et al. (2001), Zink et al. (2000b)	AF027827–AF27836, AF027838–AF027848, AF027849–AF027859, AF518771, AF518872
	Callipepla squamata species group	Scaled quail	Zink and Blackwell (1998b), Zink et al. (2001)	AF028750, AF028751, AF028753, AF028754, AF028756, AF028757, AF028759, AF028760, AF028762, AF028763, AF028765, AF028766, AF028768, AF028769, AF028771, AF0287712, AF028752, AF028755, AF028758, AF028761, AF028764, AF028767, AF028770, AF028773
	Pipilo fuscus species group	Canyon towhee	Zink et al. (2001), Zink and Dittman (1991), Zink et al. (1998c)	AF298595–298622

(Continued)

TABLE 12.1 (Continued)

List of North American Desert Taxa as Used by Riddle and Hafner (2006) and GenBank Numbers for Sequences Used for Molecular Dating

Group	Clade	Common Name	Original Source	GenBank Accession Numbers
Mammals	*Neotoma lepida* species group	Desert woodrat	Planz (1992); Patton et al. (2008)	DQ781162–DQ781256*
	Onychomys species	Grasshopper mouse	Riddle (1995), Riddle and Honeycutt (1990)	U21614–U21616, U21633–U21635, U21648–U21650
	Peromyscus eremicus species group	Cactus mouse	Riddle et al. (2000c)	AY009173–AY009237
	Dipodomys merriami species group	Merriam's kangaroo rat	Riddle et al. (2000a), Jaeger et al. (2005), Alexander and Riddle (2005)	AY926430, AY926437, AY926438, AY926439, AY926443, AY926445, AY926450
	Chaetodipus nelsoni species group	Nelson's pocket mouse	Riddle et al. (2000b), Jaeger et al. (2005)	AY009240, AY009246, AY009249
	Chaetodipus arenarius	Little desert pocket mouse	Riddle et al. (2000a)	AY010230–AY10197
	Chaetodipus penicillatus species group	Desert pocket mouse	Lee et al. (1996), Riddle et al. (2000b), Jezkova et al. (2009)	AB456529–AB456507
	Chaetodipus baileyi species group	Bailey's pocket mouse	Riddle et al. (2000b)	AY009253–AY009316, AY009238–AY009252
	Ammospermophilus species	Antelope squirrel	Riddle et al. (2000a), Jaeger et al. (2005)	AY010167–AY010196

Note: Some unpublished sequences were also used for molecular dating (provided by J. Patton).

TABLE 12.2

How to Conduct PACT and mPACT

Step	Goal	Input Data	Algorithms	Programs	Process
1	Construct taxon-area cladogram	- Phylogenetic tree - Distributions of each lineage	N/A	N/A	Replace taxon names on each phylogenetic tree with areas of distribution.
2	*Molecular dating of diversification events*	*- Taxon–area cladograms* *- Genetic data* *- Fossil data (when available)*	*- Coalescence* *- Molecular clocks*	*- Mesquite* *- BEAST* *- BEAST*[a]	*Infer age of divergence for nodes on each taxon–area cladogram using molecular dating techniques.*
3	Construct general area cladogram (GAC)	*- Dated* taxon area cladograms	*- mPACT*	N/A	Follow PACT protocols for construction of a GAC, except: *i) identify congruent events by both pattern as well as age, and ii) record number of congruent events for each branch. After the GAC has been constructed, convert the number of events (E) to 1/E for each branch.*
4	Ancestral area reconstruction	- GAC	- Parsimony *- Maximum Likelihood*	*- MacClade* - *Mesquite* - *LaGrange*	Infer ancestral states using the preferred optimization method(s). Bayesian approaches, such as those used in BayesTraits and s-DIVA/RASP are currently not compatible with mPACT because they evaluate numerous topologies and at present only one GAC is produced using the mPACT protocol.
5	Analyze nodes - BE from IS events *- General from unique events*	- GAC with ancestral areas	N/A	N/A	Count the number of nodes for each type of event: biotic expansion (BE), *in situ* (IS), *general, and unique. Dispersal from vicariance may differ in that dispersal would be unique BE events whereas vicariance would be general BE events.*

[a] Instances where the mPACT approach differs from the PACT approach are depicted in bold-italics. In cases where algorithms or programs indicate N/A, these tasks are done by hand as indicated under "process."

includes the following steps: (1) common elements (Y) between two cladograms are combined (Y + Y = Y); (2) novel elements (N) in a cladogram are retained into the output GAC (Y + N = YN); (3) superficially similar events that occur at different nodes are not combined (Y(Y– = Y(Y–))), not (Y(Y– = Y)); (4) common elements are retained even in the presence of novel elements (Y + YN = YN); and (5) multiple novel elements are retained, but until further information is provided may remain unresolved (YN′ + YN = YN′N).

12.2.2.3   Step 3: GAC Optimization

The GAC was then analyzed using the parsimony-based Delayed Transformation (DELTRAN) optimization with Mesquite version 2.5 (Maddison and Maddison 2008) to infer ancestral areas. In cases where strict parsimony optimization cannot distinguish between two character states at a node, DELTRAN optimization favors convergences by retaining the pre-existing ancestral state and transformation occurs at the last possible moment, as shown in Figure 12.7A. The alternative approach, Accelerated Transformation (ACCTRAN), favors reversals by transforming ancestral states at the earliest possible moment, as shown in Figure 12.7B. Although ACCTRAN can also discover extinction–recolonization dispersal events, we used DELTRAN optimization because it favors an explanation of a primitively widespread distribution and subsequent dispersal (Wiley 1986, 1988a,b; Agnarsson and Miller 2008; Eckstut et al. 2011).

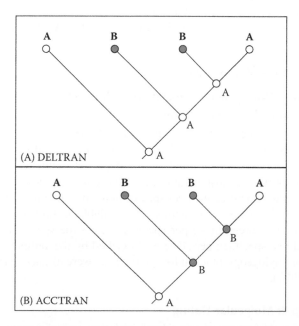

FIGURE 12.7   Comparison of parsimony ancestral state reconstructions in a hypothetical area cladogram. Bolded "A" and "B" at the tree tips represent areas where lineages occur. Regular "A" and "B" by nodes represent reconstructed ancestral states using (A) Delayed Transformation (DELTRAN) and (B) Accelerated Transformation (ACCTRAN).

12.2.2.4 Step 4: GAC Interpretation

The PACT evolutionary biogeographic data can be used to test the species–area relationship (SAR) in an evolutionary context (Halas et al. 2005; Hoberg and Brooks 2010; Eckstut et al. 2011). With the SAR, number of species should linearly correlate with area size if diversity is at equilibrium (Arrhenius 1921; MacArthur and Wilson 1963, 1967). Eckstut et al. (2011) found that the age of islands in the Hawaiian Islands and Greater Antilles were reflected in the SARs: there was a strong linear correlation of number of diversification events and area size in the older Greater Antilles and a nonsignificant, weak correlation in the younger Hawaiian Islands because the big island of Hawaii is both the largest and youngest island. Therefore, PACT can be used to test if a region is at evolutionary equilibrium. We herein refer to SARs as diversification–area relationships in the context of PACT to reflect the nature of these plots, which represent diversification events (and not species) in the GAC.

We examined the total number of OTUs (operational taxonomic units; i.e., number of tips on the GAC), and using GAC data inferred from DELTRAN ancestral state reconstruction, we calculated the total number of lineages per area, number of biotic expansion (BE) events, and number of *in situ* (IS) events. We used OTU, BE, and IS data to construct diversification–area plots (with area in km^2). We then tested diversification–area relationships using the power-law species–area relationship (SAR; $S=cA^z$; Arrhenius 1921; MacArthur and Wilson 1963, 1967) for total lineages–area relationship and linear regressions to test biotic expansion–area and *in situ*–area relationships (Halas et al. 2005; Eckstut et al. 2011). Preliminary analysis of residuals depicted nonuniformity of data; thus, all data were log-transformed for analysis. Statistical analyses were run using the Microsoft Excel Data Analysis package (2007).

12.2.3 MODIFIED PACT ANALYSIS

12.2.3.1 Step 1: Obtain Sequence Data and Convert to Taxon–Area Cladograms

We obtained DNA sequences from National Center for Biotechnology Information's GenBank database (http://ncbi.nlm.nih.gov/) and locality information for each sequence from GenBank, the original source paper, museum records, or private records of the authors, as shown in Table 12.1. To avoid overpredicting *in situ* diversification, we selected one specimen per species or a representation from each region for interspecies studies (if the information was available; in some cases, researchers only sequenced one representative per species, even if the species is widespread), and we selected one specimen per clade as described by the authors of the original source paper for phylogeographic studies. Sequences were aligned using ClustalW2 (Larkin et al. 2007).

12.2.3.2 Step 2: Molecular Dating

To estimate the most appropriate models of evolution and prior parameters for the analyses, we used the Akaike Information Criterion (AIC; Akaike 1973, 1974) and either PALM (http://palm.iis.sinica.edu.tw/), a parallel computing cluster (Chen et al. 2009), or MEGA5 (Tamura et al. 2011). The rate of substitution and date of gene divergence were estimated using BEAST v1.4.8 (Drummond and Rambaut 2007) on either the University of Oslo Bioportal (http://bioportal.uio.no) or the BioHPC v1.4.8 parallel computing cluster at the Cornell University Computational Biology Service Unit (http://biohpc.org/default.aspx). Multiple programs and clusters were used because of availability at the time of analysis. However, several examples were tested on both sets of programs and clusters to ensure that results were comparable between the sources.

All BEAST analyses were run using the Yule Process tree prior and run three times for 10^7 Markov Chain Monte Carlo (MCMC) chains that were sampled every 1000 iterations (Drummond and Rambaut 2007). All iterations were then pooled using Logcombiner v1.4.8 and log files were analyzed in Tracer v1.4.1 (Rambaut and Drummond 2007), and effective sample sizes (ESS) were used to evaluate the estimates of posterior distributions (Drummond and Rambaut 2007). The first 10% of all analyses were discarded for burn-in. A summary of the output trees was generated with TreeAnnotator v1.4.8 (Drummond and Rambaut 2007), discarding the first 1000 trees as burn-in, and then analyzed with FigTree v1.2.1 (Rambaut 2008).

We conducted each analysis in three ways using both relaxed and strict clock models: (1) universal substitution rates; (2) substitution rates that were both half and double the universal clock rates; and (3) (when possible) fossil calibration with a lognormal prior with a standard deviation of 1 to designate the split. Fossil calibrations suffer from numerous potential issues, including incorrect placement on the tree, insufficiency of single fossil calibrations, and they can only provide minimal estimates of divergence times (e.g., Yang and Rannala 2006; Donoghue and Benton 2007; Marshall 2008). For this reason, fossil calibration was used when records were available and used by other researchers on the same system.

We tested clock-like evolution of each phylogeny with relaxed clock simulations, where ucld.stdev values closer to 0.0 were considered clock-like and had a strict clock implemented and those lineages with ucld.stdev values greater than 1.0 were not clock-like had relaxed clocks implemented. For simulations that had ucld.stdev values between 0.35–0.65, we conducted analyses using both strict and relaxed clock models. We performed likelihood ratio tests (LRTs) to evaluate which models were statistically different in terms of likelihood scores, and we used the analysis with the highest significant likelihood score and posterior probability for subsequent analyses. In instances where there was not one analysis with distinctly higher likelihood scores or posterior probabilities, we included all analyses with similar scores to estimate divergence times and we incorporated all ranges of possible dates.

We acknowledge that molecular dating, especially with only one gene, can give erroneous dates, and that diversification dates may range for different taxa in response to the same biogeographic event because of varying levels of gene flow and molecular evolution rates, and the event may take place over a long period of time. Because of these potential molecular dating problems, we incorporated the 95% confidence interval of BEAST divergence dates for all nodes and placed these dates into general, but biogeographically meaningful, categories: late (0.011–0.126 mya), mid (0.126–0.781 mya), or early (0.781–2.58 mya) Pleistocene; late (2.58–3.6 mya) or early (3.6–5.3 mya) Pliocene; and late Miocene (5.3–11.6 mya). When there was overlap in dates, the node ages would be widened to include all possible dates because inclusion of all ranges maximizes information in the GAC (Folinsbee and Evans 2012).

12.2.3.3   Step 3: GAC Construction

When constructing GACs, we followed the same rules as described by Wojcicki and Brooks (2004, 2005) and detailed in the previous section ("Standard PACT Analysis"), with the exception that (i) we hypothesized congruent events by age of diversification as estimated with molecular dating, and (ii) we distinguished general from unique events by recording every congruent event (E) on each branch (e.g., if three clades exhibit a diversification at area A in the late Pleistocene, $E = 3$ for branch A), as shown in Figure 12.4. For unique branches, only one character was used to describe each branch. This approach facilitates the distinction of unique from general events by providing information for relative branch lengths (i.e., clade-unique dispersal would likely only happen once on the GAC, whereas a vicariant event or concordant dispersal would occur numerous times). Situations were excluded in which there were multiple equivocal input taxon–area cladogram placements on the GAC from analysis.

12.2.3.4   Step 4: GAC Optimization

Once we constructed the GAC, we conducted ancestral area reconstruction using both likelihood and the parsimony-based delayed transformation, or DELTRAN, optimizations. We used traditional parsimony in Mesquite version 2.5 (Maddison and Maddison 2008) because the GAC contains polytomies (which prohibits optimization using DELTRAN), and in cases where there was still ambiguity with regard to node state, we implemented DELTRAN (Wiley 1986, 1988a,b).

Maximum Likelihood ancestral area reconstructions favor shorter branches to reconstruct ancestral nodes because it is inferred that the descendants are more similar to the ancestor and that the probability of change is less likely on a short branch than on a long branch. Therefore, we calculated branch lengths as $1/E$, where E is the number of events inferred for each branch. We designated this value as $1/E$ because that weights the ancestral areas toward general events that more likely result from congruent diversification mechanisms (e.g., vicariance), whereas unique events are more likely to be the result of idiosyncratic dispersal and should not represent the ancestral state. However, there are polytomies present in the GAC, which means that traditional ancestral area reconstruction programs and models (e.g., DIVA, Ronquist 1997; Mesquite, Maddison and Maddison 2008; and Dispersal–Extinction–Cladogenesis, Ree and Smith 2008) cannot be used to infer Maximum Likelihood ancestral area reconstructions. In order to conduct Maximum Likelihood reconstruction, we first used Mesquite version 2.5 (Maddison and Maddison 2008) to build the GAC with branch lengths and then all instances of polytomies were instead converted to bifurcations with a branch length of zero. Subsequently, the saved Nexus file was imported into the APE package (Paradis et al. 2004) in R version 2.15.2 (R Development Core Team 2008). We then used the ancestral character estimation (ACE) analysis within APE to reconstruct areas using three models for discrete characters: equal rates (ER), symmetric (SYM), and all rates different (ARD). Likelihoods for each model were compared, and significance was tested using a likelihood test using chi-squares to determine which model was most suitable. In cases where there were ambiguities at internal nodes (equal probabilities), a delayed transformation was favored to remain consistent with favoring biotic expansion.

12.2.3.5 Step 5: GAC Interpretation

We interpreted the mPACT GAC similar to how we interpreted the standard PACT GAC. We examined the total number of OTUs (operational taxonomic units) and we calculated the total number of lineages per area, number of BE events, and number of IS events. We used OTU, BE, and IS data to construct diversification–area plots for each of the two mPACT datasets (mPACT with DELTRAN and mPACT with Maximum Likelihood optimization) for each region. We then tested diversification–area relationships using the SAR (Arrhenius 1921; MacArthur and Wilson 1963, 1967) and linear regressions to test biotic expansion–area and *in situ*–area relationships (Halas et al. 2005; Eckstut et al. 2011). Preliminary analysis of residuals depicted nonuniformity of data and all data were log-transformed for analysis. Statistical analyses were run using the Microsoft Excel Data Analysis package (2007).

12.3 RESULTS

12.3.1 LIKELIHOOD ANCESTRAL AREA RECONSTRUCTION MODEL SELECTION

The three reconstruction methods (ER, SYM, and ARD) yielded the following likelihoods, respectively: -40.20254, -38.31762, and -36.31635. The likelihood test showed that the ARD model was more suitable than either ER ($p < 0.01$) or SYM ($p = 0.045$) models at $p < 0.05$ for Maximum Likelihood ancestral area reconstruction of the mPACT dataset.

The Secondary BPA tree yielded a tree with seven operational taxonomic units (areas; OTUs) and six nodes; PACT with 18 OTUs and 16 nodes, as shown in Figure 12.8; and mPACT with 30 OTUs and 25 nodes, as shown in Figure 12.9 and Table 12.3. For the PACT analysis (using DELTRAN optimization), 31.25% of nodes were *in situ* nodes, 31.25% were *in situ*/biotic expansion nodes, 37.5% were ambiguous, and there were no nodes showing full biotic expansion, as shown in Figure 12.10 and Table 12.3. mPACT analysis using DELTRAN ancestral area reconstruction had 48% ambiguous nodes, 28% *in situ*, 20% *in situ*/biotic expansion, and 4% full biotic expansion, as shown in Figure 12.11 and Table 12.3. Finally, mPACT analysis using Maximum Likelihood ancestral area reconstruction had 8% ambiguous nodes, 40% *in situ* nodes, and 52% *in situ*/biotic expansion nodes, as shown in Figure 12.12 and Table 12.3.

12.3.2 DIVERSIFICATION–AREA RELATIONSHIPS

Table 12.4 and Figure 12.11 show the data that were collected for these analyses. Diversification–area relationships for total lineages were weak and statistically not significant for all three analyses, as shown in Figure 12.13A (PACT-DELTRAN: $c = 0.06$, $z = 0.48$, $R^2 = 0.22$, $p = 0.69$; mPACT-DELTRAN: $c = 0.09$, $z = 0.45$,

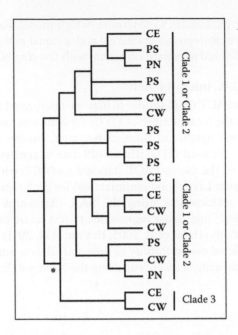

FIGURE 12.8   General area cladogram (GAC) derived from standard PACT analysis of the North American warm deserts. The placement of Clade 3 differs between the PACT GAC and Riddle and Hafner's (2006) Secondary BPA GAC (as indicated with a * at the alternate nodes).

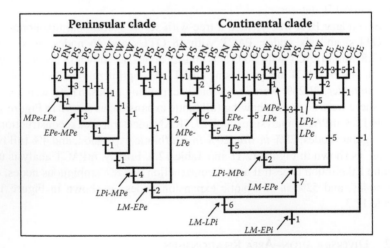

FIGURE 12.9   North American warm desert mPACT tree. Tick marks indicate number of lineages that experienced the event (degree of generality). Node ages are indicated as follows: *LM* = Late Miocene (11.6–5.3 mya), *EPi* = Early Pliocene (5.3–3.6 mya), *LPi* = Late Pliocene (3.6–2.58 mya), *EPe* = Early Pleistocene (2.58–0.781 mya), *MPe* = Mid Pleistocene (0.781–0.126 mya), and *LPe* = Late Pleistocene (0.126–0.011 mya).

TABLE 12.3

General Area Cladogram (GAC) Node Analysis

GAC Node Parameters	Secondary BPA	PACT: DELTRAN	mPACT: DELTRAN	mPACT: LnL
OTUs (tree tips)	7	18	30	30
Total nodes	6	16	25	25
In situ nodes	N/A	5	7	10
In situ/biotic expansion	N/A	5	5	13
Nodes				
Full biotic expansion nodes	N/A	0	1	0
Ambiguous nodes	N/A	6	12	2

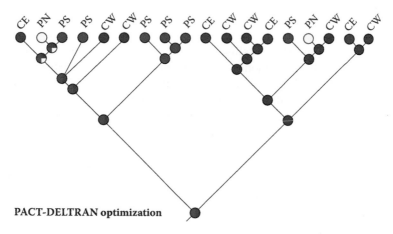

PACT-DELTRAN optimization

FIGURE 12.10 North American warm desert standard PACT tree with DELTRAN optimization.

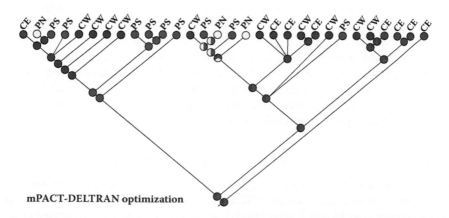

mPACT-DELTRAN optimization

FIGURE 12.11 North American warm desert mPACT tree with DELTRAN optimization.

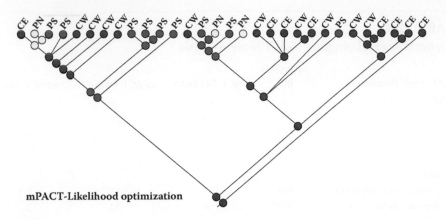

mPACT-Likelihood optimization

FIGURE 12.12   North American warm desert mPACT tree with Maximum Likelihood optimization.

$R^2 = 0.37$, $p = 0.50$; mPACT-Likelihood: $c = 0.36$, $z = 0.20$, $R^2 = 0.16$, $p = 0.79$). Additionally, all analyses for biotic expansion– and *in situ*–area relationships were also not significant. For biotic expansion–area relationships, PACT-DELTRAN and mPACT-DELTRAN yielded more similar results compared to mPACT-Likelihood, as shown in Figure 12.13B (PACT-DELTRAN: $y = 0.36x–1.44$, $R^2 = 0.49$, $p = 0.40$; mPACT-DELTRAN: $y = 0.43x–1.74$, $R^2 = 0.50$, $p = 0.38$; mPACT-Likelihood: $y = 0.11x + 0.32$, $R^2 = 0.16$, $p = 0.75$). However, all results differed for *in situ*–area relationships, as shown in Figure 12.13C; (PACT-DELTRAN: $y = 27x–1.09$, $R^2 = 0.13$, $p = 0.71$; mPACT-DELTRAN: $y = 0.16x–0.46$, $R^2 = 0.10$, $p = 0.82$; mPACT-Likelihood: $y = –0.02x + 0.71$, $R^2 < 0.01$, $p = 0.86$).

12.4   DISCUSSION

Our newly developed method, mPACT, was able to tease apart evolutionary complexity in the North American warm deserts in a way that previous methods could not. This study revealed that methods that exclude time (Secondary BPA and PACT) underpredicted the amount of pseudo-congruent diversification patterns. Moreover, parsimony-based ancestral area reconstruction produced more ambiguities and reconstructed less intuitive ancestral areas compared to likelihood-based reconstruction that incorporated branch lengths. We were also able to reveal interesting, novel patterns of diversification–area relationships in this region that can lead to interesting ideas regarding the evolutionary nature of areas of endemism.

12.4.1   REVEALING THE PREVALENCE OF PSEUDO-CONGRUENCE IN ANALYSES

The incorporation of a temporal component into analysis produced a more complex GAC compared to either standard PACT analysis or secondary BPA analysis (30 OTUs compared to 18 and 7, respectively), as shown in Table 12.3. Increased GAC complexity indicates that pseudo-congruence played a role in underestimating the

TABLE 12.4
Diversification–Area Relationships; PACT vs. mPACT

Area of Endemism	Area Size (km²)	PACT: Total Lineages DT	PACT: BE Events DT	PACT: BE Out Events DT	PACT: BE In Events DT	PACT: IS Events DT	mPACT: Total Lineages		mPACT: BE Events		mPACT: BE Out Events		mPACT: BE In Events		mPACT: IS Events	
							ML	DT	ML	DT	ML	DT	ML	DT	ML	DT
PS	48,500	8	3	0	1	2	17	13	8	3	4	1	4	2	8	4
PN	48,000	2	1	0	1	0	5	3	5	1	2	0	3	1	2	0
CW	386,500	14	5	5	0	8	21	15	11	7	9	4	2	3	17	4
CE	479,000	4	3	0	3	0	9	11	6	3	0	2	6	1	1	2

Total counts indicate resolved nodes as well as operational taxonomic units (OTUs). DT indicates the parsimony-based Delayed Transformation (DELTRAN) optimization, whereas ML indicates Maximum Likelihood-based optimization. BE in and BE out indicates biotic expansion into and out of, respectively, the designated region. IS indicates in situ diversification events. Ambiguous nodes and events were discarded for this analysis.

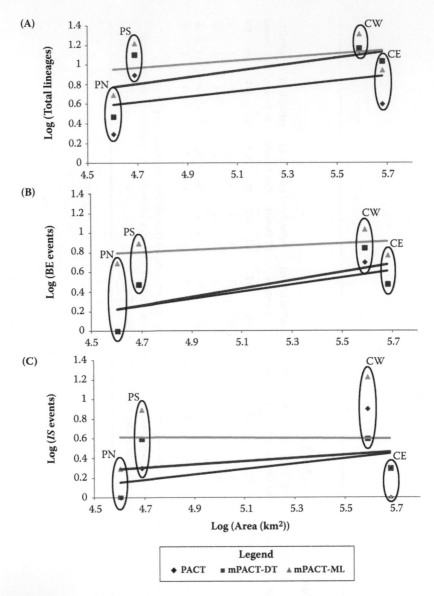

FIGURE 12.13 mPACT vs. PACT diversification–area relationship results. (See text for explanation.)

total number of diversification events prior to the incorporation of temporally calibrated area cladograms.

The input taxon–area cladograms show several instances of pseudo-congruent patterns among taxa, and there are event pseudo-congruent patterns within types of organisms (e.g., reptiles, rodents, or plants), as shown in Figure 12.6 and Table 12.1. Continental East/Continental West divergences occurred in the

Chaetodipus nelsoni, *Onychomys*, and *Kinosternon flavescens* groups (two rodents and a reptile, respectively) during the Pliocene (5.3–2.58 mya), whereas this split is estimated to have occurred in the *Peremicus eremicus* group (rodents) during the early Pleistocene (2.58–0.781 mya) and the *Uta stansburiana* group (reptiles) during the mid-to-late Pleistocene (0.781–0.011 mya). Continental West/Peninsular divergences occurred during the early Pliocene (5.3–3.6 mya) in the reptile *Sauromalus*, the Pliocene to mid-Pleistocene (3.7–0.781 mya) in the rodent *C. baileyi* group, and the late Pleistocene (0.126–0.011 mya) in the plant *Lophocereus schotti*. Finally, there were also pseudo-congruent patterns in Peninsular North/Peninsular South divergences, which was also found in mammals and reptiles using hierarchical approximate Bayesian computation (Leaché et al. 2007), although these divergences in general were younger than in the continental divergences. *Sauromalus* experienced Peninsular North/Peninsular South divergence during the late Pliocene–early Pleistocene (3.6–0.781 mya), whereas the *C. baileyi* group dispersed and subsequently diverged from south to north in the mid-Pleistocene (0.781–0.126 mya), the *C. arenarius* and *P. eremicus* groups diverged in the mid–late Pleistocene (0.781–0.011 mya) and the *U. stansburiana* group diverged in the late Pleistocene (0.126–0.011 mya).

These patterns show that temporally and mechanistically distinct historical events (e.g., mountain uplift or glacial–interglacial events) can frequently produce similar biogeographic patterns among taxa. In this case, pseudo-congruent patterns can be produced by vicariant events at one point in time (e.g., as the result of tectonic uplift) and dispersal events at a different point in time (e.g., resulting from climate change). In older events circa the Quaternary–Neogene transition tectonic uplift events (particularly during the late Pliocene to early Pleistocene, 3.6–0.781 mya), vicariance as a result of mountain and plateau uplift was likely a driving force of diversification patterns, whereas dispersal was more likely prevalent during the Quaternary glacial–interglacial cycles (2.58 mya–present). Consequently, these results underscore the importance of incorporating explicit tectonic versus orbital forcing climate information and models when testing biogeographic hypotheses.

12.4.2　LIKELIHOOD-BASED ANCESTRAL AREA RECONSTRUCTION

With regard to ancestral area reconstruction, the parsimony-based DELTRAN optimization produced ambiguity in both standard PACT and mPACT trees, with 37.5% of PACT nodes and 48% of mPACT nodes, as shown in Figures 12.10 and 12.11, respectively. These GACs are highly complex and thus difficult to resolve with parsimony. However, by incorporating general events to create branch lengths and subsequently conducting likelihood ancestral state optimization, we were able to reduce the number of ambiguities in mPACT node reconstruction (8% of nodes), as shown in Figure 12.12. With the Maximum Likelihood optimization, ancestral nodes were weighted more heavily toward general events (i.e., diversification events that were represented in numerous taxa compared to unique events that were only represented by one taxon). For this reason, there were differences in the ability of the optimization to resolve nodes that DELTRAN could not, and both resolved ambiguous nodes differed between trees, as shown in Figures 12.11

and 12.12. The mPACT-Likelihood tree, as shown in Figure 12.12, takes into account the amount of general events, and because the Continental West has more general events, it weights these basal nodes to ambiguous between Continental East or Continental West.

12.4.3 Diversification–Area Relationships

All diversification–area relationships were weak and statistically not significant, as shown in Figure 12.13 and Table 12.4. However, all analyses showed similar patterns, with the Peninsular South (48,500 km²) and Continental West (386,500 km²) yielding the highest amount of diversification, and the Peninsular North (48,000 km²) and Continental East (479,000 km²) yielding the least amount of diversification.

The equilibrium theory of island biogeography (ETIB; MacArthur and Wilson 1963, 1967) suggests that a SAR should be linear when richness is perfectly correlated with area, based on the mathematical relationship ($S = cA^z$) described by Arrhenius (1921). Under the ETIB model, we would expect the Peninsular North and South to yield the fewest diversification events and the Continental West and East to yield the most diversification events. The lack of a linear correlation between diversification events and area size (because the Peninsular South and Continental West had the most amount of diversification) could indicate that the Peninsular North and Continental East have not yet reached stable equilibrium (perhaps as a result of biotic instability during paleoclimatic oscillations), or this could have implications for the Continental West and Peninsular South as strong sources of biodiversity in the North American warm deserts.

Substantial debate has occurred on what constitutes an "area of endemism" (e.g., Platnick 1991; Riddle 1998; Humphries and Parenti 1999; Parenti and Ebach 2009; Crother and Murray 2011). Currently, operational approaches exist to diagnose areas of endemism using modern distribution data and grid-based approaches (e.g., Rosen 1988; Linder 2001; Szumik et al. 2002; Szumik and Goloboff 2004; Morrone 2014; Oliveira et al. 2015). Riddle and Hafner (2006) suggested that areas of endemism represent areas that more stably produce and maintain biodiversity and thus an evolutionary approach should be incorporated into area of endemism diagnosis, and current diagnostic techniques insufficiently describe the areas of endemism using the grid-based method Parsimony Analysis of Endemicity (PAE; Rosen 1988; Morrone 2014), which predicts areas of endemism only based on current distribution data.

Riddle and Hafner (2006) corroborated the proposal that the Continental West, Continental East, Peninsular South, and Peninsular North were all areas of endemism, but the data here indicate that these may not all be equal with regard to biodiversity production and stability. For example, even though the Continental East is the largest of the regional warm deserts, the Continental West and Peninsular South biotas had more diversification events, as shown in Figure 12.13. Moreover, the Peninsular North was unique among the areas of endemism because it only appeared three times on the GAC (as opposed to eight times for the Continental East and Peninsular South and nine times for the Continental West), and all were instances of general events (divergence events occurring in 2, 3, and 5 taxa), as shown in Figure 12.9. Interestingly, all of the analyzed events in the Peninsular North occurred during Pleistocene times and did not represent any ancestral areas, as shown in Figures 12.9 and 12.11.

Savage (1960) suggested that herpetofaunal distributions in the Baja California peninsula of modern species were shaped by Pleistocene events, wherein the cape region was likely a refuge during glacial maxima and there were invasions from the northeast, continental mainland and expansion from the cape during interglacial cycles. The general Peninsular North events and the overwhelming number of Peninsular North/Peninsular South divergences during the Pleistocene (see discussion section "Revealing the Prevalence of Pseudo-congruence in Analyses" above for examples) may be consistent with Savage's (1960) hypothesis. The Peninsular North may have been less stable through time than the Peninsular South, which may have been an area where organisms were better able to persist. Alternatively, perhaps the Peninsular North is a region that was inaccurately diagnosed as an area of endemism by the method PAE because of the large number of species that are currently distributed there.

Exploring the evolutionary dynamics of areas of endemism is something that was largely overlooked in the past, but has been addressed more recently (Crother and Murray 2011; Murray and Crother 2015). Riddle and Hafner (2006) conducted an analysis in which they reconstructed the evolutionary history of the diagnosed areas of endemism in the North American warm deserts, but their analysis lacked detail regarding analysis of the differences among these areas of endemism. Furthermore, even though methods exist to diagnose areas of endemism, experimental approaches to testing the evolutionary dynamics of areas of endemism are limited, and more sufficient analyses are needed to adequately test these hypothetical ideas (Riddle and Hafner 2006; Crother and Murray 2011).

12.4.4 mPACT Caveats

Whereas the mPACT protocol provides a unique means to explore the evolution of biodiversity across time and space, there are four inherent issues with the mPACT algorithm, as shown in Table 12.5. These issues include: (i) inability of likelihood ancestral area reconstruction to resolve complex mPACT GACs, (ii) assumption of appropriate molecular dating techniques and node ages, (iii) equivocal possibilities when generating the GAC, and (iv) the lack of a software program to conduct mPACT analyses.

Preliminary analysis of more complex mPACT GACs (81 OTUs representing 17 areas) showed that applicable likelihood-based ancestral state reconstructions are unable to resolve ancestral states for large, complex GACs because most ancestral states are reconstructed as equivocal. Most likelihood-based approaches require fully bifurcating tree topologies (e.g., DIVA, Ronquist 1997; Dispersal–Extinction–Cladogenesis, Ree and Smith 2008; Mesquite, Maddison and Maddison 2008), and full bifurcation is uncommon for mPACT GACs because of the PACT rules $Y + N + N = YNN$ and $YN' + YN = YN'N$. The inability of current models and software to resolve more complex GACs means that the Maximum Likelihood algorithms are

TABLE 12.5

Comparative Inference Types, Benefits, and Disadvantages of Secondary BPA, PACT, and mPACT

Algorithm	Inference Type	Benefits	Disadvantages
Secondary BPA	– Historical	– Allows inference into general patterns of area relationships	– Drastically oversimplifies amount of regional diversification – Restricted to historical inference
PACT	– Historical – Ecological	– Reveals more realistic degree of diversification than Secondary BPA – Can distinguish biotic expansion from *in situ* diversification events	– Pseudo-congruence – Cannot distinguish types of biotic expansion (vicariance, dispersal) – Limited options for ancestral area reconstruction – Laborious and time intensive; currently no software program available
mPACT	– Historical – Ecological	– Reduces pseudo-congruence – Facilitates distinction of dispersal from vicariance	– Likelihood-analyses may not be able to be conducted on excessively large, complex GACs – Assumes node ages (and thus, molecular dating techniques) are accurate – Equivocal possibilities impact GAC structure; currently no way to incorporate – Laborious and time intensive; currently no software program available

Note: All three assume accuracy of input cladograms.

not capable of dealing with the amount of complexity in large-scale analyses (e.g., variability of branch lengths and number of regions). Although the accuracy and performance of likelihood- compared with parsimony-based phylogenetic methods have been debated (e.g., Saitou and Imanishi 1989; Yang 1996; Swofford et al. 2001; Kolaczkowski and Thornton 2004), Maximum Likelihood ancestral state reconstruction approaches are commonly used and may reveal more information about ancestral states than parsimony-based approaches (e.g., Cunningham et al. 1998; Mooers and Schluter 1999; Royer-Carenzi et al. 2013); a Google Scholar search for "likelihood ancestral state reconstruction" yielded 136 publications since 2000, and 112 (82.35%) of those were published since 2009. With growing interest in methods to integrate ecology and evolution, more refined mPACT-appropriate Maximum Likelihood ancestral area reconstruction approaches will likely become available.

As with all phylogenetic-based analyses, the product of the mPACT GAC directly reflects the quality of the input data. Both the input tree structure and lineage ages (from divergence dating) are derived from a variety of algorithms that have benefits as well as inherent potential flaws. The genetic data may be inappropriate for phylogenetic analysis—for example, if a gene is assumed to be selectively neutral but further research reveals selective pressures (e.g., mitochondrial DNA under certain conditions; Ballard and Kreitman 1995)—or there may be error in algorithms or fossil placement on trees for divergence dating (Donoghue and Benton 2007). Under these scenarios, the mPACT GAC would be inaccurate because the input trees would be inaccurate.

Appropriateness and accuracy of input data are not a fault of the mPACT algorithm, but are something that potential users should be aware of when interpreting GACs. For that reason, we remained conservative in the analysis. For example, any degree of overlap in timing of divergence between taxon–area cladograms are considered the "same" and then the GAC date range would encompass all dates (e.g., if node A on two trees yield the results 1.5–3 and 2–4 mya, the GAC age at node A would be 1.5–4 mya). There is frequently error in molecular dating estimates (e.g., Graur and Martin 2004) and our approach minimizes the number of unique events in the GAC but may over predict the degree of generality for nodes. However, we also note that percentage of overlap itself may be an important variable to consider in future analyses, particularly when you take into account comparison of different historical events that occur on different timescales. For example, mountain uplifting and vicariance may be a more gradual process and thus a smaller percentage of overlap may indicate a congruent diversification event. However, the shift of glacial–interglacial cycles and resulting dispersal is comparatively more abrupt, and requiring a higher percentage of overlap to identify congruence would be more appropriate for glacial–interglacial timeframes.

The use of taxon–area cladograms with dissimilar depths (e.g., using overall younger or older taxon–area cladograms) also can confound the construction of the mPACT GAC. For very young taxon–area cladograms, there are multiple possible placements within the tree, and each different GAC structure is then equally likely (equivocal). In this analysis, we excluded several diversification events because there are equivocal results (i.e., there are two or more possible placements for one diversification event on the GAC). Alternatively, older taxon–area cladograms can also be problematic because they can overestimate the total number of unique events. In our results, we found only two instances of exclusively Neogene-based divergence events (one general, one unique). These two events likely do not accurately reflect the number of deep Neogene divergences in the North American warm deserts. There are several lineages of North American desert organisms with deep Neogene divergences. For example, divergences in North American heteromyid rodents (Dipodomyinae and Perognathinae) are estimated to date back to 20 mya (Hafner et al. 2007). Using several deep lineages may provide more insight into general deep Neogene biodiversity formation as opposed to partitioning into subsets of taxa. For example, the taxon–area cladograms used in Riddle and Hafner (2006) and here focus mainly on within species or within species-groups. Thus, given both the issues of using young and old taxon–area cladograms, the use of taxon–area cladograms of approximately the same timescale would reduce uncertainties and potential inaccuracies in the mPACT GAC.

Finally, there is currently no mPACT software available for GAC construction. Lack of software makes mPACT and PACT analyses laborious, time intensive, and subject to human error. Labor, time, and potential error make mPACT analyses less desirable for many scientists and diminish the possible impact of mPACT on the scientific community. Development of mPACT software would be ideal for conducting mPACT analysis, because a user-friendly mPACT computer software would facilitate the implementation of the mPACT algorithm by the scientific community and reduce the potential for human error in generating GACs. Additionally, such software would provide more options for conduction of the mPACT algorithm based on user preference and philosophy regarding how to incorporate equivocal results, allow users to identify general events based on different node ages, and facilitate an enhanced understanding of how different input taxon–area cladograms affect the overall GAC.

12.5 CONCLUSIONS

The mPACT algorithm reveals more diversity in the North American warm deserts than was observed using either Secondary BPA or PACT analysis, and relieves several of the issues and concerns associated with PACT analysis (e.g., by reducing the impact of pseudo-congruence, revealing modes of diversification—either unique or general, and facilitating likelihood-based optimization). Incorporating a temporal component both reduces pseudo-congruence and facilitates inference of general or unique diversification events to distinguish possible vicariant and dispersal events. Therefore, this approach does not assume predominantly vicariance or dispersal for data analysis, and therefore does not inhibit progress in comparative biogeography based on those two assumptions listed by Rosen (2016).

Our conservative approach to considering dates with any overlap as the "same" may still overpredict pseudo-congruent events. Use of likelihood-based ancestral area reconstruction helps resolve ambiguities that were apparent in parsimony-based DELTRAN optimizations. However, despite the benefits of the mPACT protocols, mPACT is a much more liberal approach than PACT and may overestimate regional biodiversity dynamics. Even though mPACT still suffers from potential issues, as shown in Table 12.5, it is nonetheless an exciting new approach that can be used as an alternative to the more conservative PACT algorithm should scientists desire the opportunity to integrate taxon- and area-based biogeographic methods.

ACKNOWLEDGMENTS

We thank D. Thompson, J. Rodríguez-Robles, M. Lachniet, and T. Jezkova for helpful discussion and review of this manuscript, and J. Patton for providing some *Neotoma* sequences. This research was financially supported by the National Science Foundation under Cooperative Agreement No. EPS-0814372 through an NSF EPSCoR Climate Change Graduate Fellowship (2010–2013).

REFERENCES

Agnarsson, I. and J. A. Miller. 2008. Is ACCTRAN better than DELTRAN? *Cladistics* 24:1032–1038.
Akaike, H. 1973. Information theory and an extension of the maximum likelihood principle. In *Proceeding of the Second International Symposium on Information Theory*, ed. B. N. Petrov and F. Caski, 267–281. Budapest: Akademiai Kiado.

Akaike, H. 1974. A new look at the statistical model identification. *IEEE Transactions on Automatic Contributions* 19:716–723.

Alexander, L. F. and B. R. Riddle. 2005. Phylogenetics of the New World rodent family Heteromyidae. *Journal of Mammalogy* 86:366–379.

Arrhenius, O. 1921. Species and area. *Journal of Ecology* 9:95–99.

Avise, J. C. 2000. *Phylogeography: The History and Formation of Species*. Cambridge: Harvard University Press.

Ballard, J. W. O. and M. Kreitman. 1995. Is mitochondrial DNA a strictly neutral marker? *Trends in Ecology and Evolution* 10:485–488.

Brooks, D. R. 1990. Parsimony analysis in historical biogeography and coevolution: Methodological and theoretical update. *Systematic Biology* 39:14–30.

Brooks, D. R. and D. McLennan. 2002. *The Nature of Diversity: An Evolutionary Voyage of Discovery*. Chicago: University of Chicago Press.

Chen, S. H., S. Y. Su, C. Z. Lo, et al. (2009). PALM: A paralleled and integrated framework for phylogenetic inference with automatic likelihood model selectors. *PLoS One* 4:e8116.

Cowie, R. H. and B. S. Holland. 2008. Molecular biogeography and diversification of the endemic terrestrial fauna of the Hawaiian Islands. *Philosophical Transactions of the Royal Society, London B Biological Sciences* 363:3363–3376.

Croizat, L., G. Nelson, and D. E. Rosen. 1974. Centers of origin and related concepts. *Systematic Zoology* 23:265–287.

Crother, B. I. and C. M. Murray. 2011. Ontology of areas of endemism. *Journal of Biogeography* 38:1009–1015.

Cunningham, C. W. and T. M. Collins. 1994. Developing model systems for molecular biogeography: Vicariance and interchange in marine invertebrates. In *Molecular Ecology and Evolution: Approaches and Applications*, ed. B. Schierwater, B. Streit, G. P. Wagner, and R. DeSalle, 405–433. Basel: Birkhauser Verlad.

Cunningham, C. W., K. E. Omland, and T. H. Oakley. 1998. Reconstructing ancestral character states: A critical reappraisal. *Trends in Ecology and Evolution* 13:361–366.

Cunningham, C. W. 1999. Some limitations of ancestral character-state reconstruction when testing evolutionary hypotheses. *Systematic Biology* 48:665–674.

de Queiroz, A. 2005. The resurrection of oceanic dispersal in historical biogeography. *Trends in Ecology and Evolution* 20:68–73.

Donoghue, M. J. and B. R. Moore. 2003. Toward an integrative historical biogeography. *Integrative and Comparative Biology* 43:261–270.

Donoghue, P. C. and M. J. Benton. 2007. Rocks and clocks: Calibrating the tree of life using fossils and molecules. *Trends in Ecology and Evolution* 22:424–431.

Drummond, A. J., S. Y. Ho, M. J. Phillips, and A. Rambaut. 2006. Relaxed phylogenetics and dating with confidence. *PLoS Biology* 4:e88.

Drummond, A. J. and A. Rambaut. 2007. BEAST: Bayesian evolutionary analysis by sampling trees. *BMC Evolutionary Biology* 7:214.

Eckstut, M. E., C. D. McMahan, B. I. Crother, J. M. Ancheta, D. A. McLennan, and D. R. Brooks. 2011. PACT in practice: Comparative historical biogeographic patterns and species–area relationships of the Greater Antillean and Hawaiian Island terrestrial biotas. *Global Ecology and Biogeography* 20:545–557.

Folinsbee, K. E. and D. C. Evans. 2012. A protocol for temporal calibration of general area cladograms. *Journal of Biogeography* 39:688–697.

Graham, M. R., J. R. Jaeger, L. Prendini, and B. R. Riddle. 2013. Phylogeography of the Arizona hairy scorpion (*Hadrurus arizonensis*) supports a model of biotic assembly in the Mojave Desert and adds a new Pleistocene refugium. *Journal of Biogeography* 40:1298–1312.

Graur, D. and W. Martin. 2004. Reading the entrails of chickens: Molecular timescales of evolution and the illusion of precision. *Trends in Genetics* 20:80–86.

Hafner, J. C., J. E. Light, D. J. Hafner, et al. 2007. Basal clades and molecular systematics of heteromyid rodents. *Journal of Mammalogy* 88:1129–1145.

Halas, D., D. Zamparo, and D. R. Brooks. 2005. A historical biogeographical protocol for studying biotic diversification by taxon pulses. *Journal of Biogeography* 32:249–260.

Hennig, W. 1966. *Phylogenetic Systematics*. Urbana: University of Illinois Press.

Hoberg, E. P. and D. R. Brooks. 2010. Beyond vicariance: Integrating taxon pulses, ecological fitting and oscillation in evolution and historical biogeography. In *The Biogeography of Host-Parasite Interactions*, ed. S. Morand and B. Krasnov, 7–20. Oxford: Oxford University Press.

Humphries, C. J. and L. R. Parenti. 1999. *Cladistic Biogeography: Interpreting Patterns of Plant and Animal Distribution*. Oxford: Oxford University Press.

Jaeger, J. R., B. R. Riddle, and D. F. Bradford. 2005. Cryptic Neogene vicariance and Quaternary dispersal of the red-spotted toad (*Bufo punctatus*): Insights on the evolution of North American warm desert biotas. *Molecular Ecology* 14:3033–3048.

Jezkova, T., J. R. Jaeger, Z. L. Marshall, and B. R. Riddle. 2009. Pleistocene impacts on the phylogeography of the desert pocket mouse (*Chaetodipus penicillatus*). *Journal of Mammalogy* 90:306–320.

Kolaczkowski, B. and J. W. Thornton. 2004. Performance of maximum parsimony and likelihood phylogenetics when evolution is heterogeneous. *Nature* 431:980–984.

Larkin, M. A., G. Blackshields, N. P. Brown, et al. 2007. Clustal W and Clustal X version 2.0. *Bioinformatics* 23:2947–2948.

Leaché, A. D., S. C. Crews, and M. J. Hickerson. 2007. Two waves of diversification in mammals and reptiles of Baja California revealed by hierarchical Bayesian analysis. *Biology Letters* 3:646–650.

Lee, T. E., B. R. Riddle, and P. L. Lee. 1996. Speciation in the desert pocket mouse (Chaetodipus penicillatus Woodhouse). *Journal of Mammalogy* 77:58–68.

Lieberman, B. S. and N. Eldredge. 1996. Trilobite biogeography in the Middle Devonian: Geological processes and analytical methods. *Paleobiology* 22:66–79.

Lim, B. K. 2008. Historical biogeography of New World emballonurid bats (tribe Diclidurini): Taxon pulse diversification. *Journal of Biogeography* 35:1385–1401.

Linder, H. P. 2001. On areas of endemism, with an example from the African Restionaceae. *Systematic Biology* 50:892–912.

MacArthur, R. H. and E. O. Wilson. 1963. An equilibrium theory of insular zoogeography. *Evolution* 17:373–387.

MacArthur, R. and E. O. Wilson. 1967. *The Theory of Island Biography*. Princeton: Princeton University Press.

Maddison, W. P. and D. Maddison. 2008. Mesquite: A modular system for evolutionary analysis. Version 2.5. http://mesquiteproject.org.

Marshall, C. R. 2008. A simple method for bracketing absolute divergence times on molecular phylogenies using multiple fossil calibration points. *American Naturalist* 171:726–742.

Microsoft. 2007. Microsoft Excel. Redmond, Washington.

Mooers, A. Ø. and D. Schluter. 1999. Reconstructing ancestor states with maximum likelihood: Support for one- and two-rate models. *Systematic Biology* 48:623–633.

Morrone, J. 2009. *Evolutionary Biogeography: An Integrative Approach with Case Studies*. New York: Columbia University Press.

Morrone, J. J. 2014. Parsimony analysis of endemicity (PAE) revisited. *Journal of Biogeography* 41:842–854.

Murphy R. and G. Aguirre-Léon. 2002. The nonavian reptiles. In *A New Island Biogeography of the Sea of Cortes*, ed. T. Case, M. Cody, and E. Ezcurra, 181–220. New York: Oxford University Press.

Murray, C. M. and B. I. Crother. 2015. Entities on a temporal scale. *Acta Biotheoretica* 1–10.

Nason, J. D., J. L. Hamrick, and T. H. Fleming. 2002. Historical vicariance and postglacial colonization effects on the evolution of genetic structure in *Lophocereus*, a Sonoran Desert columnar cactus. *Evolution* 56:2214–2226.

Nelson, G. and N. Platnick. 1981. *Systematics and Biogeography: Cladistics and Vicariance*. New York: Columbia University Press.

Nelson, G. and P. Y. Ladiges. 1996. Paralogy in cladistic biogeography and analysis of paralogy-free subtrees. *American Museum Novitates* 3167:1–58.

Oliveira, U., A. D. Brescovit, and A. J. Santos. 2015. Delimiting areas of endemism through kernel interpolation. *PLoS One* 10:e0116673.

Page, R. D. M. 1994. Maps between trees and cladistic analysis of historical associations among genes, organisms and areas. *Systematic Biology* 43:58–77.

Paradis, E., J. Claude, and K. Strimmer. 2004. APE: Analyses of phylogenetics and evolution in R language. *Bioinformatics* 20:289–290.

Parenti, L. R. 2007. Common cause and historical biogeography. In *Biogeography in a Changing World*, ed. M. C. Ebach and R. S. Tangney 61–71. London: CRC Press.

Parenti, L. R. and M. C. Ebach. 2009. *Comparative Biogeography: Discovering and Classifying Biogeographical Patterns of a Dynamic Earth*. Berkeley: University of California Press.

Patton, J. L., D. G. Huckaby, and S. T. Álvarez-Castañeda. 2008. The evolutionary history and a systematic revision of woodrats of the *Neotoma lepida* group. *University of California Publications in Zoology* 135:1–472.

Petren, K. and T. J. Case. 1997. A phylogenetic analysis of body size evolution and biogeography in chuckwallas (*Sauromalus*) and other iguanines. *Evolution* 51:206–219.

Petren, K. and T. Case. 2002. Updated mtDNA phylogeny for *Sauromalus* and implications for the evolution of gigantism. In *A New Island Biogeography of the Sea of Cortes*, ed. T. Case, M. Cody, and E. Ezcurra, 574–579. New York: Oxford University Press.

Pirie, M. D., A. M. Humphreys, A. Antonelli, C. Galley, and H. P. Linder. 2012. Model uncertainty in ancestral area reconstruction: A parsimonious solution? *Taxon* 61:652–664.

Planz, J. 1992. Molecular phylogeny and evolution of the American woodrats, genus *Neotoma* (Muridae). PhD Dissertation, University of North Texas.

Platnick, N. I. 1991. On areas of endemism. *Australian Systematic Botany* 4(1):xi–xii.

Platnick, N. I. and G. Nelson. 1978. A method of analysis for historical biogeography. *Systematic Zoology* 27:1–16.

Pyron, R. A. and F. T. Burbrink. 2010. Hard and soft allopatry: Physically and ecologically mediated modes of geographic speciation. *Journal of Biogeography* 37:2005–2015.

R Development Core Team. 2008. *R: A Language and Environment for Statistical Computing*. R Foundation for Statistical Computing, Vienna, Austria. ISBN 3-900051-07-0, URL http://www.R-project.org.

Radtkey, R. R., S. M. Fallon, and T. J. Case. 1997. Character displacement in some *Cnemidophorus* lizards revisited: A phylogenetic analysis. *Proceedings of the National Academy of Sciences USA* 94:9740–9745.

Rambaut, A. 2008. FigTree v. 1.1.2. Institute of Evolutionary Biology, University of Edinburgh, Edinburgh, United Kingdom.

Rambaut, A. and A. Drummond. 2007. Tracer v1.4. http://beast.bio.ed.ac.uk/Tracer.

Ree, R. H. and I. Sanmartín. 2009. Prospects and challenges for parametric models in historical biogeographical inference. *Journal of Biogeography* 36:1211–1220.

Ree, R. H. and S. A. Smith. 2008. Maximum likelihood inference of geographic range evolution by dispersal, local extinction, and cladogenesis. *Systematic Biology* 57:4–14.

Reeder, T. W., C. J. Cole, and H. C. Dessauer. 2002. Phylogenetic relationships of whiptail lizards of the genus *Cnemidophorus* (Squamata: Teiidae): A test of monophyly, reevalua-
tion of karyotypic evolution, and review of hybrid origins. *American Museum Novitates*
3365:1–61.

Richards, C. L., B. C. Carstens, and L. Knowles. 2007. Distribution modelling and statistical
phylogeography: An integrative framework for generating and testing alternative biogeographical hypotheses. *Journal of Biogeography* 34:1833–1845.

Riddle, B. R. and R. L. Honeycutt. 1990. Historical biogeography in North American arid
regions: An approach using mitochondrial-DNA phylogeny in grasshopper mice (genus
Onychomys). *Evolution* 44:1–15.

Riddle, B. R. 1995. Molecular biogeography in the pocket mice (*Perognathus* and
Chaetodipus) and grasshopper mice (*Onychomys*): The late Cenozoic development of a
North American aridlands rodent guild. *Journal of Mammalogy* 76:283–301.

Riddle, B. R. 1998. The historical assembly of continental biotas: Late Quaternary range-
shifting, areas of endemism, and biogeographic structure in the North American mammal fauna. *Ecography* 21:437–446.

Riddle, B. R., D. J. Hafner, L. F. Alexander, and J. R. Jaeger. 2000a. Cryptic vicariance in the
historical assembly of a Baja California Peninsular Desert biota. *Proceedings of the
National Academy of Sciences USA* 97:14438–14443.

Riddle, B. R., D. J. Hafner, and L. F. Alexander. 2000b. Phylogeography and systematics of the
Peromyscus eremicus species group and the historical biogeography of North American
warm regional deserts. *Molecular Phylogenetics and Evolution* 17:145–160.

Riddle, B. R., D. J. Hafner, and L. F. Alexander. 2000c. Comparative phylogeography of
Baileys' pocket mouse (*Chaetodipus baileyi*) and the *Peromyscus eremicus* spe-
cies group: Historical vicariance of the Baja California Peninsular Desert. *Molecular
Phylogenetics and Evolution* 17:161–172.

Riddle, B. R. and D. J. Hafner. 2006. A step-wise approach to integrating phylogeographic and
phylogenetic biogeographic perspectives on the history of a core North American warm
deserts biota. *Journal of Arid Environments* 66:435–461.

Riddle, B. R., M. N. Dawson, E. A. Hadly, et al. 2008. The role of molecular genetics in
sculpting the future of integrative biogeography. *Progress in Physical Geography*
32:173–202.

Riddle, B. R. and D. J. Hafner. 2010. Integrating pattern with process at biogeographic boundaries: The legacy of Wallace. *Ecography* 33:321–325.

Riddle, B. R, T. Jezkova, M. E. Eckstut, V. Oláh-Hemmings, and L. N. Carraway. 2014.
Cryptic divergence and revised species taxonomy within the Great Basin pocket
mouse, *Perognathus parvus* (Peale, 1848), species group. *Journal of Mammalogy*
95:9–25.

Ronquist, F. 1997. Dispersal-vicariance analysis: A new approach to the quantification of
historical biogeography. *Systematic Biology* 46:195–203.

Rosen, B. R. 1988. From fossils to earth history: Applied historical biogeography. In *Analytical
Biogeography: An Integrated Approach to the Study of Animal and Plant Distributions*,
ed. A. Myers and P. Giller, 437–481. London: Chapman and Hall.

Rosen, D. E. 2016. Assumptions that *inhibit* scientific progress in comparative biology. In
Assumptions Inhibiting Progress in Comparative Biology, ed. B. I. Crother and L. R.
Parenti, 1–4. Boca Raton: CRC Press.

Royer-Carenzi, M., P. Pontarotti, and G. Didier. 2013. Choosing the best ancestral character
state reconstruction method. *Mathematical Biosciences* 242:95–109.

Saitou, N. and T. Imanishi. 1989. Relative efficiencies of the Fitch-Margoliash, maximum-
parsimony, maximum-likelihood, minimum-evolution, and neighbor-joining methods
of phylogenetic tree construction in obtaining the correct tree. *Molecular Biology and
Evolution* 6:514–525.

Savage, J. M. 1960. Evolution of a peninsular herpetofauna. *Systematic Zoology* 9:184–212.

Serb, J. M., C. A. Phillips, and J. B. Iverson. 2001. Molecular phylogeny and biogeography of *Kinosternon flavescens* based on complete mitochondrial control region sequences. *Molecular Phylogenetics and Evolution* 18:149–162.

Siddall, M. E. and A. G. Kluge. 1997. Probabilism and phylogenetic inference. *Cladistics* 13:313–336.

Swofford, D. L., P. J. Waddell, J. P. Huelsenbeck, et al. 2001. Bias in phylogenetic estimation and its relevance to the choice between parsimony and likelihood methods. *Systematic Biology* 50:525–539.

Szumik, C. A., F. Cuezzo, P. A. Goloboff, and A. E. Chalup. 2002. An optimality criterion to determine areas of endemism. *Systematic Biology* 51:806–816.

Szumik, C. A. and P. A. Goloboff. 2004. Areas of endemism: An improved optimality criterion. *Systematic Biology* 53:968–977.

Tamura, K., D. Peterson, N. Peterson, G. Stecher, M. Nei, and S. Kumar. 2011. MEGA5: Molecular evolutionary genetics analysis using likelihood, distance, and parsimony methods. *Molecular Biology and Evolution* 28:2731–2739.

Upton, D. E. and R. W. Murphy. 1997. Phylogeny of the side-blotched lizards (Phrynosomatidae: *Uta*) based on mtDNA sequences: Support for a midpeninsular seaway in Baja California. *Molecular Phylogenetics and Evolution* 8:104–113.

Wells, S. and D. Haragan. 1983. *Origin and Evolution of Deserts*. Albuquerque: University of New Mexico Press.

Wiens, J. J. and M. J. Donoghue. 2004. Historical biogeography, ecology and species richness. *Trends in Ecology and Evolution* 19:639–644.

Wiens, J. J. 2012. Perspective: Why biogeography matters: Historical biogeography vs. phylogeography and community phylogenetics for inferring ecological and evolutionary processes. *Frontiers in Biogeography* 4:129–135.

Wiley, E. O. 1986. Methods in vicariance biogeography. In *Systematics and Evolution*, ed. P. Hovenkamp, 283–306. Utrecht: University of Utrecht Press.

Wiley, E. O. 1988a. Parsimony analysis and vicariance biogeography. *Systematic Zoology* 37:271–290.

Wiley, E. O. 1988b. Vicariance biogeography. *Annual Review of Ecology and Systematics* 19:513–542.

Wojcicki, M. and D. R. Brooks. 2004. Escaping the matrix: A new algorithm for phylogenetic comparative studies of coevolution. *Cladistics* 20:341–361.

Wojcicki, M. and D. R. Brooks. 2005. PACT: An efficient and powerful algorithm for generating area cladograms. *Journal of Biogeography* 32:755–774.

Yang, Z. 1996. Phylogenetic analysis using parsimony and likelihood methods. *Journal of Molecular Evolution* 42:294–307.

Yang, Z. and B. Rannala. 2006. Bayesian estimation of species divergence times under a molecular clock using multiple fossil calibrations with soft bounds. *Molecular Biology and Evolution* 23:212–226.

Yoder, A. D. and M. D. Nowak. 2006. Has vicariance or dispersal been the predominant biogeographic force in Madagascar? Only time will tell. *Annual Review of Ecology and Systematics* 37:405–431.

Zink, R. M. and D. L. Dittmann. 1991. Evolution of brown towhees: Mitochondrial DNA evidence. *Condor* 93:98–105.

Zink, R. M., R. C. Blackwell, and O. Rojas-Soto. 1997. Species limits in the Le Conte's thrasher. *Condor* 99:132–138.

Zink, R. M. and R. C. Blackwell. 1998a. Molecular systematics and biogeography of arid-land gnatcatchers (genus *Polioptila*) and evidence supporting species status of the California gnatcatcher (*Polioptila california*). *Molecular Phylogenetics and Evolution* 9:26–32.

Zink, R. M. and R. C. Blackwell-Rago. 2000. Species limits and recent population history in the curve-billed thrasher. *Condor* 102:881–886.

Zink, R. M., D. L. Dittmann, J. Klicka, and R. C. Blackwell-Rago. 1999. Evolutionary patterns of morphometrics, allozymes, and mitochondrial DNA in thrashers (genus *Toxostoma*). *Auk* 116:1021–1038.

Zink, R. M. and R. C. Blackwell. 1998b. Molecular systematics of the scaled quail complex (genus *Callipepla*). *Auk* 115:394–403.

Zink, R. M., S. J. Weller, and R. C. Blackwell. 1998c. Molecular phylogenetics of the avian genus *Pipilo* and a biogeographic argument for taxonomic uncertainty. *Molecular Phylogenetics and Evolution* 10:191–201.

Zink, R. M., R. C. Blackwell-Rago, and F. Ronquist. 2000a. The shifting roles of dispersal and vicariance in biogeography. *Philosophical Transactions of the Royal Society, London B Biological Sciences* 267:497–503.

Zink, R. M., G. F. Barrowclough, J. L. Atwood, and R. C. Blackwell-Rago. 2000b. Genetics, taxonomy, and conservation of the threatened California gnatcatcher. *Conservation Biology* 14:1394–1405.

Zink, R. M., A. E. Kessen, T. V. Line, and R. C. Blackwell-Rago. 2001. Comparative phylogeography of some aridland bird species. *Condor* 103:1–10.

13 Raising Cain
On the Assumptions That Inhibit Scientific Progress in Comparative Biogeography

Lynne R. Parenti
National Museum of Natural History,
Smithsonian Institution

CONTENTS

13.1 INTRODUCTION

Stanley Adair Cain (1902–1995) was a botanist, ecologist, and biogeographer who moved readily among leadership roles in academia, the federal government, and conservation biology throughout a career that spanned seven decades. Cain is one of the founders of the modern field of conservation biology in the United States (Thomas 1995; Evans 1996). He helped establish the Nature Conservancy and served as its president; his academic appointments included Professor of Botany at the University of Tennessee and Charles Lathrop Pack Professor of Conservation at the University of Michigan's School of Natural Resources, where he started the Department of Conservation (Evans 1996). He served in the US federal government as Assistant Secretary of the Interior during a leave of absence from Michigan in the mid-1960s. He was elected a member of the National Academy of Sciences in 1970.

Even with these establishment credentials, Cain was something of a rebel. He and fellow botanist Léon Croizat (1958) "... were among the first scientists to challenge vocally the dispersal explanation as the main process in biogeography and promote vicariance as an equally important process" (Crisci 2001:160). Cain (1944) countered the long-distance dispersal biogeography of the Modern Synthesis of Ernst Mayr (1943, 1946) and others in his major thesis, *Foundations of Plant Geography*. The book focused on the geography of plants yet also reviewed the general principles that apply to all biological distributions. This highly respected text was well reviewed (Godwin 1945) and is well cited, particularly by botanists (e.g., Kruckeberg and Rabinowitz 1985; Crisci 2001; Crisci et al. 2003). Yet even though *Foundations of Plant Geography* appeared before Croizat (1952) began publishing his intensive writings on biogeography, Cain is much less visible among modern biogeographers. None of Cain's writings was reproduced in the enormous *Foundations of biogeography: classic papers with commentaries* (Lomolino et al. 2004). One could have been: "Criteria for the indication of center of origin in plant geographical studies," published in the botanical journal *Torreya* (Cain 1943) and reprinted, with edits, as Chapter 14 in Cain (1944). We may easily overlook a paper with such a neutral title published in a specialty journal, but this one challenged the orthodoxy by concluding that there was no straightforward method to identify the center of origin of a group. More important, Cain (1943:151) warned that the lack of clarity of thought in biology and "... the assumptions arising from deductive reasoning have so thoroughly permeated the science of geography and have so long been a part of its warp and woof that students of the field can only with difficulty distinguish fact from fiction."

Cain wrote in an accessible style reminiscent of that of Donn Eric Rosen (1929–1986) who spoke and published extensively on biogeography (e.g., Rosen 1974, 1978, 1979; Nelson and Rosen 1981). It is natural that of Rosen's (2016) 33 assumptions that inhibit scientific progress in comparative biology, eight explicitly address biogeography; I relabel these B1 through B8 in Table 13.1.

Like Croizat (1964:605), Rosen and Cain agreed that life and Earth evolve together, not separately, although all three may not have used the same language or methods to describe and interpret the relationship between the biotic and abiotic components

---

TABLE 13.1

Assumptions about Biogeography That *Inhibit* Scientific Progress in Comparative Biology (See Chapter 1)

B1. All distribution patterns result from vicariance.

B2. All distribution patterns result from dispersal.

B3. Centers of origin can be found.

B4. Historical geology tests and can reject biological theories of area relationships.

B5. Widespread taxa and two taxon systems are informative about the relationships of areas.

B6. Geographic hybridization and biotic mixing make vicariance analysis impossible.

B7. A knowledge of geologic processes is necessary for a vicariance analysis of areas of endemism.

B8. Some organisms are better indicators of biotic history than others depending on their ecology and means of dispersal.

of the world. Rosen worked with a set of methods and a theory of biogeography that drew on the cladistics of Willi Hennig (1966) and the vicariance of Croizat (1952, 1958), as outlined in Brundin (1966), Croizat et al. (1974), Nelson and Platnick (1981), Nelson and Rosen (1981) and elsewhere, although this characterization of methods was not sanctioned by Croizat (1982; for example). Cain's (1944) method is less formal although equally robust. He called it "interpretive plant geography … which derives its distinction as a field of study from synthesis and integration" (Cain 1944:3). It relies on maps of distributions and interpretation of the shape and sizes of biotic areas and the formation of area characteristics, called areography (see also Rapoport 1981), among other considerations.

We may classify Cain as an ecologist, Rosen a systematist; in biogeography, they met on middle ground: congruence between distribution patterns of biology and geology or geography. Elsewhere (Parenti 2007) I have commented on Rosen's view of comparative biogeography from the perspective of one of his graduate students. Here I take the opportunity offered by a focus on Rosen's inhibiting assumptions to demonstrate why rejecting them aligns with the general principles of biogeography that Cain formulated. To be fair, we should view Cain's work in the context of the state of geology and biogeography in 1944: the continents were permanent, land bridges may have connected some of them, and organisms became distributed via chance dispersal from centers of origin (Wolfson 1986). Of particular interest are Rosen's inhibiting assumptions numbered B1 (on vicariance), B2 (on dispersal), B3 (on centers of origin), B7 (on knowledge of geology and its relationship to vicariance), and B8 (on the relative importance of ecology and dispersal) in Table 13.1, timeless topics of biogeography. The three other assumptions, B4–B6, focus on the details of a vicariance biogeographic analysis, *sensu* Nelson and Platnick (1981), which were not relevant to Cain. I do not attempt an exhaustive review of Cain's classic contribution to biogeography, but, instead, address some topics that are of particular interest to modern biogeographers of all philosophies. Note that Stanley A. Cain should not be confused with his contemporary, British biologist A. J. Cain (1921–1999).

13.2 CONGRUENCE

The principal aim of comparative biogeography is to discover congruence, within and among biological groups, and between patterns of biological distribution and Earth or geological history (e.g., Rosen 1978, 1984; Nelson and Platnick 1981; Ebach and Humphries 2002; Parenti and Ebach 2009; Wiley and Lieberman 2011). Complete congruence between the relationships of the species in three clades of four species each (T1–T4, T5–T8, and T9–T12) and of the four areas in which they live (A–D) is demonstrated in Figure 13.1 (from Rosen 1984:fig. 10). Species T1–T4 are related in the hierarchy (T1(T2(T3,T4))) and areas A–D as (A(B(C,D))). Likewise, species T5–T8 and species T9–T12 are related in the same way and live in the same areas of endemism.

Relationships among the areas of endemism in this hypothetical case are inferred from the relationships among the taxa (T1–T12) that live in the areas. Geologists could also propose relationships among the areas. In either case, for biogeography, the areas are recognized by their unique biota. The area hierarchy (A(B(C,D))) is analogous to a phylogeny and specifies, for example, that areas B, C, and D share a

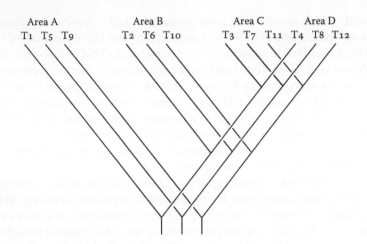

FIGURE 13.1 Overlapping biological and geological cladograms demonstrate congruence between the relationships of three clade with four species each, T1–T4, T5–T8, and T9–T12 and four areas (A–D). (Rosen, D.E. Hierarchies and history. In *Evolutionary Theory: Paths into the Future.* ed. J.W. Pollard, 77–97, 1984. Chichester: John Wiley & Sons Ltd. Reproduced with permission of John Wiley & Sons.)

history that is not shared by area A. Rosen (1984:87) considered that such a "...corroboration of nature's hierarchical structure would be compelling." Discovery of complete congruence such as that in this hypothetical example is rare, but at least partial, informative congruence (involving at least three areas and taxa) is discovered routinely in comparative biogeographic analyses (*sensu* Parenti and Ebach 2009).

Cain (1944:7) also addressed the relationship between phylogeny (as distinct from taxonomy) and geography: "Taxonomy attains a logical basis when the data of comparative morphology can be arranged in a geographical pattern that coincides with the probable phylogeny of the group and the history of the floras and climaxes in which it has been involved. The advantage, of course, is reciprocal for geography." Although Cain (1944:478) did not compare phylogenies with geological hierarchies, he integrated biology and Earth history, most often as climate change, for example here to describe the ecological phenomenon of succession or the evolution of a clisere: "A series of climaxes following one another in any area as a result of climatic change. Contiguous climaxes move together in a common direction because of a widespread climatic change that induces regional parallelism." "Regional parallelism" is congruence to Cain (1944). The history of the floras, not the history of individual species, is key.

13.3   DISPERSAL AND MIGRATION

Plants and animals move naturally throughout their native habitat or area of endemism. What to call this movement and how to assess its role in forming distributions and distribution patterns has always occupied biogeographers (e.g., Platnick 1976; Heads 2014). Cain (1944) distinguished between two types of movement of plants: dispersal and migration. Dispersal refers to the "Transport of diaspores. It does not constitute migration, but is a necessary antecedent to it" (Cain 1944:479). Migration occurs "... when the dissemination of a diaspore is followed by establishment of the organism in a new area..." (Cain 1944:484). This may lead to range expansion.

Individuals of a species live throughout the limits of their range, as diagrammed in Figure 13.2A. This is the species' area of endemism (Ebach and Humphries 2002). Cain (1944:20) stated the obvious about endemism: "Organisms do not live everywhere." And because "Ranges are limited by tolerance" (Cain 1944:19), "... an organism can live and develop only within certain circumscribed limits..." (Cain 1944:89).

How did organisms become distributed? Organisms have all "... become dispersed within limits that are set not by their capacity for dispersal but by other considerations, internal and external" (Cain 1944:88). Cain rejected the relative dispersal ability of organisms to establish ranges in favor of range limits that are established by environmental conditions because of the parallels that are exhibited by major distribution patterns.

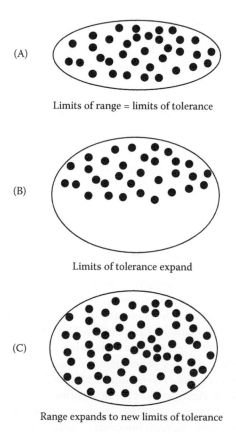

(A)

Limits of range = limits of tolerance

(B)

Limits of tolerance expand

(C)

Range expands to new limits of tolerance

FIGURE 13.2   Schematic diagram to illustrate a hypothetical sequence of events in the range expansion of a species. (A) Individuals (solid circles) live throughout the limits of their species' range, which equals the limits of their tolerance (oval). (B) Limits of tolerance expand, through events of Earth history (oval becomes larger). (C) Species is able to expand its range by movement of individuals throughout the new limits of tolerance. No differentiation occurs.

An area of endemism—a range limited by tolerance—may expand or contract as environmental or other Earth history conditions change. If the limits of tolerance of a species expand, as in Figure 13.2B, then individuals of that species may expand their range to the new limits of tolerance, as in Figure 13.2C.

On the possibility of organisms migrating into new areas with different environments (or climates), Cain (1944:21) was not encouraging:

> "When further environmental changes make life increasingly difficult ... and when there is no opportunity of further migration, they are doomed to extinction unless fortuitous evolution makes them better fitted for the new conditions."

Naturalist Harry Greene (2013:3) described this habit for animals:

> "... although many people believe animals relocate when their habitats are destroyed, most organisms have nowhere to go. They would rather die than move."

Of long-distance dispersal, Cain was equally critical:

> "All too frequently the assumption of long-distance dispersal is merely a careless and easy way out of a difficult problem and it leads to fanciful and even ridiculous conclusions" (Cain 1944:305–306).

Why was Cain so dismissive of long-distance dispersal? Because of congruence:

> "Modern distributional patterns and areas show so many specific parallels under both continental and oceanic conditions that, because of the element of chance involved in occasional random long-distance dispersals, this theory cannot reasonably be generally applied" (Cain 1944:284–285). See also Cain (1944:161 ff., 243 ff.).

Again, parallels.

13.4   MIGRATION AND VICARIANCE

For a range to expand—for organisms to migrate—something has to happen. For Cain (1944:89), climate change [Earth history] effects migration: "... a migratory route is in its essence a matter of the continuity of suitable environmental conditions."

The disruption of an ancestral biota by events of Earth history is inferred to subdivide it into descendent biotas in a process called vicariance (e.g., Croizat et al. 1974). Cain (1944:fig. 5) illustrated this phenomenon for allopatric species of American sycamores in the genus *Platanus*, reproduced here in Figure 13.3. These are what Cain called vicariads or vicarious species in vicarious areas, the "Mutually exclusive areas belonging to closely related species or subspecies" (Cain 1944:490). Here, vicariance drove speciation: "Several botanists ... have commented on the similarity between certain areas of chaparral in southern California and in southern Arizona on the two sides of the desert. This similarity could result in recent time only from migration across or around the desert—this seems unlikely—or from the fact that the vegetational type at an earlier time extended across the area previous to the development of desert conditions" (Cain 1944:96). In the modern biogeographic

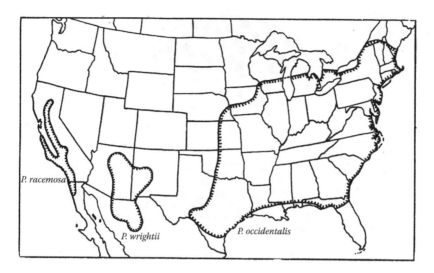

FIGURE 13.3   Allopatric distribution of the American sycamores, genus *Platanus*. From west to east (A) *P. racemosa* (Californian), (B) *P. wrightii* (southwestern), and (C) *P. occidentalis* (eastern). (Modified from Cain, S.A. 1944. *Foundations of Plant Geography*. New York: Harper and Brothers.)

idiom, change in Earth history is the driver of species diversity. No change in Earth history maintains species diversity.

Cain (1944) endorsed the general principles of plant biogeography of Good (1931:152) who argued for the ceaseless succession of floras: "…great movements of floras have taken place in the past and are still continuing." These large-scale migrations lead to cosmopolitanism, the widespread distribution of an ancestral biota or biotas, followed by disruption of biotas and the continuous, inevitable "…replacement of one flora by another." I illustrate this schematically in Figure 13.4. Individuals of a species distributed throughout their range may be subdivided by events of Earth history; this habitat differentiation, as in Figure 13.4A, leads to differentiation into two populations. Continued split of the habitat leads to biotic differentiation and ultimately speciation, as in Figure 13.4B (see also Nelson and Platnick 1984:fig. 1). A breakdown of the ecological or physical barriers that separate the species may lead to the movement of one species into the range of the other to form an area of sympatry, as in Figure 13.4C. Croizat et al. (1974) considered such sympatry to be evidence of dispersal (here migration).

These processes of dispersal (migration) and vicariance (splitting) are continuous and form the complex distribution patterns we see today. Taxa that are thriving Cain (1944) recognized as being in harmony with the environment. Taxa that are in relative disharmony, perhaps no longer undergoing speciation, are the relictual taxa or epibiotics. Examples of epibiotics in the North American freshwater fish fauna are relicts such as the living gars (Lepisosteiformes), the bowfin (Amiiformes), and the paddlefish (Polypteriformes), all of which are members of once more widespread lineages that have gone extinct elsewhere throughout their ranges (Parenti 2008).

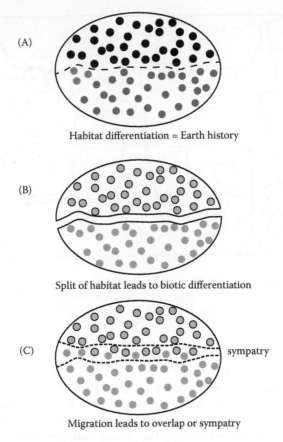

(A)

Habitat differentiation = Earth history

(B)

Split of habitat leads to biotic differentiation

(C)   sympatry

Migration leads to overlap or sympatry

FIGURE 13.4 Schematic diagram to illustrate a hypothetical sequence of events in the differentiation of a species (as in Figure 13.2C). (A) Habitat differentiation initiates isolation and differentiation. (B) Split of the habitat leads to biotic differentiation. (C) Subsequent movement of the members of one species into the range of the other results in an area of sympatry. This movement is facilitated by habitat differentiation in the area of overlap to maintain individuals of both species.

13.5   CENTER OF ORIGIN

Cain has been associated with the concept of a center of origin since his influential and often-quoted paper exposed the criteria used to recognize such areas in plant geography (Cain 1943; above). He appreciated the general interest in where a species might have come from, but detailed why this question could not be answered with the supposed certainty with which some had identified a center of origin, such as by location of the greatest number of species. A simple and as common a mechanism as extinction, paired with emigration, could mislead (Cain 1943:137): "Through emigration and extinction due to climatic and physiographic changes the variety of types may be reduced in one region so that a secondary center comes to contain more variety." In sum, "… we find ourselves confronted by many 'ifs'" (Cain 1944:204).

The concept of a center of origin in biogeography was reviewed by Croizat et al. (1974). A center of origin is rejected in panbiogeography (Croizat 1964), vicariance biogeography (Nelson and Platnick 1981), cladistic biogeography (Humphries and Parenti 1999), and comparative biogeography (Parenti and Ebach 2009), but has a home in the biogeography of Hennig (1966). For Hennig, the relationship between phylogeny and biogeography could be explained by dispersal from a center of origin—identified as the location of the most "basal" taxon—outwards, corresponding to the distribution of increasing apomorphy (e.g., Hennig 1966:fig. 40). One problem with this explanation—called the Progression Rule—for the distribution of a group is that it is not presented as a testable hypothesis (see Parenti and Ebach 2013). The paradox is that the (phylogenetic) hypothesis is the evidence for the (dispersal) hypothesis. It cannot be rejected. It can only be replaced by a new phylogenetic hypothesis, which may specify a new center of origin and new dispersal routes. Further, in practice, the Progression Rule is applied to just one clade at a time. There is no comparison among clades to identify distribution patterns; there is no search for biological congruence. Despite this, the Progression Rule, with its dispersal from a center of origin, persists in much modern biogeography (see Nelson 2004; Heads 2005; Mooi, Chapter 14), particularly that associated with the molecular dating of lineages.

13.6 TIMING AND GEOLOGY

That there could be congruence between biological and geological relationships was bolstered in the late 1960s to early 1970s by general acceptance of a theory of continental drift and its mechanisms of plate tectonics and sea-floor spreading (Hallam 1973; Romano and Cifelli 2015). The continental drift hypothesis was proposed formally by Alfred Wegener (1915) who was so challenged by "hostile and bitter opponents" among geophysicists that "… the theory was soon shelved and forgotten" (Roman and Cifelli 2015:915) only to be corroborated decades later. Yet biologists did not wait for geologists to confirm continental rearrangement to apply it as a general mechanism of biological distribution (see Parenti 2008). Biological distributions, such as that of the Gondwanan seed-fern fossil *Glossopteris* on all of the southern continents, provided some of the strongest and most compelling support for Wegener's theory even without the approval of geophysicists. At least since Alfred Russel Wallace (letter to H. W. Bates, January 4, 1858, in Marchant 1916) compared the biota of the Indonesian islands of Bali and Lombok and postulated an ancient west Pacific continent, we have understood that biological distributions are powerful predictors of geology.

Cain (1944) entertained the theory of a mobile earth as one possible explanation for biological distribution decades before it was accepted by the scientific community at large. He (1944:fig. 42) mapped the distribution of the five genera of the aster family Stylidiaceae using two projections, as in Figure 13.5: "By way of contrast with the ordinary Mercator projection, Map A is based on a reconstruction by H. B. Baker (… 1932) to illustrate his hypothesis of continental displacement." The genus *Phyllachne* comprises four species, one in southern South America and three in New Zealand; its distribution is outlined by a track, labeled 5 in Figure 13.5A and

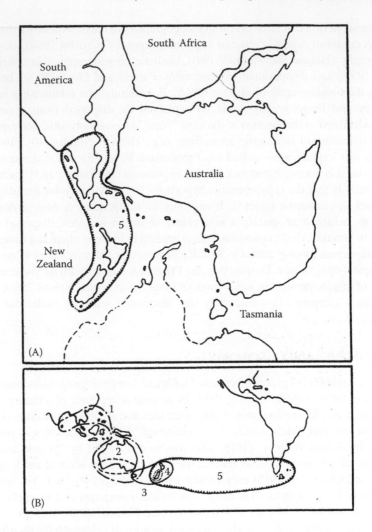

FIGURE 13.5 Distribution of the Stylidiaceae (Asterales). (A) Distributional range of the genus *Phyllachne* (genus 5) that lives in southern South America and New Zealand drawn on a map of closely arranged southern continents. (B) Distributional range of the five genera of Stylidiaceae: *Stylidium* (1), *Levenhookia* (2), *Fostera* (3), *Oreostylidium* (4), and *Phyllachne* (5) drawn on a modern map. (Modified from Cain, S.A. 1944. *Foundations of Plant Geography.* New York: Harper and Brothers.)

B. Cain (1944) used a straightforward method—mapping ranges on current and past geology—to interpret the history of a biotic distribution.

Note that Cain (1944) did not choose a classic reconstruction of Gondwana as rendered by Wegener in his book, the fourth and last edition of which was published in 1929 (Wegener 1929). In Wegener's Gondwana, the southern tip of South America and Australia/New Zealand are separated by Antarctica (e.g., Wegener 1966:fig. 23). Baker's (1932) map, in Figure 13.5A, places New Zealand close to southern South America, as it is in the rendition of an alternative Gondwana of McCarthy et al. (2007), which was based on a summary of plant and animal distributions. Cain (1944) gave no hint as to why he chose Baker's Gondwana over Wegener's. One guess is that Baker's more accurately reflected the concordance of the distribution of the trans-Pacific *Phyllachne* and the southern landmasses.

On the widespread and disjunct distribution of the genus *Magnolia*, he was equivocal on the mechanism of distribution (Cain 1944:288): "It is apparent that land surfaces for migration have been amply available (whether we accept the land-bridge hypothesis or some form of continental displacement) and that the *Magnolia* species of the regions now so widely disjunct are descendants of the same phylogenetic stock." Cain seemed almost reluctant to identify a mechanism of distribution and appeared more concerned with patterns of disjunctions: "Irmscher [1922], making an extensive inquiry into the distribution of families and genera of flowering plants in connection with the Wegenerian continental-drift hypothesis, came to the conclusion that the data can be regarded as a support for the theory. Our present concern, however, is not with this conclusion but with the data on major discontinuities" (Cain 1944:245).

The balance between biogeographic mechanisms as explanations of data tipped back and forth, towards vicariance if hierarchies of taxa and areas matched up, as in Figure 13.1, and over to dispersal if they did not. Even after the theory of plate tectonics became convention, it was explicitly rejected as a general mechanism of biotic distribution by many biogeographers who argued that although geography moved, it moved too long ago to have affected many distributions (e.g., Briggs 1984; Darlington 1965; McDowall 1971; see Rosen 1985). Therefore, chance dispersal was endorsed as the distribution mechanism despite, for example, the agreement between distribution and past continental arrangement as demonstrated by Cain (1944) for the Stylidiaceae which, one might argue, favored vicariance as a mechanism.

There is no way to arbitrate between vicariance and dispersal. These are hypotheses, not observations. Those who argue that taxa known from relatively young fossils could not have been affected by relatively old geological events and, therefore, obviously dispersed, are making "… two assumptions, both wrong at some level: (1) that fossils can tell us how old a taxon is and (2) that the ages of geologic events have been correctly assigned" (Rosen 1985:636) [Rosen (1975) and Patterson and Rosen (1977) also considered the minimum age of a taxon]. I add a third assumption, also wrong at some level, (3) that molecular sequence data can tell us the maximum or absolute age of a taxon. Three decades on, these assumptions about geology and the age of taxa are still part of the "warp and woof" of biogeography.

The age of the oldest known fossil is an estimate of the minimum age of the taxon, not the maximum (Rosen 1985; Nelson and Ladiges 2009; Heads 2015; among others). Likewise, fossil-calibrated molecular clock ages for taxa are minimum ages; they are not maximum ages or absolute ages. When fossil ages are considered to be maximum ages, they yield relatively young clade age estimates. This molecular clock method has been used to reject older vicariance hypotheses in favor of more recent long-distance dispersal explanations, for example, of plants moving across the southern oceans in a "Green Web" (de Queiroz 2014). This inferred widespread dispersal of plants has been rejected by discovery and analysis of an assemblage of Patagonian fossils, which supports the hypothesis of Gondwanan vicariance (Wilf and Escapa 2014). Of equal interest, even if Gondwanan vicariance is rejected as an explanation for a distribution pattern, this does not mean long-distance dispersal is the only alternative. Earth history did not stop at the end of Gondwanan break-up; some other, perhaps younger, vicariance event or geological formation may be used to interpret distribution patterns, especially those across the Indo-Pacific (see Heads 2015 on ratite birds, for example). As above, Cain (1944) also chose an alternate Gondwana.

Finally, as biologists we cannot depend upon geological hypotheses to tell us how taxa may have become distributed. There is no reason to reject a hypothesis of distribution—either dispersal or vicariance—because it is not concordant with the geological event postulated to have been its cause. As part of the history of Gondwana, a large portion of the New Zealand biota is hypothesized to have evolved along with the separation of New Zealand from eastern Gondwana starting some 80 mya (Fleming 1979). This explanation has been rejected recently because of geological theories that New Zealand was completely inundated during the Late Oligocene to early Miocene marine transgression, about 35–25 mya, and, therefore, it is assumed, must have been colonized "... by water-borne and air-borne waifs, strays and pioneers that arrived during the past 22 million years" (Landis et al. 2008:174). "Goodbye Gondwana," say some (McGlone 2005).

New Zealand biologists and geologists have come forward with studies to counter the complete inundation hypothesis. Geologists have discovered and described "... rocky shorelines of Oligocene age ... [that] ... provide unequivocal evidence for the existence and extent of land during the period of maximum marine transgression ..." (Lee et al. 2014:195). Endemic New Zealand frogs of the genus *Leiopelma*, have been estimated, using mitochondrial genome data, to have diverged from their closest living relative, the North American *Ascaphus truei* over 150 mya, given as support for the vicariance hypothesis (Carr et al. 2015). Biology is a powerful predictor of geology.

13.7　CONCLUSIONS

Like Donn Rosen, Stanley Cain had a comparative approach to biogeography that did not depend on a particular distribution mechanism (vicariance or dispersal) but instead used the empirical data of plant and animal distributions, and their relationships, to reveal patterns to be explained. Both were critical of biogeographic analyses that treat assumptions about age and dispersal ability as more informative than biological distribution patterns, the raw data of biogeography. And both dismissed attempts to identify the center of origin of a group when these too required numerous assumptions and gave ambiguous results. They lived and worked before the popular use of molecular clocks to arbitrate between vicariance and dispersal explanations, but that approach would likely not have changed theirs.

They used geological hypotheses to shed light on their biological hypotheses—Rosen's (1978:187) "reciprocal illumination"—but not to reject biology. I close with an observation from Cain (1944:155):

"Incidentally, any critical coordination between the geographical data and paleoecology, geology, or paleoclimatology must be regarded as a piece of luck. It should be emphasized that the first task of floristic geography—and one that has always been its forte—is the accumulation of distributional data and their organization on maps. This static aspect of plant or animal geography leads naturally to interpretations of vegetational dynamics, but it is important that hypotheses grow from observations rather than that hypotheses form the framework of the science."

ACKNOWLEDGMENTS

The Herbert R. and Evelyn Axelrod Chair in Systematic Ichthyology, Division of Fishes, Department of Vertebrate Zoology, National Museum of Natural History, Smithsonian Institution, supported the preparation of this paper. Comments of two reviewers improved the text. I thank Paulette Goldweber, John Wiley & Sons, for permission to reproduce the image in Figure 13.1.

REFERENCES

Baker, H. B. 1932. *The Atlantic Rift and Its Meaning*. Ann Arbor: Edwards Brothers, privately printed.
Briggs, J. 1984. Freshwater fishes and biogeography of Central America and the Antilles. *Systematic Zoology* 33:428–435.
Brundin, L. 1966. Transantarctic relationships and their significance as evidenced by chironomid midges. *Kungliga Svenska Vetenskapsakademiens Handlinger* 11:1–472.
Cain, S. A. 1943. Criteria for the indication of center of origin in plant geographical studies. *Torreya* 43:132–154.
Cain, S. A. 1944. *Foundations of Plant Geography*. New York: Harper and Brothers.
Carr, L. M., P. A. McLenachan, P. J. Waddell, N. J., Gemmell and D. Penny. 2015. Analyses of the mitochondrial genome of *Leiopelma hochstetteri* argues against the full drowning of New Zealand. *Journal of Biogeography* 42:1066–1076.
Crisci, J. V. 2001. The voice of historical biogeography. *Journal of Biogeography* 28:157–168.
Crisci, J. V., L. Katinas and P. Posadas. 2003. *Historical Biogeography. An Introduction*. Cambridge: Harvard University Press.
Croizat, L. 1952. *Manual of Phytogeography or an Account of Plant Dispersal throughout the World*. The Hague: W. Junk.
Croizat, L. 1958. *Panbiogeography or an Introductory Synthesis of Zoogeography, Phytogeography, and Geology; with Notes on Evolution, Systematics, Ecology, Anthropology, etc*. Caracas: Published by the author.
Croizat, L. 1964. *Space, Time, Form: The Biological Synthesis*. Caracas: Published by the author.
Croizat, L. 1982. Vicariance/vicariism, panbiogeography, "vicariance biogeography," etc.: A clarification. *Systematic Zoology* 31:291–304.
Croizat, L., G. Nelson and D. E. Rosen. 1974. Centers of origin and related concepts. *Systematic Zoology* 23:265–287.
Darlington, P. J. 1965. *Zoogeography, the Geographical Distribution of Animals*. New York: John Wiley & Sons.

de Queiroz, A. 2014. *The Monkey's Voyage: How Improbable Journeys Shaped the History of Life*. New York: Basic Books.

Ebach, M. C. and C. J. Humphries. 2002. Cladistic biogeography and the art of discovery. *Journal of Biogeography* 20:427–444.

Evans, F. C. 1996. Resolution of respect. Stanley Adair Cain 1902–1995. *Bulletin of the Ecological Society of America*, April, 1996:80–81.

Fleming, C. A. 1979. *The Geological History of New Zealand and Its Life*. Auckland: Auckland University Press.

Godwin, H. 1945. Rejuvenation of plant geography. *Nature* 155:286–287.

Good, R. D'O. 1931. A theory of plant geography. *The New Phytologist* 30:149–171.

Greene, H. W. 2013. *Tracks and Shadows. Field Biology as Art*. Berkeley: University of California Press.

Hallam, A. 1973. *A Revolution in the Earth Sciences. From Continental Drift to Plate Tectonics*. Oxford: Clarendon Press.

Heads, M. 2005. Towards a panbiogeography of the seas. *Biological Journal of the Linnean Society* 84:675–723.

Heads, M. 2014. *Biogeography of Australasia. A Molecular Analysis*. Cambridge: Cambridge University Press.

Heads, M. 2015. Panbiogeography, its critics, and the case of the ratite birds. *Australian Systematic Botany* 27:241–256.

Hennig, W. 1966. *Phylogenetic Systematics*. Urbana: University of Illinois Press.

Humphries, C. J. and L. R. Parenti. 1999. *Cladistic Biogeography: Interpreting Patterns of Plant and Animal Distributions*, Second edition. Oxford: Oxford University Press.

Irmscher, E. 1922. Pflanzenverbreitung und Entwicklung der Kontinente. Studien zur genetischen Pflanzengeographie. *Mitteilungen aus dem Institut für Allgemeine Botanik Hamburg* 5:17–235.

Kruckeberg, A. R. and D. Rabinowitz. 1985. Biological aspects of endemism in higher plants. *Annual Review of Ecology and Systematics* 16:447–479.

Landis, C. A., H. J. Campbell, J. G. Begg, D. C. Mildenhall, A. M. Paterson and S. A Trewick 2008. The Waipounamu erosion surface: Questioning the antiquity of the New Zealand land surface and terrestrial fauna and flora. *Geological Magazine* 145:173–197.

Lee, D. E., J. K. Lindqvist, A. G. Beu, J. H. Robinson, M. A. Ayress, H. E. G. Morgans and J. K. Stein. 2014. Geological setting and diverse fauna of a Late Oligocene rocky shore ecosystem, Cosy Dell, Southland. *New Zealand Journal of Geology and Geophysics* 57:195–208.

Lomolino, M. V., D. F. Sax and J. H. Brown (eds.). 2004. *Foundations of Biogeography. Classic Papers with Commentaries*. Chicago: University of Chicago Press.

Marchant, J. 1916. *Alfred Russel Wallace – Letters and Reminiscences*, Vol. 1. London: Cassell and Company Ltd.

Mayr, E. 1943. The zoogeographic position of the Hawaiian Islands. *The Condor* 45:45–48.

Mayr, E. 1946. History of the North American bird fauna. *The Wilson Bulletin* 58:3–41.

McCarthy, D., M. C. Ebach, J. J. Morrone and L. R. Parenti. 2007. An alternative Gondwana: Biota links South America, New Zealand and Australia. *Biogeografía* 2:2–12.

McDowall, R. M. 1971. Fishes of the family Aplochitonidae. *Journal of the Royal Society of New Zealand* 1:31–52.

McGlone, M. S. 2005. Goodbye Gondwana. *Journal of Biogeography* 32:739–740.

Nelson, G. 2004. Cladistics: Its arrested development. In: *Milestones in Systematics: The Development of Comparative Biology*, eds. D. M. Williams and P. L. Forey, 127–147. London: Taylor & Francis.

Nelson, G. and D. E. Rosen (eds.). 1981. *Vicariance Biogeography. A Critique*. New York: Columbia University Press.

Nelson, G. and N. I. Platnick. 1981. *Systematics and Biogeography: Cladistics and Vicariance.* New York: Columbia University Press.

Nelson, G. and N. I. Platnick. 1984. *Biogeography.* Burlington, North Carolina: Carolina Biological Supply Company.

Nelson, G. and P. Y. Ladiges. 2009. Biogeography and the molecular dating game: A futile revival of phenetics? *Bulletin de la Société géologique de France* 180:39–43.

Parenti, L. R. 2007. Common cause and historical biogeography. In *Biogeography in a Changing World. Fifth Biennial Conference of the Systematics Association, UK*, eds. M. C. Ebach and R. S. Tangney, 61–82. Boca Raton: CRC Press.

Parenti, L. R. 2008. Life history patterns and biogeography: An interpretation of diadromy in fishes. *Annals of the Missouri Botanical Garden* 95:232–247.

Parenti, L. R. and M. C. Ebach. 2009. *Comparative Biogeography. Discovering and Classifying Biogeographical Patterns of a Dynamic Earth.* Berkeley: University of California Press.

Parenti, L. R. and M. C. Ebach. 2013. Evidence and hypothesis in biogeography. *Journal of Biogeography* 42:1801–1808.

Patterson, C. and D. E. Rosen. 1977. Review of ichthyodectiform and other Mesozoic teleost fishes, and the theory and practice of classifying fossils. *Bulletin of the American Museum of Natural History* 158:85–172.

Platnick N. I. 1976. Concepts of dispersal in historical biogeography. *Systematic Zoology* 25:294–295.

Rapoport, E. H. 1981. *Areography: Geographical Strategies of Species.* New York: Fundación Bariloche, Pergamon Press.

Roman, M. and R. L. Cifelli. 2015. 100 years of continental drift. *Science* 350:915–916.

Rosen, D. E. 1974. The phylogeny and zoogeography of salmoniform fishes and the relationships of *Lepidogalaxias salamandroides. Bulletin of the American Museum of Natural History* 153:265–326.

Rosen, D. E. 1975. A vicariance model of Caribbean biogeography. *Systematic Zoology* 24:431–464.

Rosen, D. E. 1978. Vicariant patterns and historical explanation in biogeography. *Systematic Zoology* 27:159–188.

Rosen, D. E. 1979. Fishes from the uplands and intermontane basins of Guatemala: Revisionary studies and comparative geography. *Bulletin of the American Museum of Natural History* 162:267–376.

Rosen, D. E. 1984. Hierarchies and history. In *Evolutionary Theory: Paths into the Future*, ed. J. W. Pollard, 77–97. Chichester: John Wiley & Sons Ltd.

Rosen, D. E. 1985. Geological hierarchies and biogeographic congruence in the Caribbean. *Annals of the Missouri Botanical Garden* 72:636–659.

Rosen, D. E. 2016. Assumptions that *inhibit* scientific progress in comparative biology. In *Assumptions Inhibiting Scientific Progress in Comparative Biology, UK*, eds. B. I. Crother and L. R. Parenti, 1–4. Boca Raton: CRC Press.

Thomas, R. McG, Jr. 1995. Obituary: Stanley Adair Cain, 92, led the field. *New York Times.* April 2, 1995.

Wegener, A. 1915. *Die Enstehung der Kontinente und Ozeane.* Braunschweig: F. Vieweg & Sohn.

Wegener, A. 1929. *Die Enstehung der Kontinente und Ozeane, 4.* Braunschweig: F. Vieweg & Sohn.

Wegener, A. 1966. The origin of continents and oceans. Translation of the fourth edition of *Die Enstehung der Kontinente und Ozeane*, by J. Biram. New York: Dover Publications, Inc.

Wiley, E. O. and B. S. Lieberman. 2011. *Phylogenetics. Theory and Practice of Phylogenetic Systematics.* New Jersey: Wiley-Blackwell.

Wilf, P. and I. H. Escapa. 2014. Green web or megabiased clock? Plant fossils from Gondwanan
    Patagonia speak on evolutionary radiations. *New Phytologist* (2014):1–8. doi 10.1111/
    nph.13114.
Wolfson, A. 1986. Bird migration and the concept of continental drift. In *Plate Tectonics and
    Biogeography. Special Issue of Earth Sciences History, Journal of the History of the
    Earth Sciences Society* 4(2), ed. A. E. Levition and M. L. Aldrich, 182–186. New York:
    History of the Earth Sciences Society.

14 Evidence, Pattern and Assumptions
Reintroducing Rosen's Empiricism and Skepticism to Systematics and Biogeography

Randall D. Mooi
The Manitoba Museum

CONTENTS

14.1   INTRODUCTION

... you may imagine our vexation in discovering that within this book there is no hint
of the controversy, no inkling that modern systematics is undergoing epistemological
surgery, no trace even of a restlessness of spirit among the author's contemporaries.

Rosen and Schuh (1975:504)

Few, if any, phylogenies are incompatible with the proposed mechanisms. Hence,
current theories of process matter little, if at all, to how we discover nature's hierarchy
or recover its singular history.

Rosen (1982:84)

The revolutions in both systematics and biogeography that took place in the latter
third of the twentieth century have Donn Rosen's fingerprints all over them. Not
only did he contribute directly through his own publications (Rosen 1974, 1975b,
1978a, 1979, 1985a,b), but through promotion and debate of these revolutionary
ideas via meetings, symposia, lectures, book reviews, discussion groups, and student mentoring (Nelson et al. 1987). His influence is also apparent by his appearance in the acknowledgments of important publications of that time (e.g., Nelson
1978a,b, 1979; Platnick and Nelson 1978). The enormity of the paradigm shift from
1966 (Greenwood et al. 1966) to 1985 (Rosen 1985a,b) is difficult to comprehend for
recent generations of biologists that have grown up within the newer paradigm. For
Rosen and his contemporaries, however, it meant digging from under a mountain of
entrenched orthodoxy while espousing unpopular alternatives considered subversive
by the establishment. Confidence and open-mindedness are requisites to question
fundamentals, and by all accounts Rosen had these qualities in abundance (Nelson
et al. 1987) and encouraged them in his colleagues. In fact, he was mystified by
anyone who would "keep to himself ... new ideas that might be viewed as fundamental to our science" (Rosen 1981:3). These character traits permitted him freedom
to examine critically the assumptions underlying the biological species concept and
evolutionary taxonomy in systematics, as well as dispersal from centers of origin in
biogeography. He did this not only through theory, but empirically, from the principle
that "general patterns demand general explanations" (Rosen 1978a:186). He found
these patterns in congruent morphological characters and in congruent area cladograms; pattern preceded explanation and required a cladistic approach to be testable.

Rosen's (2016, fig. 1.1) list of "Assumptions that inhibit scientific progress in
comparative biology, was produced only a year and a half before his early death at
age 57, and succinctly encapsulates his published record of challenge to orthodoxy
(see also Crother, this volume). The systematic and biogeographic communities scru-
tinized many of these same assumptions over the last 20 years of his life. Moving
from the ancestors, origins, narratives, and mechanisms of the Neo-Darwinian synthesis to the relationships, apomorphies, empiricism, and patterns of cladistics has
presented some challenges—new ideas are hard work. Distanced from the immedi-
acy of these revolutions, focus has wandered from the fundamental elements of that
paradigm shift: relationships over origins, pattern before process, empirical evidence
before narrative. It has once again become acceptable to embrace consensus over

self-criticism and individual explanation over congruence (e.g., Waters et al. 2013 for the former, de Queiroz 2005, 2014 for the latter). Given this trend, perhaps we would do well to re-examine systematic and biogeographical practices through the critical and skeptical lens of Rosen's inhibiting assumptions.

Mapping and optimizing characters and biogeographic distributions on existing cladograms is *de rigueur* in today's comparative biology landscape (e.g., Grande et al. 2013; Dornburg et al. 2015), but the rationale for these practices and their theoretical implications are rarely considered. This contribution will look at a selection of Rosen's inhibiting assumptions and their applicability to recent contributions in systematics and biogeography that employ mapping and optimization. Scrutiny of these examples indicates that notions of evidence and the relationship between pattern and process need to be reconsidered in the spirit of the critiques of the 1970s encouraged and presented by Rosen and his contemporaries.

14.2 EVIDENCE, TOPOLOGY AND CHARACTER MAPPING—ROSEN GENERAL INHIBITING ASSUMPTION 5: REDEFINING THE PROBLEM CAN SOLVE IT

… ambiguous or inconsistently distributed characters still plague fish systematics and are used again and again to support some proposed relationship.

Rosen (1985a:55)

The mapping of morphological characters onto nodes of a molecular tree results in an empirically empty procedure for synapomorphy discovery.

Assis and Rieppel (2011:94)

What to do with inconsistent characters? Hennig (1966) introduced to English-speaking systematists the concept of reciprocal illumination—a system of checking, correcting, and rechecking whereby incongruence is identified and resolved resulting in stronger hypotheses of character homology and relationships among taxa. But cladistic approaches have changed considerably over the last 50 years and these cycles of testing are rarely applied; we are left comparing topologies and the methods by which they were produced rather than the evidence (the characters) they summarize. As Rosen predicted (general assumption 5), this redefinition of the problem has not solved the issue of incongruent characters, rather, it has changed the questions we ask. This has shifted the focus from data → evidence (What evidence do these data provide?—empirical) to data → topology (What topology can we derive from these data?—methodological).

In part, this stems from how we think about characters and resolve issues of homology. There is generally recognized a two-step process to identify homology (Patterson 1982; de Pinna 1991), an *a priori* assessment based on various similarity criteria (Remane 1952) and an *a posteriori* evaluation via congruence with other characters. Because of the nature of many of the characters we use or how we code them, congruence has become the major player in determining homology, even the final verdict. However, Hennig's concept of *reciprocal* illumination implies that these tests should inform one another (Mooi and Gill 2016). Grant and Kluge (2004) suggested that similarity testing and congruence testing are decoupled and complementary. Hence, in some instances characters might exhibit a degree of structural and/or developmental complexity "to defensibly choose among competing hypotheses of homology" (Grant and Kluge 2004:28). This explains why character systems such as hair in mammals, the Weberian apparatus in Otophysa, and asymmetry in Pleuronectiformes are so compelling as evidence for monophyly.

In contrast, parts of organisms can be indistinguishable (e.g., nucleotides) or coded in such a way as to make them seem indistinguishable (e.g., meristic characters) regardless of their actual historical identities. In these instances, "separate tests of homology are inert" (Grant and Kluge 2004:28) and congruence is the only option for choosing among competing hypotheses for these types of characters. As these types of characters and codings have come to dominate most studies, similarity tests have lost influence, indeed, they are unavailable, making further investigation of congruence of evidence via reciprocal illumination moot. Instead, the focus has shifted to congruence of topologies, redefining the problem yet again and taking discussion one additional step removed from the evidence (characters).

As an example, Thacker (2009) introduced the Gobiiformes as comprised of Kurtidae ((Apogonidae + Pempheridae) + Gobioidei) based on four mtDNA genes and some morphology (but see Mooi and Gill 2010). Using several nuclear genes, Near et al. (2013) and Betancur-R. et al. (2013a) independently removed pempherids from Thacker's gobiiforms to be very distantly related and changed the relationships among remaining included taxa to ((Kurtidae + Apogonidae) + Gobioidei). Despite these substantial changes to membership and internal relationships, Betancur-R. et al. (2013a) suggested that their results "provide partial support for Gobiiformes *sensu* Thacker." Such a conclusion can only arise from a superficial comparison of the topologies—some of the taxa are shared in some combination. But these radically different topologies, each boasting high support values, cannot indicate congruence among the characters that supported either of these trees. What is the evidence for one topology over another? What are the character distributions?

Although the flatfishes (Pleuronectiformes) can be defined by some of the more convincing morphological synapomorphies of any group of fishes (asymmetry, including migration of eyes to one side of the head; Chapleau 1993; Friedman 2008, 2012), several molecular studies have suggested that the group is not monophyletic (Campbell et al. 2013; Near et al. 2013; Betancur-R. et al. 2013a). Yet, other molecular findings have supported flatfish monophyly (Campbell et al. 2014b; summarized in Betancur-R. and Ortì 2014). Resolution among competing topologies has not been sought through examination of characters, but rather by examination of methodology (e.g., Betancur-R. et al. 2013b; Betancur-R. and Ortì 2014). Methodological issues are certainly significant for explaining discrepancies among pleuronectiform molecular topologies, and many enumerated by Campbell et al. (2014a) are broadly applicable. For example, the notions that more data necessarily mean more evidence and the reliance on support values to choose among topologies are problematic (see also Simmons et al. 2004; Simmons and Freudenstein 2011). Inadvertent incorporation of symplesiomorphies introduced via methodological artifact can also result in critical errors and flawed conclusions (Kück and Wägele 2015). Until the focus is turned from the results (topologies) and how they are obtained (methods) back to the evidence itself (distribution of apomorphies), debate will continue without solid foundation. Campbell et al. (2014a:152) concluded that "evidence for flatfish monophyly is not broadly, cleanly or strongly preserved in living flatfish genomes." And, reasonably, they also rejected "using the compelling shared body plan of flatfishes to differentially weigh conflicting results" because this "invalidates the value of conducting these studies" (Campbell et al. 2014a:151). In other words, they argued against mapping molecules on morphological topologies to provide evidence. Mooi and Gill (2010) similarly argued that morphology should not be mapped onto molecular-based topologies to reinterpret incongruence as evidence. Such re-scored characters "cannot lend support to the inference of monophyly, because such 'synapomorphies' are empirically empty: they can never be shown to be wrong" (Assis and Rieppel 2011:97). Neither molecule nor morphology has a monopoly on phylogenetic "truth"; a topology cannot be used to resolve initial homology assessments, but only to identify incongruence.

Mooi and Gill (2010) called this mapping "pseudomorphology" to emphasize that the contribution of this approach is limited to pointing out incongruence; these topological reinterpretations are not resolutions. Hilton et al. (2015:867) objected to this characterization, stating that: "this process ... is, in principle, no different than examining and comparing the morphology of fishes in light of hypotheses of relationships based on different morphological studies, historical or current. Data are data, and interpretations of these data can be—and should be—evaluated in the context of *all* data available." However, data are indeed only data—what we want is evidence (interpreted data) and that means homology. Mapping data without homology estimation onto a topology does not contribute evidence, it only points to incongruence. Hilton et al. (2015:867) suggested that "novel hypotheses regarding the comparative anatomy of fishes (e.g., hypotheses of homology, i.e., synapomorphy) have come from morphological investigation of relationships initially proposed by molecular analysis (e.g., a paracanthopterygian relationship of *Stylephorus*; Miya et al., 2007; Grande et al., 2013)." But both of the cited studies mapped morphological evidence onto conflicting molecular topologies and only point out incongruence. Neither provided alternative homology interpretations that overturn the initial hypotheses (i.e., those of Olney et al. 1993).

Grande et al. (2013:385, 389) were explicit in their approach: "New interpretations of homologies of published characters are proposed based on topological and phylogenetic data" and "Morphological characters were mapped onto a topology based on our molecular likelihood phylogeny." Do these re-mapped morphological characters fit as well or better on the preferred molecular tree? Of the 17 characters that potentially speak to the relationships of *Stylephorus* on their tree (Grande et al. 2013:fig. 2, their character numbering), one is inapplicable (char. 1), one must be interpreted as a loss (char. 8), ten are also found in Lampriformes (the competing hypothesis) (chars. 2, 6, 11, 12, 15, 17, 20, 22, 24, 25), one is homoplastic among other unrelated derived paracanthopterygians (char. 19), three involve features highly modified in *Stylephorus* where homology is difficult to interpret (chars. 5, 21, 26), and the one unique synapomorphy supporting paracanthopterygian relationship is of questionable homology by admission (char. 14, p. 399). And because the paper did not examine the morphological characters that most convincingly place the taxon among lampriforms (see Olney et al. 1993), it did not evaluate the hypothesis "in the context of all data" as preferred by Hilton et al. (2015).

Borden et al. (2013:419) followed a similar mapping strategy where the contribution of morphological characters was evaluated by "congruence with a phylogenetic hypothesis derived from the analysis of DNA sequence data." They concluded (p. 446) that the characters "provide additional morphological support for the monophyly of paracanthopterygians and their intrarelationships, albeit with non-trivial amounts of homoplasy." Indeed, the only characters that might be interpreted as unequivocally congruent are autapomorphies for *Stylephorus* and for the Gadiformes. The preferred molecular tree only serves to discover incongruence, not resolve homology. What is happening at the character level in the molecular tree? Why not permit the morphology to speak for itself and map molecular data?

If such mapping is seen to perform a service—that of pointing out incongruence—there remains some way to go before this becomes a real contribution to understanding characters and their homology. Just as mapping pleuronectiform asymmetry on conflicting molecular trees or molecular data on conflicting morphological trees based on asymmetry would be unsatisfying approaches, mapping is unable to clarify the history of the many characters among paracanthopterygians that have remained opaque to homology determination. For example, the PU2 neural spine in the teleost caudal skeleton appears to develop independently in several groups, a situation recognized by earlier workers. Rosen (1985a) hoped that ontogeny might provide the key to understanding this problematic character; techniques introduced for studying caudal development in other taxa might be helpful (Wiley et al. 2015). Grande et al. (2013) were aware of these issues, but used re-mapped PU2 states (and other characters) as reinterpreted evidence. Changing the original interpretation of the homology of a character based only on congruence with other data shortchanges the new methods that Hilton et al. (2015) espouse and casts aside a basic principle of Hennigian systematics: reciprocal illumination (Mooi and Gill 2016). To merely map these characters is empirically empty and results only in apparent congruence "born out of utilitarianism" (Kluge 2005:39)—pseudomorphology. To map, see conflict and then re-examine characters via ontogeny or other evidence and provide some new understanding of homology is a different matter—a solid foundation for the "new and vibrant science" of Hilton et al. (2015:868).

Britz et al. (2014) provide an example of the perils of mapping morphology onto molecular trees. Due to contradictory results among molecular and morphological datasets, there has been considerable discussion regarding the relationships of *Paedocypris*, an enigmatic miniaturized cypriniform fish. Rüber et al. (2007) and Britz and Conway (2009) suggested that the genus was related to a restricted group of other miniaturized species within the Cyprinidae and Cypriniformes. In contrast, Mayden and Chen (2010) placed *Paedocypris* as the sister group to all other cypriniforms based on six nuclear genes with apparent high support. They then mapped ("optimized", p. 156) the morphology onto the resulting tree to argue for convergent evolution of miniaturization. Britz et al. (2014) employed splits spectrum analysis, a technique pioneered by Wägele and Mayer (2007), to examine all available molecular data for quality of information. Not unlike the conclusions of Campbell et al.

(2014a) regarding evidence from the flatfish genome, Britz et al. (2014) concluded that there was insufficient signal from molecular markers to make any clear phylogenetic inference regarding the relationships of *Paedocypris*. They also thoroughly reviewed the morphological evidence, including that provided by Mayden and Chen (2010), and determined that it best supported *Paedocypris* as embedded within the cyprinoids of the Cypriniformes. The erosion of reciprocal illumination and the emphasis on congruence through mapping and reinterpretation rather than treating characters as independent homology statements has hindered progress in systematics; "... highlighting ambiguous and contradictory results may point to interesting aspects of ... evolution" (Campbell et al. 2014a:151).

The value of highlighting contradictory results through mapping is illustrated by an important study by McMahan et al. (2015). They pointed out that (p. 528), "Most prior work has asked the question 'why do morphological data not recover molecular relationships?' We ask the reverse 'why do molecular data not recover morphological relationships?' We argue that consideration of both these questions is an important and crucial step in attempting to understand incongruence between these two groups of hypotheses." Rather than mapping and reinterpreting characters to fit a preferred tree, they explored the distribution of apomorphic characters on competing topologies and took an important step towards incorporating Hennig's reciprocal illumination into molecular methods. Identifying apomorphic characters and asking how they fit on one tree compared to another examines evidence. Investigating two cases of incongruent molecular and morphological hypotheses of squamate phylogeny, they found that the molecular apomorphies mapped more consistently on the competing morphological topology than they did on the molecular-based topologies! Discovering that the evidence (apomorphies) fits better on a topology other than that derived from the original dataset is rather startling and should be of great concern for systematists using these methods. McMahan et al. (2015:528) suggested that molecular and morphological hypotheses are incongruent "... not because the molecular data are correct or that the morphology is rife with homoplasy, but because there may be an inherent analytical problem with the molecular data." They are uncertain as to what that problem might be, although use of distant outgroups might contribute. This might also be tied, in part, to the systematic errors identified by Kück and Wägele (2015) where it appears plesiomorphy is influencing the outcome of some of these methods rather than synapomorphy alone. Perhaps this should come as no surprise, as at the introduction of numerical methods it was explicit that they use "all states, derived and primitive" (Kluge 1976:43). Regardless, as McMahan et al. (2015) noted, the methods need to be looked at critically to see why this is happening and what impact it has had on systematics.

Citing Mooi and Gill (2010), Near et al. (2015:8) stated that "Some ichthyologists are hesitant to accept taxonomic suggestions based on molecular phylogenetic analyses, preferring morphological evidence for all proposals regarding classification." The issue, though, is not a preference for one kind of evidence over another, but a preference for classifications based on clear statements of homology. There should be a focus on individual characters as statements of homology. There is no hesitation to map morphology onto molecular topologies and point out conflict, but where is the complementary search for conflict in molecular data? Britz et al. (2014),

Campbell et al. (2014a), Kück and Wägele (2015) and McMahan et al. (2015), among others, have all presented good reasons to examine characters as statements of homology in and of themselves rather than only in the context of topology. Like Wägele and Mayer (2007:17), "We are convinced ... that the search for reliable evidence is good practice in science."

The goal of systematics is not a topology or a taxonomy, but discovery of homology. As Rosen suggested, redefining the problem does not solve it. Nelson and Platnick (1981:199) had this useful perspective on the relation of character distributions and topology: "... as long as there is conflict among positive occurrences, there is a problem that may be investigated: namely, of the conflicting occurrences, which are real and which not? This residual problem cannot be solved, except perfunctorily, through the use of a clustering procedure. Its solution is possible only through the study of organisms and new knowledge of, or new insight into, their real characteristics." If we keep this as the aim of systematics, and follow the empiricism of Rosen for whom "the glycerin dish was the crucible in which truth had to leave its residues" (Nelson et al. 1987:543), our science might be better served.

14.3 FOSSILS AND TAXON AGE—ROSEN INHIBITING EVOLUTIONARY ASSUMPTION 7: FOSSILS SPECIFY THE AGE OF THEIR INCLUDING TAXON

... in all probability such ancestors have been dead for many tens or hundreds of millions of years, and that even in the fossil record they are not accessible to us.

Nelson (1969a:27)

Much paleontological effort is directed toward discovery of older and older fossils; as a result they are found with some regularity.

Nelson (1978a:331)

How much older than its oldest fossil can a lineage be?

Heads (2014c:287)

Traditionally, fossils were given a special place for understanding the history of life as either ancestors or indicators of ancestral birthplace. One of the major revolutions in systematics of the late 1960s and the 1970s was to recognize that ancestors are inaccessible and that fossils are taxa like any other except that they happen to be extinct: "... fossilized descendants are neither more nor less significant than Recent descendants" (Nelson 1969a:26). As summarized by Nelson (1994:137): "A cladogram differs from other types of phylogenetic trees in placing all organisms, both fossil and recent, in terminal positions, implying that ancestral taxa are artifacts." Fossils do, however, provide an alluring time dimension and the notion that they can provide dates of taxon origin has proven difficult to discourage in biogeography.

Today, paleontologists are clear that fossils are not indicative of maximum age. For example, when discovery of *Lunataspis*, a horseshoe crab from the late Ordovician (~445 Ma), extended the known age of xiphosurids by almost 100 My,

Rudkin et al. (2008:7) wrote, "The new minimum age will provide an important temporal benchmark..." (see also Rudkin and Young 2009). And the discovery of older minimum ages is the norm (Nelson's epilogue, above; see also Heads 2005:682). For a fish example, through the 1970s, the oldest esocoid fossils were from the Oligocene (34–23 Ma), pushed to the Paleocene (66–56 Ma) in the 1980s and to the late Cretaceous (83–66 Ma) by the 1990s (Wilson et al. 1992), more than doubling the minimum age known fewer than 20 years prior. Recent papers also push the age of several percomorph taxa well into the Cretaceous from previous Eocene or more recent minimum ages (e.g., Carnevale and Johnson 2015). Hence, Rosen's inhibiting assumption that fossils specify clade age should have no relevance today. Yet, there is ample evidence that workers assign fossil age as a proxy for maximum age despite theoretical and empirical caveats and the (detrimental) impact this has on entire research programs. For example, Herrera et al. (2015:104) discounted particular biogeographic scenarios as "actively misleading" where a proposed region of origin did not include the oldest known fossils; whatever one thinks of centers of origin (more on this below), the conclusion of Herrera et al. certainly equates clade age with fossil age. Renema et al. (2008:655) turn minimum ages into maximum ages by stating, "Estimated minimum divergence ages of extant cowrie lineages suggest that the majority of extant species originated in the Miocene."

Most modern workers lament "Unfortunately, fossils only give a lower bound (i.e., minimal age)" (Christin et al. 2008:S1). But maximum fossil ages are seen as so desirable that sophisticated molecular clock models have been developed that ostensibly provide 95% credibility intervals for maximum taxon ages, and best practices have been established to justify these fossil calibrations (Parham et al. 2012). It is a regrettable inconvenience that these models and rules have not managed to halt the discovery of fossils that fall well outside those confidence intervals; there is no reason to suspect that the actual ages of taxa will follow the models. As noted in Parham et al. (2012:352), there is no practical way to estimate model parameters and they are little more than educated guesswork based on ambiguous assumptions. Near et al. (2013) provided a 95% upper bound for the age of esocoids of 87.5 Ma, about 10 My older than the oldest known fossil. Applying lognormal prior probability curves to the oldest known fossils of esocoids of 30 Ma in the 1970s would never have bounded the 76 My old fossils discovered in the 1990s that, of course, are still representative of a minimum age and not of an actual upper bound.

This bias of molecular clocks toward young ages is widespread. The selection of prior probability curves is based on steep priors and results in estimates of maximum clade ages that are not that different from the minimum ages provided by the empirical data from fossils (Heads 2012). For example, Near et al. (2013) rarely provided maximum clade ages more than 10 My older than the minimum age and usually far younger, except in cases where older fossils of near-relatives forced the maximum age to be at least as old as the minimum age of that relative. With their veneer of statistical respectability, calibrated young ages provide taxa that are confidently "revealed," "shown," and "demonstrated" to be too young for particular biogeographic explanations (summarized in Heads 2005, 2012, 2014a–c). This confidence seems misplaced given that there can be no empirical data on maximal clade ages. Yet, the use of minimum and soft maximum age constraints is advocated as a "pragmatic solution" until priors are "fully justified" and "evidence-based" (Warnock et al. 2012:157). But given that ancestors are not accessible and an ancestor–descendent relationship is impossible to demonstrate, evidence of maximal clade age will not be forthcoming. Rather than formulate research programs that rely on unknowable/indeterminate maximum ages, Nelson and Ladiges (2009) caution that we should recognize the limits of our present knowledge and, when dealing with molecular timescales, follow the advice of Graur and Martin (2004:85): "demand uncertainty." Others have taken the opposite approach and provided an explanation for the distribution of cichlids resting on the statistical certainty of maximum age based on an absence of fossils (Friedman et al. 2013). Never mind that individual case histories will not lack for explanation whatever the distribution and/or hypothesized age, and that it is empirical evidence of common distributional patterns that are paramount (Sparks and Smith 2005 for cichlids; also see below), to rest a biogeographic explanation on absence of fossils suggests uncertainty might be prudent.

That fossils specify clade age is, in practice, an inhibiting assumption that is still with us. But the only empirical evidence available provides minimum taxon age. A taxon can be shown, revealed, and demonstrated to be *too old* for certain narratives, but it can only ever be a methodological inference that a taxon is too young. This has important implications for biogeography.

14.4   ANCESTORS AND ORIGINS IN BIOGEOGRAPHY—ROSEN INHIBITING EVOLUTIONARY ASSUMPTION 5: IT IS IMPORTANT TO SEARCH FOR ANCESTORS, AND BIOGEOGRAPHY INHIBITING ASSUMPTION 3: CENTERS OF ORIGIN CAN BE FOUND

Despite its disadvantages, the quest for centers of origin continues to be a dominant theme of modern zoogeography.

Croizat et al. (1974:270)

In short, biogeography today, sadly, has become an arena for indiscriminate invention rather than a stage for the disciplined interpretation of data.

Rosen (1975a:69)

The phylogeographers have confused technical advances with conceptual advances.

Heads (2005:680)

Dispersal from a center of origin has been a dominant explanatory narrative in biogeography since its inception as a science (Humphries and Parenti 1999, and others), although its relative dominance in recent times over a perceived opposite ideology, vicariance, has been debated (de Queiroz 2014; cf. Heads 2014c). Regardless, this polarity of dispersal and vicariance explanations is not the most important; for historical biogeography, the critical dichotomies are common pattern vs. no pattern, general explanation vs. unique explanation, and relationships vs. origins.

The initial Linnaean explanation of organismal distribution of origin and dispersal from a single, mountainous island gradually became more sophisticated as biogeographical regions with unique biotas were identified and similarities among areas were recognized (see histories in Nelson 1978b; Humphries and Parenti 1999; Parenti and Ebach 2009). Sclater (1858:131) capped the pre-Darwinian period with the perceptive remark that "little or no attention is given to the fact that two or more of these geographical divisions may have much closer relations to each other than to any third." This inattention persisted for over 100 years, and emphasis remained on origin and dispersal until the revolutions inspired by Hennig (1966), Brundin (1966), and Croizat (1964), formulated largely by Nelson and Platnick (Nelson 1969a,b, 1973, 1974; Platnick and Nelson 1978) and applied by, for example, Rosen (1974, 1975b, 1978a, 1979).

Perhaps this inattention to area relationships can be traced to Darwin, although the responsibility cannot be laid at his feet alone (Brady 1989). It seems a paradox that Darwin (1859), who so successfully used patterns of distinctive though similar biota in different geographic regions to support evolution, fell to independent case-by-case arguments to explain those distributions. Darwin turned to individual explanations to defeat attempts at falsification of evolution from special creation. Disjunct distributions on a stable Earth provided the strongest case for special creation; extraordinary dispersal provided a counter-argument if its plausibility could be defended. His "how-possibly" explanations (O'Hara 1988) were designed to remove objections to dispersal events that could otherwise maintain what Darwin (1859:349) saw as "the simplicity of the view that each species was first produced within a single region" that so "captivates the mind"—one birthplace, one center of origin, one tree of life. It was so captivating, and his concentration on this case-by-case explanation so complete, that he missed the fact that disjunct distributions are frequently repeated. Having done so, he never found patterns that might point to general explanations.

Instead, Darwin (1859:457) felt that "when we better know the many means of migration, then, by the light geology now throws ... we shall surely be enabled to trace in an admirable manner the former migrations of the inhabitants of the whole world" from particular centers of origin. But Darwin never found these traces nor the centers of origin, and those that followed encountered similar difficulties. Almost a century later, Darlington (1957:236) found centers of origin and dispersal of birds "very difficult to trace and understand." Cain (1943:132) noted the criteria to identify centers of origin "have been largely accepted without question, despite the lack of substantiating data" and he outlined various contradictions among them (see Parenti, this volume). Brady (1989:115) mused on this predicament: "Consider the possibility that biologists, following Darwin's imagined future, would insist on tracing the routes of migration from geological evidence when traces of such routes were not there to be found." We can not only consider it, but we can examine the results—this describes the bulk of biogeographical endeavour over 150 years. And, as Brady (1989:125) warned, "We do not need another century of searching the data of observation for what it does not contain, while overlooking the relations that it does."

14.4.1   THE PERSISTENCE OF CENTERS OF ORIGIN

Despite the upheaval in systematics, paleontology, and biogeography publicly recorded in *Systematic Zoology* through the 1970s and 1980s, most biogeographical studies continued with an emphasis on origin. As an example from fishes, esocoid origins shifted from Eurasia (Gilbert 1976) to North America (Briggs 1986) dependent upon discovery of oldest fossils. Wilson et al. (1992:839) noted that similarly aged fossils on both continents suggested that the group radiated when Eurasia and North America were still joined concluding, strangely, that "pikes were as likely to be North American as European in origin on the evidence of fossils." Yet Wilson (1980:311) had unwittingly provided an interpretation consistent with vicariance biogeography: "... modern *Esox* species in North America may represent survivors of an ancient fauna rather than recent immigrants." Patterson (1981) used congruent area cladograms derived from modern taxa to hypothesize current distribution as the result of the splitting of a widespread Laurasian taxon. Despite this, Briggs (1995:112) continued to claim, "The presence of the oldest fossils and four of the five living species in North America indicates a probable origin in that area with subsequent dispersals to Asia." This reliance on fossils and discovery of origin in historical biogeography persists, if it is not dominant. Under the title, "The historical biogeography of coral reef fishes: global patterns of origination and dispersal," Cowman and Bellwood (2013:220, 221) lamented the "lack of independent evidence of origination of fishes in the IAA [Indo-Australian Archipelago]" and advised "solid independent evidence, preferably fossil, is urgently required." And typical of most biogeographic studies, Thacker (2015:9) noted that though gobies "do not have a detailed enough fossil record to directly include extinct taxa in the analysis ... what fossils do exist support the hypothesis of origin in the Eocene Tethys Sea."

Identifying centers of origin implies, of course, identifying ancestral distributions or ancestral areas, and this in turn would require the identification of ancestors. Given that ancestors are inaccessible (see the previous section), it seems clear that their distributions would be equally inaccessible. Hence, Rosen included as an inhibiting assumption that centers of origins could be discovered. But attempts to find origins and ancestral distributions fill many journal pages. One could go so far as Nelson (1970:375, 376) and suggest that these "logical deceptions" result in a "proliferation of data and literature irrelevant to science, much to its detriment if not its ultimate demise." Or, we could hold centers of origin and ancestral areas to Brady's (1979:617) standard of "fruitfulness" whereby, "A theory which is popular enough to be applied often can be credited with a power to generate activity, but this is desirable only if some profit is derived from it. Mere activity, in itself, seems to have no intrinsic value other than paying the bills. Activity within the context of a community given to reflection and self-criticism, however, might be though a different matter." Center of origin theory certainly passes the test of popularity and activity generation; its success on the remainder is examined through a few examples.

14.4.2 Moving Taxa or Moving Tectonics?

The Indo-Australian Archipelago (IAA) or some portion of it has been recognized as a biogeographic realm or center of biodiversity since Sclater and Wallace over 150 years ago. Most investigations have focused on processes of speciation or accumulation of diversity rather than taxon or area relationships. An example of the former approach is that of Renema et al. (2008) who used fossils and a molecular clock to examine areas of high species richness over time "to distinguish between the movement of a single hotspot across the globe and the successive origination and extinction of hotspots" (p. 657). They provided maps of the number of species in various genera of benthic foraminifera suggesting that areas of high biodiversity shifted from "hotspots" of the West Tethys (about 40 Ma), to the Arabian (about 20 Ma), and to the IAA (Recent). Unfortunately, there was no analysis that took into account cladistic relationships of the taxa nor any attempt to identify historical areas of endemism that might have provided evidence of common patterns and relationships among these areas. However, even with the rather crude estimate of relationship provided by genus-level species richness, Renema et al. (2008:655) noted that "One of the most striking features of these three hotspots is that each in turn marks the location of a major collision between tectonic plates." And despite the application of minimum fossil ages as dates of origin, they concluded that representatives of IAA taxa were present in the region before its development as a "hotspot"; tectonic activity occurred with component taxa in place. Their maps showed that benthic foraminifera were widespread and already occurred in each of the identified hotspots from the earliest period under examination. There seems to be little reason to suspect dispersal as a major factor given that tectonic events provide "opportunities for isolation and disruption of genetic connectivity … and accumulation of diversity as a result of the juxtaposition of communities by accretion of tectonic terranes" (p. 656). They contradicted the general inference of young taxon age based on fossil-calibrated molecular clocks (p. 656): "The strong correlation between the presence of hotspots and major tectonic events suggests that the primary drivers may operate over time scales beyond those traditionally used to examine diversity."

Although their emphasis was on speciation, species richness and centers of origin, it was patterns of distribution of *in situ* taxa that prompted Renema et al. (2008) to conclude, "The critical role of tectonic events emphasizes the importance of abiotic factors in shaping the world's biotic realm. They drive and underpin the birth, life, and senescence of biodiversity hotspots" (p. 657). Although presented as novel without acknowledging its previous manifestations or guises, this basic concept has a long history dating to Wildenow in 1798 and Humboldt and Bonpland in 1805 (see Humphries and Parenti 1999:19, 20), and was perhaps most simply stated by Croizat (1964:605): "… earth and life evolve together." Renema et al. (2008:657) suggested that it is paleontological and molecular data interpreted in an ecological context that has "enabled us to understand the true antiquity of hotspots and their component species." To the contrary, a fair reading of the history and development of historical biogeography would suggest, instead, that it is in spite of paleontological and molecular data interpreted by ecology that taxon antiquity and relationship to tectonics might finally be realized. There is, perhaps, no clearer example of how Rosen's list of assumptions have inhibited progress in science; fossils interpreted as clade age and fixation on dispersal from centers of origin as explanations for distribution have held back a simple but critical concept from general recognition for over 200 years. De Candolle (1820:383, translation from Nelson 1978b:281) had recognized the impediment placed on biogeography when combining ecological and historical concepts: "The confusion of these two classes of ideas is one of the causes that have most retarded the science, and that have prevented it from acquiring exactitude."

But the small change in emphasis as formulated by Renema et al. (2008) has had little impact on subsequent approaches and interpretations. Their concept of hopping hotspots, that is, shifts of areas of tectonic activity across the globe influencing regional diversity, was reinterpreted by Dornburg et al. (2015) under an ancestral area/dispersal paradigm—it is the biodiversity that does the hopping to create the hotspots. Using squirrelfishes and soldierfishes (Holocentridae) as a test case, they asked (p. 147), "Do lineages use common pathways to colonize newly forming hotspots?" The implication is that taxa are the active players and geological formations are merely destinations; that Earth and life evolve together is lost in this formulation. However, given that several taxa (foraminifera, mangroves, various gastropods—Renema et al. 2008) pre-existed in the not-yet biodiversity hotspots (i.e., the taxa exhibit ancient, widespread distributions), why would we not expect other contemporaneous taxa such as holocentrids to have occurred in these regions prior to tectonic activity? Why would we not look for common patterns of area relationships first, rather than immediately invoke a causal mechanism?

The simple answer is that the ancestral areas/dispersal approach examines individual taxon histories as opposed to a cladistic biogeography approach that identifies common patterns of relationship among areas of endemism ("hotspots", at least potentially) (cf. Crisp et al. 2011 for the former; Ebach et al. 2003, Parenti and Ebach 2009 for the latter). Dornburg et al. (2015) provide an entry to examine briefly several topics from the perspective of Rosen's era and modern practices in biogeography: areas of endemism, methods, no pattern vs. patterns or origins vs. relationships, and process explanations.

14.4.3 Areas of Endemism

Although none of Rosen's inhibiting assumptions directly concerns the nature of areas of endemism, it is clear from his publications that he considered endemism to be an important element of biogeography (Rosen 1978a, 1979). The identification of areas of endemism as historically relevant biogeographic units receives far less attention than it deserves. Regardless of philosophical approach, poorly delimited areas make for weak biogeographic analyses and hypotheses (Harold and Mooi 1994). Platnick (1991) suggested that biogeographic studies should prefer taxa that are maximally endemic, that is, those with the largest number of species exhibiting the smallest ranges in the region of interest. Instead, most workers do not select study organisms with biogeography in mind and apply whatever taxa are at hand regardless of distribution. Distributions of these taxa are then overlain onto areas that have been predetermined by species richness, zones of biodiversity, arbitrary levels of endemism, ecology, geography, or even tradition. Such areas are unlikely to be based on parameters that are historically meaningful for the taxa under study. Both Renema et al. (2008) and Dornburg et al. (2015) employed predetermined areas. For the former, taxa are not identified specifically and there are no cladograms involved in the analysis so there can be no real attempt at defining areas based on intrinsic parameters; the included areas are based on maximum diversity rather than particular distributions and relationships. Dornburg et al. employed the areas defined by Briggs and Bowen (2012) without examining the efficacy or applicability of these for holocentrids. Although ancestral areas/dispersal programs are designed to map movement of taxa to and from areas, not delimit them, identifying historically relevant areas should be a concern. Because the algorithms will reconstruct ancestral areas however the original regions are defined, employing areas that have some historical meaning to the taxa under investigation should be paramount.

The IAA, a focus of both Renema et al. and Dornburg et al., is known to be a geological composite (e.g., Hall 1998) and is likely to exhibit several, even competing, taxon and area histories; neither of the studies addresses how that issue might be reflected in the distributions or relationships of taxa. Although Dornburg et al. (2015) offer holocentrids as an exemplar lineage for investigating the evolution of biodiversity hotspots, the distributions they map onto their consensus chronograms indicate otherwise. The majority of taxa are widespread over several identified areas. Two of the included areas, Eastern Atlantic and IAA, have no endemic holocentrid taxa associated with them—there is no reason to suspect that these areas have historical relevance for this fish family. To include areas in an analysis even though no study taxa are endemic to them has been interpreted as "the triumph of hope over evidence" (Platnick 1991:xii). In this instance, the taxon of choice has little, if anything, to contribute to understanding the biogeography of the area that is the focus of the study.

Why would expectations be so high for a taxon to provide insight for an area it can't define? Perhaps, as Platnick (1991:xi) has suggested, by the time we have done fieldwork to gather samples, cladistically analyzed relationships, mapped distributions, and have calculated and plotted taxon ages, the taxon in question "gains a heavy burden of anticipation for biogeographically decisive resolution." In addition, for funding agencies and employers or supervisors, there is the imperative to demonstrate "fruitfulness," or at least generate activity and pay the bills (Brady 1979).

14.4.4   METHODS IN RECONSTRUCTION OF ANCESTRAL AREAS

The number of biogeographers who confidently drew dispersal routes on fixed continent maps ten or more years ago and now just as confidently draw dispersals of the same organisms on continental drift maps must cause us to seriously question the procedures of biogeographers.

Edmunds (1975:251)

Comparing the maps in Thacker (2015:fig. 2) with those in Darlington (1957), one is left to wonder at how far biogeography has come over 60 years. Considering the fundamental and revolutionary changes in phylogenetics, paleontology, and biogeography that have taken place since Hennig (1966), Brundin (1966), and Nelson's influential 1969 lecture (Williams and Ebach 2004), along with better data and analyses, it is astounding that many "modern" results could be slipped into the appropriate chapters of Darlington and require only a change to pagination. One reason for this is that estimates of maximum clade age based on the minimum ages provided by fossils frequently suggest taxa are young. These young taxa are assumed to be beyond the influence of classical vicariance explanations such as continental drift; in effect, the taxa are presumed to have evolved on a stable Earth. Darlington's centers of origin and dispersal routes were also drawn on a stable Earth (1957:606, 607): "... animal distribution now is fundamentally a product of movement of animals, not movement of land ... Even if drift did occur, it was probably long ago, and existing distributions of animals and plants probably would not show it; they are probably too recent."

Although Waters et al. (2013:496) advocated banning panbiogeography from evolutionary journals to protect readers from what they identified as perhaps an "attractive notion several decades ago," they were quite willing to maintain as "mainstream evolutionary biology" the antediluvian notions of centers of origin and dispersal that have a long history of inhibiting scientific progress. What seem new and shiny are ideas that actually date back to Linnaeus, if not further (Nelson 1983; Heads 2005). Ancestral area biogeographers have "confused technical advances with conceptual advances" (Heads 2005:680). Computers apply molecular data to sophisticated models based on eighteenth-century theory to identify inaccessible centers of origin and to plot untraceable dispersal events. This has been said before, and better, yet regrettably without a notable impact on the field (Heads 2005, 2014b,c; Ladiges et al. 2012; Nelson and Ladiges 2001).

A popular implementation of ancestral area reconstruction is based on time-calibrated trees and likelihood dispersal–extinction–cladogenesis (DEC). Although users frequently provide extensive lists of caveats regarding the assumptions required by these methods (1.5 pages worth in Cowman and Bellwood 2013), there seems little reticence in making declarative statements regarding origin, dispersal, migration, invasion, and all manner of evolutionary processes. Yet, the results rely heavily on particular settings and the inclusion or exclusion of taxa (either fossil or recent). Dornburg et al. (2015) showed that results using extant and fossil species of holocentrids differed "fundamentally" from those sampling only extant species. Holocentrids are considered to have a "very good fossil record," although this is only in comparison to most taxa where the fossil record is poor (Heads 2005:682). For the latter, ancestral reconstruction is problematic given the role fossils are deemed to play in identifying centers of origin, and time calibration would rely extensively on fossils from other taxa. Even for holocentrids, if a fossil were found in another area (all known fossils are from the Mediterranean West Tethys—a suspicious distribution that is argued as real rather than artifact by Dornburg et al.) or if a fossil of different age were discovered, neither being an unreasonable expectation, there would be substantial changes to the reconstruction. A minor point specific to the holocentrid study is that Dornburg et al. (2015) included Lessepsian migrants in the reported distributions (although inconsistently and incorrectly); how inclusion of recent anthropogenic introductions might impact the analysis was not discussed. Two important caveats not often mentioned: (1) the program will provide a reconstruction regardless of the quality of the input data; (2) the program will provide a reconstruction regardless of the reality of the parameters of the model. Nonetheless, the most undermining element of this approach still looms: centers of origin, ancestors, and ancestral areas are unknowable.

Despite the apparent complexity of the DEC model, its assumptions and limitations are substantial. Crisp et al. (2011:71) called for further complexity and sophistication for models to be more realistic, despite the fact that "validation with independent empirical data on crucially important parameter values ... are difficult to obtain, especially in an historical context." Advice on all this comes from Croizat (1964:527), an unexpected source: "... never figure out details until you are reasonably sure about fundamentals." It might also be worth considering Mishler's (2005:69) point of view: "More-complicated models ... are fundamentally attempts to compensate for marginal data."

The application of time-calibrated phylogenies and assumption-burdened models such as DEC will always identify a center of origin and long-distance dispersal (LDD) routes. These results are often described as having shown (e.g., Crisp et al. 2011; Herrera et al. 2015) or revealed (e.g., de Queiroz 2014) ancestral areas and LDD. Because the conclusions that taxa are too young or that they dispersed, or that they originated in any particular place are a product of method, they cannot overturn empirical observation.

It should be emphasized that process explanations for individual histories and origins do not provide any empirical evidence against the observation of common patterns of distribution or common patterns of area history (Parenti and Ebach 2013a). Process and history explanations such as dispersal and taxon age are methodological inferences that have little, or perhaps nothing, to say about general patterns because they do not employ them nor look for them (Ebach et al. 2003).

14.4.5 PATTERN VS. NO PATTERN; RELATIONSHIPS VS. ORIGINS

Few, if any, phylogenies are incompatible with the proposed mechanisms.

Rosen (1982:84)

If we lose the distinction between the detection of pattern and its explanation by a process hypothesis, we lose the reason for our inquiry, not merely historically, but logically.

Brady (1985:125)

... the revolution in biogeography has been stalled by attempts to generate explanations rather than to discover patterns.

Parenti (2007:63)

Gill and Mooi (in press) noted that biogeography has yet to recognize fully in itself the same mistake that for so long plagued systematics: "thinking in terms of origins rather than relationships" (Patterson 2011:124). Much of current historical biogeography is structured around this mistake, with emphasis on centers of origin, origins of biodiversity, and species origins instead of focusing on relationships among areas of endemism. Ebach et al. (2003) felt that biogeography had, in a sense, been "high-jacked" by the modern synthesis and its centers of origin. It became an evolutionary biology process-dominated research program that explores unique histories, changing from its original formulation as a systematic, pattern-based program exploring relationships among historically relevant areas.

Although this "pattern vs. process" dichotomy has dominated discussions of biogeographical theory and philosophy (e.g., Crisp et al. 2011), it is, in some ways, a false one. The dichotomy would be better described as "pattern vs. no pattern." Darwin (1859) provided a substantial boost to the "no pattern" approach with an emphasis on centers of origin and dispersal. He did not see patterns because he wasn't looking for them; he was too busy arguing single origins and unique dispersal explanations for particular taxon distributions to defeat special creation (Brady 1989). But the modern biogeographer has no such excuse. There is ample evidence of pattern in biogeography, recognized early on by Buffon, Humboldt, and de Candolle, and described in Wallace's line among other patterns (Humphries and Parenti 1999). But an entire research program—ancestral area/center of origin—proceeds as though no such patterns exist, employs methods that will never discover them, and actively works to discount them.

De Queiroz (2014:82) suggested that distributions were chaotic and "run all over the place," although this is more a reflection of approach than any reality in nature. Uniquely derived center of origin and dispersal explanations for each taxon would indeed suggest "no pattern," if not actual chaos. Although contrary to the broad scientific principle that "general patterns demand general explanations" (Rosen 1978a:186), this "no pattern" approach appears workable because the application of a computer algorithm provides a result. It has been argued that dispersal explanations are unscientific because they can be tailored to fit any distribution (Platnick and Nelson 1978). This seems difficult to deny in the face of origin/dispersal papers that emphasize the necessity of multiple explanations instead of general ones; rather than a biodiversity area being the result of one of several "center of" theories, it is argued that *all* "center of" theories are applicable (Bowen et al. 2013; Dornburg et al. 2015). Perhaps this is evidence of chaos. Platnick and Nelson (1981:119) pointed out that "there is no way for us to actually determine whether the disorder that we perceive exists in nature, or only in our own hypotheses."

Heads (2014c) has suggested that chaos is the ultimate conclusion of dispersal biogeography because every distribution pattern has its own explanation. He argued that this "nihilistic approach to distributions and the interpretation of distribution as 'shaped by miracles' [de Queiroz 2014:281] effectively short-circuit a science of biogeography … The approach requires little work, because there is no need to understand the geology of the area, which is often complex, or to compare the distribution with a large number of others in the same area to assess whether or not it conforms to a standard pattern" (Heads 2014c:288). It seems explanations for distributions are limited only by our imaginations (or, for the less imaginative, by the number of distributions). It is more than uncharitable to suggest that the dispersalist approach is less work—it is different work. But is that work fruitful in the sense of Brady (1979)? Is it generating activity "within the context of a community given to reflection and self-criticism"? Or is it generating activity because it provides results and "'business as usual' is profitable business"? Most biogeographers work within a center of

origin/dispersal paradigm and do not question its orthodoxy. "It is probably true …
that any community of co-workers which spends its energy carrying out practices
dictated by communal belief and shared technique, will believe that such energy is
fruitfully spent" (Brady 1979:617).

Crisp et al. (2011) have defended individual case history, process-based explanations over pattern-based approaches on the grounds that the latter are not scientific. They accomplished this by changing the usual biogeographical definition
of pattern—repeating relationships of areas among multiple clades (repeating areagrams)—to one meaning a single taxon phylogeny with distributions optimized to
infer ancestral areas. Crisp et al. (2011) rejected the latter approach, one that is actually process-based, on many of the same grounds that would be raised by any practi-
tioner of pattern-based biogeography (under the original definition). Strangely, Crisp
et al. sought improvement on the single taxon phylogeny ancestral area approach by
adding fossils as estimates of clade age and further process-burdened explanations.
They are left only with narrative explanations for the distributions of single clades.
Crisp et al. did not address truly pattern-based methods.

The schism between process and pattern, or no pattern and pattern, is not only
methodological, but also philosophical. Fundamentally, all biogeographers are trying to understand what lives where and why. But one approach, based on no pattern,
looks for origins of taxa through individual histories; the other, based on pattern,
looks for the relationships of areas through congruence. As Parenti and Ebach
(2013b) suggested, this leaves little common ground. They were hopeful that debate
such as that between themselves and de Bruyn et al. (2013) "encourages the good
health of the field of biogeography." It is difficult to be optimistic when the patient is
being torn asunder by opposing doctors of philosophy. If biogeographers are serious
about contributing general conclusions with broader applications for understanding
the distribution of life on Earth, they must move from unique, independent narratives
and invest in comparative studies.

Is there sufficient reflection and self-criticism in biogeography? That several
of Rosen's inhibiting assumptions provide the foundation for a popular research
program suggests not. That this program considers centers of origin discoverable
without acknowledging the healthy literature in opposition is disappointing. That
its explanations of individual histories take precedence over discovery of general
patterns is not progressive, but adheres to practices entrenched for well over a century. That the program allows as admissible all possible explanations of distribution
means we have no way of choosing among them. Brady (1979:617) provided a reason
to temper over-exuberant embrace of any mainstream approach, and one with perhaps particular relevance to biogeography: "Consensus is a witness of convincing
power but not, unless we owe the flat-earthers an apology, of truth."

14.5 CONCLUSION

> … ideas and beliefs have a history; and, in the search for that history … be candid
> with students so that they may not wander in a world of make-believe and pretense—
> however reputable and orthodox that world might seem.

Croizat et al. (1974:277)

Rosen valued challenges to orthodoxy and embraced the conflict that might arise; this was an important aspect of scientific development (Rosen 1981). This challenge is embodied in his list of assumptions that, from his perspective, inhibited scientific progress (Rosen 2016). That several of these assumptions remain relevant indicates that many elements of the systematic and biogeographic orthodoxy of 30 and more years ago have not developed in directions that Rosen might have expected. Consensus approaches support optimizing and mapping of characters onto topologies to reinterpret homoplasy as homology. Similarly, consensus advocates optimizing and mapping of distributions onto topologies to infer centers of origin and dispersal routes. Despite community consensus, these approaches have considerable limitations and fall short of their presumed outcomes.

Consensus contends fossils are "emerging again as being crucially important" and future, ideal biogeographical models will be "more … sophisticated and realistic" (Crisp et al. 2011:71). This emphasis on "the search for … the ultimate source of data, or the ultimate clustering algorithm, that by itself would guarantee a worthwhile result, inaugurate a modern age of systematics [and biogeography], and in the process relieve us all of a heavy burden—reading the systematic [and biogeographic] literature of the past and reaching an informed judgement of its relevance" (Nelson 1979:20). Systematics and biogeography did not commence with the use of computers and molecules; these fields have had long and productive histories that should be acknowledged to avoid making the mistakes of the past. The "promise that the study of cladistic congruence between the earth and its life will be the next 'revolution in the earth sciences'—an integrated natural history of geological and biological systems" (Rosen 1981:5) remains unrealized for the reason, in part, that this history has been largely ignored. Rosen's list is grounded in that history that has lost its influence and in the fact that systematists and biogeographers have become less skeptical of consensus and more satisfied with "business as usual" (Brady 1979:617).

That establishing empirical pattern takes precedence over proposing explanatory process has yet to be generally appreciated in biogeography. Present consensus approaches relying on fossil-calibrated molecular clocks, centers of origin, and long distance dispersal dismiss common pattern for individual case histories. Consensus approaches have generated a great deal of activity, but, to reiterate, "Mere activity, in itself, seems to have no intrinsic value other than paying the bills. Activity within the context of a community given to reflection and self-criticism, however, might be though a different matter" (Brady 1979:617). Reflection is not a luxury generally afforded in today's scientific climate. Funding agencies and employers have little patience for a biogeography that requires comparison of accumulated taxonomic revisionary work, robust morphological- and molecular-based cladograms, and detailed distributional data, bolstered by thorough geological underpinning. Instead, the more certain results from individual narratives are favored, those that are derived from application of consensus methods based on pre-Darwinian concepts where the logical, philosophical, and scientific implications are assumed to have been resolved by those that created the methods. Our system presently rewards those who, as defined by Rosen and Schuh (1975:505), are masters "… in the sense of one who has acquired a comprehensive understanding of certain facts … and a facility in using an existing paradigm or model to order these facts," but is less likely to produce scientists "... whose wisdom will lay bare the essences and solutions to existing differences of viewpoint."

There seems to be a disconnect between the researcher, the data, and the search for pattern. For Rosen, the glycerin dish was the crucible of truth (Nelson et al. 1987) and discovery of common patterns, that is congruence, among character traits or areagrams generated general hypotheses of relationship. Today, pattern is cryptic and the crucible of truth resides in the guts of a computer, where algorithms produce patterns that researchers, trained to discover them, cannot themselves find when examining the data. McMahan et al. (2015) found that, at least in some cases, there is higher congruence when these data are compared to patterns researchers have discovered (as opposed to computer algorithms). Consensus systematics prefers to have cryptic patterns redefine observable patterns by optimizing them on those cryptic trees; this is not only empirically empty (Assis and Rieppel 2011), but seems completely unsatisfying as a research program. Where has our sense of discovery gone? Are we willing to abnegate our role as systematists, that of discovering patterns, to machines that produce irretrievable patterns? This is not a Luddite manifesto, but merely a reminder that we cannot let our tools, no matter how sophisticated, either create evidence or be its interpreter. Consensus biogeography employs methods that cannot identify common patterns among taxa even if they exist. Individual narratives created by model-burdened programs that rest on inaccessible ancestral distributions and where fossils might or might not be found do not inspire confidence. We are mired in orthodoxy, with the mistaken impression that technological advances have provided theoretical and conceptual advances (Heads 2005).

If the aim of systematics and biogeography is to be fruitful in the sense of Brady (1979), we need to challenge "business as usual." Rosen did this along with a handful of colleagues, chief among them Gary Nelson, Norman Platnick, and Colin Patterson. But challenging orthodoxy has its risks. Science is conservative and consensus rules; it has always been difficult to be different (e.g., Croizat). Rosen, an established figure in a privileged position at a prestigious institution, could afford to take those risks and encourage others to do so. However, Grehan (2014:36) offers quite the opposite advice as a result of the power of orthodoxy; Heads (2014c:286–287) adds to this cautionary tale regarding the perils of challenging the science establishment. McGlone (2016) contested the details of that version of events and maintained that challenging, "established evolutionary theory" (i.e., long-distance dispersal and molecular clocks) deserves no institutional support. He justified this view by claiming there is "no convincing evidence" contrary to what amounts to established "business as usual." But what evidence is there, or can there be, for centers of origin, ancestors, and maximal clade ages? It is disturbing to note that the establishment continues to actively pursue extinction of its critics despite, or perhaps because of, the severe criticisms it has yet to overcome (Waters et al. 2013). But we cannot shirk our responsibility as scientists to question orthodoxy. As Brady (1985:125) noted: "Reflection may be unpopular, but science cannot do without inquiry."

This volume provides a wide variety of empirical and theoretical papers that perform that necessary inquiry to one degree or another. All, and perhaps especially this one, should be read with a critical and skeptical eye. It is through this constant challenge of ideas and approaches, not consensus, that science can make strides in understanding our world. Like Rosen (1975a:70), we should be distressed by any-one's "readiness to dismiss reasonable alternatives to his own views." Perhaps by hanging Rosen's (2016) list of progress-hindering assumptions prominently in our labs, we can become that critical and self-reflectant community that will provide the "... endless happy hours of useful contention and strife" that Rosen (1978b:373) treasured.

ACKNOWLEDGMENTS

I would like to thank Brian Crother and Lynne Parenti for their invitation to partici-pate in the symposium and this publication. This contribution, if it can be so-defined, has expanded considerably beyond the original scope of the oral presentation in Nevada. That effort was produced in collaboration with Tony Gill who, with typi-cal generosity, relinquished second authorship as the project expanded (or perhaps just because he knew better). I am privileged to count Tony as my closest friend and colleague; not only is he of generous spirit, incredible wit, and somehow always wearing a wool jumper, he is a gifted ichthyologist and conscientious scientist whose talents and knowledge have gone underappreciated. The manuscript was improved by David Williams, Gary Nelson, Lynne Parenti, and Tony Gill; I am grateful for their time. And I must also thank Brian and Lynne for their patience in permitting both the growth of this paper and my exploration of these issues. I appreciate the opportunity to focus on the relevance of Rosen's inhibiting assumptions in the milieu of today's uncritical application of concepts and methods—we need more skepticism and uncertainty.

REFERENCES

Assis, L. C. S. and O. Rieppel. 2011. Are monophyly and synapomorphy the same or differ-ent? Revisiting the role of morphology in phylogenetics. *Cladistics* 27:94–102.

Betancur-R., R., R. E. Broughton, E. O. Wiley, et al. 2013a. The tree of life and a new classifi-cation of bony fishes. *PLoS Currents Tree of Life* 2013 Apr 18. Edition 1. DOI: 10.1371/currents.tol.53ba26640df0ccaee75bb165c8c26288.

Betancur-R., R., C. Li, T. A. Munroe, et al. 2013b. Addressing gene tree discordance and non-stationarity to resolve a multi-locus phylogeny of the flatfishes (Teleostei: Pleuronectiformes). *Systematic Zoology* 62:763–785.

Betancur-R., R. and G. Ortì. 2014. Molecular evidence for the monophyly of flatfishes (Carangimorphariae: Pleuronectiformes). *Molecular Phylogenetics and Evolution* 73:18–22.

Borden, W. C., T. Grande and W. L. Smith. 2013. Comparative osteology and myology of the caudal fin in Paracanthopterygii (Teleostei: Acanthomorpha). In *Mesozoic fishes 5—Global diversity and evolution*, eds. G. Arratia, H.-P. Schultze and M. V. H. Wilson, 419–455. München: Verlag.

Bowen, B. W., L. A. Rocha, R. J. Toonen, et al. 2013. The origins of tropical marine biodiver-sity. *Trends in Ecology and Evolution* 28:359–366.

Brady, R. H. 1979. Natural selection and the criteria by which a theory is judged. *Systematic Zoology* 28:600–621.

Brady, R. H. 1985. On the independence of systematics. *Cladistics* 1:113–126.

Brady, R. H. 1989. The global patterns of life: A new empiricism in biogeography. In *Gaia and evolution (Proceedings of the Second Annual Camelford Conference on the Implications of the Gaia Thesis)*, eds. P. Bunyard and E. Goldsmith, 111–126. Cornwall: Wadebridge Ecological Centre.

Briggs, J. C. 1986. Introduction to zoogeography of North American fishes. In *The zoogeography of North American freshwater fishes*, eds. C. H. Hocutt and E. O. Wiley, 1–16. New York: John Wiley and Sons.

Briggs, J. C. 1995. *Global biogeography*. Amsterdam: Elsevier.

Briggs, J. C. and B. W. Bowen. 2012. A realignment of marine biogeographic provinces with particular reference to fish distributions. *Journal of Biogeography* 39:12–30.

Britz, R. and K. W. Conway. 2009. Osteology of *Paedocypris*, a miniature and highly developmentally truncated fish (Teleostei: Ostariophysi: Cyprinidae). *Journal of Morphology* 270:389–412.

Britz, R., K. W. Conway and L. Rüber. 2014. Miniatures, morphology and molecules: *Paedocypris* and its phylogenetic position (Teleostei, Cypriniformes). *Zoological Journal of the Linnean Society* 2014:1–60.

Brundin, L. 1966. Transantarctic relationships and their significance as evidenced by midges. *Kungliga Svenska Vetenskapsakademiens Handlinger* 11:1–472.

Cain, S. A. 1943. Criteria for the indication of center of origin in plant geographical studies. *Torreya* 43:132–154.

Campbell, M. A., W. J. Chen and J. A. Lòpez 2013. Are flatfishes (Pleuronectiformes) monophyletic? *Molecular Phylogenetics and Evolution* 69:664–673.

Campbell, M. A., W. J. Chen and J. A. Lòpez. 2014a. Molecular data do not provide unambiguous support for the monophyly of flatfishes (Pleuronectiformes): A reply to Betancur-R and Ortì. *Molecular Phylogenetics and Evolution* 75:149–153.

Campbell, M. A., J. A. Lòpez, T. P. Satoh, et al. 2014b. Mitochondrial genomic investigation of flatfish monophyly. *Gene* 551:176–182.

Carnevale, G. and G. D. Johnson. 2015. A Cretaceous cusk-eel (Teleostei, Ophidiiformes) from Italy and the Mesozoic diversification of percomorph fishes. *Copeia* 103:771–791.

Chapleau, F. 1993. Pleuronectiform relationships: A cladistic reassessment. *Bulletin of Marine Science* 52:516–540.

Christin, P.-A., G. Besnard, E. Samaritani, et al. 2008. Oligocene CO_2 decline promoted C_4 photosynthesis in grasses. *Current Biology* 18:37–43.

Cowman, P. F. and D. R. Bellwood. 2013. The historical biogeography of coral reef fishes: Global patterns of origination and dispersal. *Journal of Biogeography* 40:209–224.

Crisp, M. D., S. A. Trewick and L. G. Cook. 2011. Hypothesis testing in biogeography. *Trends in Ecology and Evolution* 26:66–72.

Croizat, L. 1964. *Space, time, form: The biological synthesis*. Caracas: Published by the author.

Croizat, L., G. Nelson and D. E. Rosen. 1974. Centers of origin and related concepts. *Systematic Zoology* 23:265–287.

Darlington, P. J. Jr. 1957. *Zoogeography: The geographical distribution of animals*. New York: John Wiley and Sons.

Darwin, C. 1859. *The origin of species*. New York: Penguin Books. [1984 reprint of the first edition]

de Bruyn, M., B. Stelbrink, T. J. Page, et al. 2013. Time and space in biogeography: Response to Parenti and Ebach (2013). *Journal of Biogeography* 40:2204–2206.

de Candolle, A. P. 1820. Géographie botanique. In *Dictionnaire des sciences naturelles*, ed. G. Cuvier, Vol. 18:359–422. Strasbourg: F. G. Levrault; Paris: Le Normant.

de Pinna, M. G. G. 1991. Concepts and tests of homology in the cladistics paradigm. *Cladistics* 7:367–394.

de Queiroz, A. 2005. The resurrection of oceanic dispersal in historical biogeography. *Trends in Ecology and Evolution* 20:68–73.

de Queiroz, A. 2014. *The monkey's voyage: How improbable journeys shaped the history of the world.* New York: Basic Books.

Dornburg, A., J. Moore, J. M. Beaulieu, R. I. Eytan and T. J. Near. 2015. The impact of shifts in marine biodiversity hotspots on patterns of range evolution: Evidence from the Holocentridae (squirrelfishes and soldierfishes). *Evolution* 69:146–161.

Ebach, M. C., C. J. Humphries and D. M. Williams. 2003. Phylogenetic biogeography deconstructed. *Journal of Biogeography* 30:1285–1296.

Edmunds, G. F. 1975. Phylogenetic biogeography of mayflies. *Annals of the Missouri Botanical Gardens* 62:251–263.

Friedman, M. 2008. The evolutionary origin of flatfish asymmetry. *Nature* 454 (10 July 2008):209–212. DOI: 10.1038/nature07108.

Friedman, M. 2012. Osteology of †*Heteronectes chaneti* (Acanthomorpha, Pleuronectiformes), an Eocene stem flatfish, with a discussion of flatfish sister-group relationships. *Journal of Vertebrate Paleontology* 32:735–756.

Friedman, M., B. P. Keck, A. Dornburg, et al. 2013. Molecular and fossil evidence place the origin of cichlid fishes long after Gondwanan rifting. *Proceedings of the Royal Society B* 280:20131733. http://dx.doi.org/10.1098/rspb.2013.1733.

Gilbert, C. R. 1976. Composition and derivation of the North American freshwater fish fauna. *Florida Scientist* 39:104–111.

Gill, A. C. and R. D. Mooi. In press. Biogeography of Australian marine fishes. In *Handbook of Australasian biogeography*, ed. M. C. Ebach. Boca Raton: CRC Press.

Grande, T., W. C. Borden and W. L. Smith. 2013. Limits and relationships of Paracanthopterygii: A molecular framework for evaluating past morphological hypotheses. In *Mesozoic fishes 5—Global diversity and evolution*, eds. G. Arratia, H.-P. Schultze and M. V. H. Wilson, 385–418. München: Verlag.

Grant, T. and A. Kluge. 2004. Transformation series as an ideographic character concept. *Cladistics* 20:23–31.

Graur, D. and W. Martin. 2004. Reading the entrails of chickens: Molecular timescales of evolution and the illusion of precision. *Trends in Genetics* 20:80–86.

Greenwood, P. H., D. E. Rosen, S. H. Weitzman and G. S. Meyers. 1966. Phyletic studies of teleostean fishes, with a provisional classification of living forms. *Bulletin of the American Museum of Natural History* 131:339–456.

Grehan, J. 2014. Into the storm: A personal retrospective on panbiogeography, part II. *Biogeografía* 7:35–44.

Hall, R. 1998. The plate tectonics of Cenozoic SE Asia and the distribution of land and sea. In *Biogeography and geological evolution of SE Asia*, eds. R. Hall and J. D. Holloway, 99–131. Leiden: Backhuys Publishers.

Harold, A. S. and R. D. Mooi. 1994. Areas of endemism: Definition and recognition criteria. *Systematic Biology* 43:261–266.

Heads, M. 2005. Towards a panbiogeography of the seas. *Biological Journal of the Linnean Society* 84:675–723.

Heads, M. 2012. Bayesian transmogrification of clade divergence times: A critique. *Journal of Biogeography* 39:1749–1756.

Heads, M. 2014a. *Biogeography of Australasia: A molecular analysis.* Cambridge: Cambridge University Press.

Heads, M. 2014b. Panbiogeography, its critics, and the case of the ratite birds. *Australian Systematic Botany* 27:241–256.

Heads, M. 2014c. Biogeography by revelation: Investigating a world shaped by miracles. *Australian Systematic Botany* 27:282–304.

Hennig, W. 1966. *Phylogenetic systematics.* Urbana: University of Illinois Press.

Herrera, N. D., J. J. ter Poorten, R. Bieler, et al. 2015. Molecular phylogenetics and historical biogeography amid shifting continents in the cockles and giant clams (Bivalvia: Cardiidae). *Molecular Phylogenetics and Evolution* 93:94–106.

Hilton, E. J., N. K. Schnell and P. Konstantinidis. 2015. When tradition meets technology: Systematic morphology of fishes in the early 21st century. *Copeia* 103:858–873.

Humphries, C. J. and L. R. Parenti. 1999. *Cladistic biogeography: Interpreting patterns of plant and animal distributions,* ed. 2. Oxford: Oxford University Press.

Kluge, A. 1976. Phylogenetic relationships in the lizard family Pygopodidae: An evaluation of theory, methods and data. *Miscellaneous Publications of the Zoological Museum, University of Michigan* 152:1–72.

Kluge, A. 2005. What it the rationale for "Ockham's razor" (a.k.a. parsimony) in phylogenetic inference? In *Parsimony, phylogeny, and genomics,* ed. V. A. Albert, 15–42. Oxford: Oxford University Press.

Kück, P. and J. W. Wägele. 2015. Plesiomorphic character states cause systematic error in molecular phylogenetic analyses: A simulation study. *Cladistics.* DOI: 10.1111/cla.12132.

Ladiges, P. Y., M. J. Bayly and G. Nelson. 2012. Searching for ancestral areas and artifactual centers of origin in biogeography: With comments on east-west patterns across Australia. *Systematic Biology* 61:703–708.

Mayden, R. and W.-J. Chen. 2010. The world's smallest vertebrate species of the genus *Paedocypris*: A new family of freshwater fishes and the sister group to the world's most diverse clade of freshwater fishes (Teleostei: Cypriniformes). *Molecular Phylogenetics and Evolution* 57:152–175.

McGlone, M. 2016. Once more into the wilderness of panbiogeography: A reply to Heads (2014). *Australian Systematic* 28:388–393.

McMahan, C. D., L. R. Freeborn, W. C. Wheeler and B. I. Crother. 2015. Forked tongues revisited: Molecular apomorphies support morphological hypotheses of squamate evolution. *Copeia* 103:525–529.

Mishler, B. 2005. The logic of the matrix in phylogenetic analysis. In *Parsimony, phylogeny, and genomics,* ed. V. A. Albert, 57–70. Oxford: Oxford University Press.

Miya, M., N. I. Holcroft, T. P. Satoh, M. Yamaguchi, M. Nishida and E. O. Wiley. 2007. Mitochondrial genome and a nuclear gene indicate a novel phylogenetic position of deep-sea tube-eye fish (Stylephoridae). *Ichthyological Research* 54:323–332.

Mooi, R. D. and A. C. Gill. 2010. Phylogenies without synapomorphies—a crisis in fish systematics: Time to show some character. *Zootaxa* 2450:26–40.

Mooi, R. D. and A. C. Gill. 2016. Hennig's auxiliary principle and reciprocal illumination revisited. In *The future of phylogenetic systematics: The legacy of Willi Hennig,* ed. D. W. Williams, 258–285. Cambridge: Cambridge University Press.

Near, T. J., A. Dornburg, R. I. Eytan, et al. 2013. Phylogeny and tempo of diversification in the superradiation of spiny-rayed fishes. *Proceedings of the National Academy of Sciences USA* 110:12738–12743.

Near, T. J., A. Dornburg, R. C. Harrington, et al. 2015. Identification of the notothenioid sister lineage illuminates the biogeographic history of an Antarctic adaptive radiation. *BMC Evolutionary Biology* (2015) 15:109. DOI: 10.1186/s12862-015-0362-9.

Nelson, G. 1969a. Origin and diversification of teleostean fishes. *Annals of the New York Academy of Sciences* 167:18–30.

Nelson, G. J. 1969b. The problem of historical biogeography. *Systematic Zoology* 18:243–246.

Nelson, G. 1970. Outline of a theory of comparative biology. *Systematic Zoology* 19:373–384.

Nelson, G. 1973. Comments on Leon Croizat's biogeography. *Systematic Zoology* 22:312–320.

Nelson, G. J. 1974. Historical biogeography: An alternative formalization. *Systematic Zoology* 23:555–558.

Nelson, G. 1978a. Ontogeny, phylogeny, paleontology, and the biogenetic law. *Systematic Zoology* 27:324–345.

Nelson, G. 1978b. From Candolle to Croizat: Comments on the history of biogeography. *Journal of the History of Biology* 11:269–305.

Nelson, G. 1979. Cladistic analysis and synthesis: Principles and definitions, with a historical note on Adanson's *Familles des Plantes* (1763–1764). *Systematic Zoology* 28:1–21.

Nelson, G. 1983. Vicariance and cladistics: Historical perspectives with implications for the future. In *Evolution, time and space: The emergence of the biosphere*, eds. R. W. Sims, J. H. Price and P. E. S. Whalley, 469–492. London: Academic Press.

Nelson, G. 1994. Homology and systematics. In *Homology: The hierarchical basis of comparative biology*, ed. B. K. Hall, 101–149. San Diego: Academic Press.

Nelson, G. and N. I. Platnick. 1981. *Systematics and biogeography: Cladistics and vicariance*. New York: Columbia University Press.

Nelson, G. and P. Y. Ladiges. 2001. Gondwana, vicariance biogeography, and the New York School revisited. *Australian Journal of Botany* 49:389–409.

Nelson, G. and P. Y. Ladiges. 2009. Biogeography and the molecular dating game: A futile revival of phenetics? *Bulletin de la Société Géologique de France* 180:39–43.

Nelson, G., J. W. Atz, K. D. Kallman and C. Lavett Smith. 1987. Donn Eric Rosen 1929–1986. *Copeia* 1987:541–547.

O'Hara, R. 1988. Homage to Clio, or towards an historical philosophy for evolutionary biology. *Systematic Zoology* 37:142–155.

Olney, J. E., G. D. Johnson and C. C. Baldwin. 1993. Phylogeny of lampridiform fishes. *Bulletin of Marine Science* 52:137–169.

Parenti, L. R. 2007. Common cause and historical biogeography. In *Biogeography in a changing world*, ed. M. C. Ebach and R. S. Tangney, 61–82. Boca Raton: CRC Press.

Parenti, L. R. and M. C. Ebach. 2009. *Comparative biogeography: Discovering and classifying biogeographical patterns of a dynamic Earth*. Berkeley: University of California Press.

Parenti, L. R. and M. C. Ebach. 2013a. Evidence and hypothesis in biogeography. *Journal of Biogeography* 40:813–820.

Parenti, L. R. and M. C. Ebach. 2013b. The explanatory power of biogeographical patterns: A reply to de Bruyn et al. *Journal of Biogeography* 40:2204–2208.

Parham, J. F., P. C. J. Donoghue, C. J. Bell, et al. 2012. Best practices for justifying fossil calibrations. *Systematic Biology* 61:346–359.

Patterson, C. 1981. The development of the North American fish fauna—a problem of historical biogeography. In *The evolving biosphere*, ed. P. L. Forey, 265–281. Cambridge: Cambridge University Press.

Patterson, C. 1982. Morphological characters and homology. In *Problems of phylogenetic reconstruction*, eds. K. A. Joysey and A. E. Friday, 21–74. London: Academic Press.

Patterson, C. 2011. Adventures in the fish trade. *Zootaxa* 2946:118–136. [edited and with an introduction by D. M. Williams and A. C. Gill]

Platnick, N. I. 1976. Concepts of dispersal in historical biogeography. *Systematic Zoology* 25:294–295.

Platnick, N. I. 1991. Commentary. On areas of endemism. *Australian Systematic Botany* 4:xi–xii.

Platnick, N. I. and G. Nelson. 1978. A method of analysis for historical biogeography. *Systematic Zoology* 27:1–16.

Platnick, N. I. and G. Nelson. 1981. The purposes of biological classification. In *PSA 1978, Proceedings of the 1978 biennial meeting of the Philosophy of Science Association, volume two, symposia*, eds. P. D. Asquith and I. Hacking, 117–129. East Lansing: Philosophy of Science Association.

Remane, A. 1952. *Die Grundlagen des Natürlichen Systems, der Vergleichehenden Anatomie und der Phylogenetik*. Leipzig: Geest and Portig.

Renema, W., D. R. Bellwood, J. C. Braga, et al. 2008. Hopping hotspots: Global shifts in marine biodiversity. *Science* 321:654–657.

Rosen, D. E. 1974. Phylogeny and zoogeography of salmoniform fishes and relationships of *Lepidogalaxias salamandroides*. *Bulletin of the American Museum of Natural History* 153:265–326.

Rosen, D. E. 1975a. Doctrinal biogeography. *The Quarterly Review of Biology* 50:69–70.

Rosen, D. E. 1975b. A vicariance model of Caribbean biogeography. *Systematic Zoology* 24:431–464.

Rosen, D. E. 1978a. Vicariant patterns and historical explanation in biogeography. *Systematic Zoology* 27:159–188.

Rosen, D. E. 1978b. Darwin's demon. *Systematic Zoology* 27:370–373.

Rosen, D. E. 1979. Fishes from the uplands and intermontane basins of Guatemala: Revisionary studies and comparative geography. *Bulletin of the American Museum of Natural History* 162:267–376.

Rosen, D. E. 1981. Introduction. In *Vicariance biogeography: A critique*, eds. G. Nelson and D. E. Rosen, 1–5. New York: Columbia University Press.

Rosen, D. E. 1982. Do current theories of evolution satisfy the basic requirements of explanation? *Systematic Zoology* 31:76–85.

Rosen, D. E. 1985a. An essay on euteleostean classification. *American Museum Novitates* 2827:1–57.

Rosen, D. E. 1985b. Geological hierarchies and biogeographic congruence in the Caribbean. *Annals of the Missouri Botanical Gardens* 72:636–659.

Rosen, D. E. 2016. Assumptions that *inhibit* scientific progress in comparative biology. In *Assumptions inhibiting progress in comparative biology*, ed. B. I. Crother and L. R. Parenti, 1–4. Boca Raton: CRC Press.

Rosen, D. E. and R. T. Schuh. 1975. A review of: Flowering plants, evolution above the species level, by G. L. Stebbins. *Systematic Zoology* 24:504–506.

Rüber, L., M. Kottelat, H. H. Tan, P. K. L. Ng and R. Britz. 2007. Evolution of miniaturization and the phylogenetic position of *Paedocypris*, comprising the world's smallest vertebrate. *BMC Evolutionary Biology* 7:38–47.

Rudkin, D. M. and G. A. Young. 2009. Horseshoe crabs—an ancient ancestry revealed. In *Biology and conservation of horseshoe crabs*, eds. J. T. Tanacredi, M. L. Botton and D. R. Smith, 25–44. Dordrecht: Springer.

Rudkin, D. M., G. A. Young and G. S. Nowlan. 2008. The oldest horseshoe crab: A new xiphosurid from Late Ordovician Konservat-Lagerstätten deposits, Manitoba, Canada. *Palaeontology* 51:1–9.

Sclater, P. L. 1858. On the general geographical distribution of the members of the class Aves. *Journal of the Linnean Society of London, Zoology* 2:130–145.

Simmons, M. P. and J. V. Freudenstein. 2011. Spurious 99% bootstrap and jackknife support for unsupported clades. *Molecular Phylogenetics and Evolution* 61:177–191.

Simmons, M. P., K. M. Pickett and M. Miya. 2004. How meaningful are Bayesian support values? *Molecular Biology and Evolution* 21:188–199.

Sparks, J. S. and W. L. Smith. 2005. Freshwater fishes, dispersal ability, and nonevidence: "Gondwana life rafts" to the rescue. *Systematic Biology* 54:158–165.

Thacker, C. E. 2009. Phylogeny of the Gobioidei and placement within Acanthomorpha, with a new classification and investigation of diversification and character evolution. *Copeia* 2009:93–104.

Thacker, C. E. 2015. Biogeography of goby lineages (Gobiiformes: Gobioidei): Origin, invasions and extinction throughout the Cenozoic. *Journal of Biogeography* 42:1615–1625.

Wägele, J. W. and C. Mayer. 2007. Visualizing differences in phylogenetic information content of alignments and distinction of three classes of long-branch effects. *BMC Evolutionary Biology* 7:147. DOI: 10.1186/1471-2148-7-147.

Warnock, R. C. M., Z. Yang and P. C. J. Donoghue. 2012. Exploring uncertainty in the calibration of the molecular clock. *Biology Letters* 8:156–159.

Waters, J. M., S. A. Trewick, A. M. Paterson, et al. 2013. Biogeography off the tracks. *Systematic Biology* 62:494–498.

Wiley, E. O., A. M. Fuiten, M. H. Doosey, et al. 2015. The caudal skeleton of the zebrafish, *Danio rerio*, from a phylogenetic perspective: A polyural interpretation of homologous structures. *Copeia* 2015:740–750.

Williams, D. M. and M. C. Ebach. 2004. The reform of palaeontology and the rise of biogeography—25 years after "ontogeny, phylogeny, paleontology and the biogenetic law" (Nelson, 1978). *Journal of Biogeography* 31:685–712.

Wilson, M. V. H. 1980. Oldest known *Esox* (Pisces: Esocidae), part of a new Paleocene teleost fauna from western Canada. *Canadian Journal of Earth Sciences* 17:307–312.

Wilson, M. V. H., D. B. Brinkman and A. G. Neuman. 1992. Cretaceous Esocoidei (Teleostei): Early radiation of the pikes in North American fresh waters. *Journal of Paleontology* 66:839–846.

Index

Printed and bound by CPI Group (UK) Ltd, Croydon, CR0 4YY

24/10/2024

01778308-0008